An Introduction to the Properties of Engineering Materials

THIRD EDITION

An Introduction to the Properties of Engineering Materials

by **K. J. PASCOE,** M.A., C.ENG., F.I.MECH.E., F.INST.P.

University Lecturer in Engineering and Fellow of St. John's College, Cambridge

VAN NOSTRAND REINHOLD COMPANY

LONDON

NEW YORK CINCINNATI TORONTO MELBOURNE

First published 1961
Second edition 1972
Third edition 1978
Reprinted 1979, 1980

Published by Van Nostrand Reinhold Company Ltd.,
Molly Millars Lane, Wokingham, Berkshire, England

Published in 1978 *by Van Nostrand Reinhold Company*
A Division of Litton Educational Publishing Inc.,
135 *West* 50*th Street, New York, N.Y.* 10020, *U.S.A.*

Van Nostrand Reinhold Limited
1410 *Birchmont Road, Scarborough, Ontario, M*1*P* 2*E*7, *Canada*

Van Nostrand Reinhold Australia Pty, Limited
17 *Queen Street, Mitcham, Victoria* 3132, *Australia*

Library of Congress Cataloguing in Publication Data

Pascoe, K. J.

An introduction to the properties of engineering materials.
Includes bibliographical references and index.
1. Materials. 2. Metals I. Title.

TA403.P.28 1978 620. 1′ 12 78–14056
ISBN: 0 442 30232 0
ISBN: 0 442 30233 9 pbk.

Typeset at the Alden Press, Oxford,
printed and bound in Great Britain
by Henry Ling Ltd, at the Dorset Press,
Dorchester, Dorset.

PREFACE

The engineering designer is always limited by the properties of available materials. Some properties are critically affected by variations in composition, in state or in testing conditions, while others are much less so. The engineer must know this if he is to make intelligent use of the data on properties of materials that he finds in handbooks and tables, and if he is to exploit successfully new materials as they become available.

He can only be aware of these limitations if he understands how properties depend on structure at the atomic, molecular, microscopic and macroscopic levels.

Inculcating this awareness is one of the chief aims of the book, which is based on a successful course designed to give university engineering students the necessary basic knowledge of these various levels. The material is equivalent to a course of about eighty to a hundred lectures.

In the first part of the book the topics covered are mainly fundamental physics. The structure of the atom, considered in non-wave-mechanical terms, leads to the nature of interatomic forces and aggregations of atoms in the three forms—gases, liquids and solids. Sufficient crystallography is discussed to facilitate an understanding of the mechanical behaviour of the crystals. The band theory of solids is not included, but the basic concepts which form a preliminary to the theory—energy levels of electrons in an atom, Pauli's exclusion principle, and so on—are dealt with. The inherent weaknesses of single crystals are explained in terms of dislocations, and the various methods of strengthening are discussed.

The next part of the book is more technological in nature. The mechanical testing of metals and the relationship of test results to engineering design form the subject matter of two chapters. Thereafter, the heat treatment of steels is considered at some length, and modifications in properties resulting from welding are considered. An introduction is given to electrochemical phenomena, providing a basic understanding of the very complex problems associated with corrosion.

Two chapters deal with inorganic and organic materials with especial emphasis on those groups of them that are widely applied in engineering practice.

A final chapter introduces some nuclear physics and shows that yet other properties have to be considered when selecting materials for use in nuclear engineering projects.

Throughout the book, actual materials are quoted by way of example only, no exhaustive list of any class of material being given.

In producing this new second edition, the first edition has been thoroughly revised and extended. Besides the two chapters on non-metallic materials mentioned above, the main additions are: a more detailed treatment of strain fields around dislocations and the forces associated with them, further treatment of corrosion and a brief discussion of fracture mechanics. Also, and by no means least, the units in the book have now been converted into SI. Additional problems have been added to several chapters.

My thanks are due to many colleagues and friends who gave me help and encouragement in writing the original book and who gave constructive criticism for this second edition.

K. J. PASCOE

Cambridge

December, 1971

PREFACE TO THIRD EDITION

The main changes in this edition are the expansion of the chapter on the relation of mechanical testing to design into three chapters, dealing respectively with fracture, fatigue and creep, presenting more up-to-date information in these fields, and the addition of a new chapter on composite materials. Minor revisions and additions have been made elsewhere and the change to SI units has been completed. Also more problems have been added to most chapters.

In adding the new material and making other changes, I am once again indebted to colleagues and friends who have checked the material and provided pictorial matter.

K. J. PASCOE

Cambridge

February, 1978

CONTENTS

ACKNOWLEDGMENTS

The publishers gratefully acknowledge permission to use, and help with, the following illustrations:

Figs. 3.2 and 3.4, J. K. Roberts, HEAT AND THERMODYNAMICS, Blackie.
Fig. 5.5, Macmillan & Co. Ltd.
Figs. 5.7 and 10.4, Dr. C. F. Tipper: G. H. Elam, DISTORTION OF METAL CRYSTALS, Oxford University Press.
Fig. 8.1, Dr. E. Gregory, METALLURGY, Blackie.
Fig. 10.5, A. F. Brown and the Institute of Metals.
Fig. 10.26, Dr. A. R. Verma and the Royal Society.
Figs. 10.29 and 11.2, Messrs. Whelan, Hirsch, Horne and Bollmann, and the Royal Society.
Fig. 12.14, Dr. D. Tabor, THE HARDNESS OF METALS, Oxford University Press.
Fig. 14.3, Dr. W. A. Wood, PROCEEDINGS OF THE INTERNATIONAL CONFERENCE ON FATIGUE OF METALS, Institution of Mechanical Engineers.
Fig. 14.4, P. J. E. Forsyth, PROCEEDINGS OF THE INTERNATIONAL CONFERENCE ON FATIGUE OF METALS, Institution of Mechanical Engineers.
Fig. 14.15, Messrs. Woodward, Gunn and Forrest, PROCEEDINGS OF THE INTERNATIONAL CONFERENCE ON FATIGUE OF METALS, Institution of Mechanical Engineers.
Figs. 15.8, 15.9, 15.10 and 15.11, Prof M. F. Ashby.
Fig. 16.6, TRANSFORMATION CHARACTERISTICS OF DIRECT-HARDENING NICKEL-ALLOY STEELS, International Nickel Co. (Mond) Ltd.
Figs. 17.6 and 17.7, Messrs. Irvine, Pickering, Heselwood and Atkins, and the Iron and Steel Institute.
Fig. 19.4, The Welding Institute.
Fig. 20.11, Prof. R. H. Mills.
Fig. 22.7, Dr. P. W. R. Beaumont.
Fig. 23.2, S. Glasstone and M. C. Edlund, THE ELEMENTS OF NUCLEAR REACTOR THEORY, MacMillan.

The publishers also acknowledge permission for the use of the following figures which are based on diagrams or data in the works detailed:

Figs. 10.10, 10.12, 11.6, Boas, AN INTRODUCTION TO THE PHYSICS OF METALS AND ALLOYS, Melbourne University Press.
Fig. 11.7, Tin Research Institute.
Fig. 11.8, Seitz, PHYSICS OF METALS, McGraw-Hill.
Fig. 11.9, Rollason, METALLURGY FOR ENGINEERS, Edward Arnold.
Figs. 15.3, 15.4, THE NIMONIC ALLOYS: DESIGN DATA, Henry Wiggin & Co. Ltd.
Figs. 16.12, 16.13, 16.19, Clark and Varney, PHYSICAL METALLURGY FOR ENGINEERS, Van Nostrand.
Figs. 16.8, 16.10, 16.22, 16.23, 17.4, TRANSFORMATION CHARACTERISTICS OF DIRECT-HARDENING NICKEL-ALLOY STEELS, International Nickel Co. (Mond) Ltd
Figs. 16.16, 17.5, THE MECHANICAL PROPERTIES OF NICKEL ALLOY STEELS, International Nickel Co. (Mond) Ltd.
Figs. 16.20, 16.24, METALS HANDBOOK, American Society of Metals.
Fig. 17.2, Heyer, ENGINEERING PHYSICAL METALLURGY, Van Nostrand.
Figs. 17.8, A.1 to A.14, and Table 6.1, METALS REFERENCE BOOK, second edition edited by Smithells, Butterworth.
Fig. 18.5, CORROSION-RESISTANT MATERIALS IN MARINE ENGINEERING, International Nickel Co. (Mond) Ltd.
Fig. 19.1, Bruckner, WELDING METALLURGY, Pitman.
Fig. 20.9, P. W. McMillan, GLASS CERAMICS, Academic Press.

The author gratefully acknowledges permission given by the Syndics of the Cambridge University Press to reproduce certain questions from Examination papers.

The sources of these questions are indicated as follows:

[MST] Mechanical Science Tripos.
[P] Preliminary Examination in Mechanical Sciences.
[S] Examination in Engineering Studies.
[E] Engineering Tripos.

CHAPTER 1

Introduction

1.1 The fundamental nature of material properties

The primary jobs of the engineer are the design, construction, and maintenance of structures, machinery, etc. In his function as a designer, he makes use of the principles of thermodynamics, electricity, and the statics and dynamics of solids and fluids, but he is always limited finally by the materials at his disposal.

For example, the maximum distance that can be spanned by a cable is dependent upon the strength/weight ratio of the material used. Also the efficiency of gas turbines increases with the temperatures attained, but the top temperature usable is limited by the availability of materials with sufficient strength at that temperature. For some applications, more than one property may be needed, such as high strength, associated with good electrical conductivity and corrosion resistance. As will be seen in Chapter 23, the nuclear-power engineer has to consider properties that are of no concern in other engineering fields. Although the range of materials is very wide and countless variations of properties are available, the ideal material for a particular application, having a combination of optimum values for various properties, does not always exist and a compromise is necessary. Economics may also play a part, so that a cheaper, though less suitable, material may be used where the capital saving may more than compensate for higher maintenance costs or a shorter life.

The properties of any class of available material can always be obtained by reference to appropriate catalogues and handbooks, and the design engineer can carry out his work merely by reference to these sources of information and to experience. He can, however, attain a much better appreciation of properties of materials and the possibilities that may be afforded by use of a different type of material or by development along certain lines if he has an understanding of the fundamentals which govern the properties.

The aim of this book is to introduce the reader to the relationships between these fundamentals and the properties. As the properties of any material depend upon the structure of the atom and the manner in which the atoms are arranged, it is necessary for a suitable course to begin with the structure of the atom and then pass on to a consideration of the behaviour of atoms in large numbers, firstly as gases and then as liquids and solids.

Of materials used by the engineer, the non-metallic building materials—concrete, stone, and brick—represent the greatest tonnage. Following them come metals, with steel ranking first, and it is mainly with these that this book is concerned. The majority of the contents may properly be called physical metallurgy, but much of the fundamental behaviour discussed is also applicable to non-metallic materials.

The reader is not expected to memorize values of the mechanical properties of the individual metals and alloys, as those data are always available from handbooks. For this reason, catalogues of data which would be of direct use to the designer are not given, and where values are quoted they are by way of illustration only.

CHAPTER 2

Atomic Structure

2.1. Classification of the elements

One of the more immediately obvious methods of subdivision of the elements is the distinction between metals and non-metals. The distinction is based on both physical and chemical properties, some of the distinguishing properties for a metal being as follows:

Ductility, that is, ability to be bent or otherwise deformed.

High conductivity for heat and electricity.

Capability of taking a polish, giving metallic lustre.

Ease of alloying, alloys differing from other forms of chemical combination in that there is no law of constant proportions.

Chemically, the oxides are basic and salts are formed with acids.

For most elements, the classification as metal or non-metal is obvious, but the two classes merge into one another, some metals possessing both metallic and non-metallic properties. Arsenic, for example, possesses many of the physical properties of a metal, but chemically it is much more like a non-metal. Such elements are called *metalloids.*

A list of all the elements arranged in alphabetical order is given in Table 2.1, together with their chemical symbols, atomic weights, atomic numbers, melting-points, boiling-points, and data concerning their structure.

Newlands (1864) pointed out that if the elements were tabulated in order of increasing atomic weight, the properties of the first seven elements re-appeared in the second seven. He named this the *law of octaves.* Mendelejeff (1869) developed a more elaborate and systematic representation of this which is known as his *periodic law.* The periodic table in the general arrangement proposed by Mendelejeff, but including all naturally occurring elements (several of which were not known in Mendelejeff's time) is shown in Table 2.2.

The elements fall into 7 periods and 9 groups. The first period contains only the lightest element—hydrogen. The second and third periods contain 8 elements each, the fourth and fifth periods contain 18 elements each, there are 32 in the sixth period, and the seventh is incomplete.

In the second and subsequent periods, the first element is a chemically

TABLE 2.1—SOME PHYSICAL PROPERTIES OF THE ELEMENTS

Element	Symbol	Atomic number	Atomic weight[1]	M.P. °C[2]	B.P. °C[2]	Crystal structure[3]	Lattice constants[4] nm		
							a	b	c
Actinium	Ac	89	277	1050	3200	f.c.c.	0·5311		
Aluminium	Al	13	26·9815	600	2520	f.c.c.	0·40496		
Americium	Am	95	[243]	995		d.c.p.h.	0·34680		1·1240
Antimony	Sb	51	121·75	631	1587	r.	0·45067	α = 57° 6·5′	
Argon	A	18	39·948	– 189	– 186	f.c.c.	0·542	(at – 233 °C)	
Arsenic	As	33	74·9216	817[5]		r.	0·41318	α = 54° 8′	
Astatine	At	85	[210]	302	337				
Barium	Ba	56	137·34	729	1898	b.c.c.	0·5013		
Berkelium	Bk	97	[247]						
Beryllium	Be	4	9·0122	1287	2472	c.p.h.	0·2286		0·3584
Bismuth	Bi	83	208·980	271	1564	r.	0·4736	α = 57° 14′	
Boron	B	5	10·811	2027	4002	t.	0·880		0·505
Bromine	Br	35	79·909	– 7	59	orth.	0·448	0·667 (at – 150 °C)	0·872
Cadmium	Cd	48	112.40	321	767	c.p.h.	0·29788		0·56167
Caesium	Cs	55	132·905	28	671	b.c.c.	0·6079	(at – 100 °C)	
Calcium	Ca	20	40·08	839	1484	f.c.c.	0·55884		
Californium	Cf	98	[249]						
Carbon (graphite)	C	6	12·01115	[6]		hex.	0·24612		0·67079
Cerium	Ce	58	140·12	798	3426	d.c.p.h.	0·3673		1·1802
Chlorine	Cl	17	35·453	– 101	34	orth.	0·624	0·448 (at – 160 °C)	0·826
Chromium	Cr	24	51·996	1860	2672	b.c.c.	0·28846		
Cobalt	Co	27	58·9332	1495	2928	c.p.h.	0·2507		0·4070
Copper	Cu	29	63·54	1084	2563	f.c.c.	0·36147		
Curium	Cm	96	[245]						
Dysprosium	Dy	66	162·50	1409	2562	c.p.h.	0·35903		0·56475
Einsteinium	E	99	[254]						
Erbium	Er	68	167·26	1522	2863	c.p.h.	0·35588		0·55874
Europium	Eu	63	151·96	817	1597	b.c.c.	0·45820		
Fermium	Fm	100	[253]						
Fluorine	F	9	18·9984	– 220	– 188	cub.	0·667	(at – 223 °C)	
Francium	Fr	87	[223]	27	667				
Gadolinium	Gd	64	157·25	1312	3266	c.p.h.	0·36360		0·57826
Gallium	Ga	31	69·72	30	2205	orth.	0·4523	0·4524	0·7661
Germanium	Ge	32	72·59	937	2834	d.	0·56575		
Gold	Au	79	196·967	1063	2853	f.c.c.	0·40785		
Hafnium	Hf	72	178·49	2227	4603	c.p.h.	0·31946		0·50511
Hahnium	Ha	105	[262]						
Helium	He	2	4·0026	– 271[7]	– 269	b.c.c.	0·4110	(at – 271 °C)[7]	
Holmium	Ho	67	164·930	1470	2695	c.p.h.	0·35773		0·56158
Hydrogen	H	1	1·00797	– 259	– 253	c.p.h.	0·357	(at – 271 °C)	0·612
Indium	In	49	114·82	157	2073	f.c.t.	0·45979		0·49467
Iodine	I	53	126·9044	114	185	orth.	0·4792	0·7271	0·9773
Iridium	Ir	77	192·2	2443	4428	f.c.c.	0·38389		
Iron	Fe	26	55·847	1536	2862	b.c.c.	0·28664		

(1) The values of atomic weights are those recommended in the Reoort of the International Commission on Atomic weights (1961). These are for the naturally occurring mixture of isotopes, except where a value is given in brackets [ ]. Each of these values denotes the mass number of the isotope of longest known half-life, which is not necessarily the most important in atomic energy work. The atomic weights are given in terms of the unified atomic mass unit which is one twelfth of the mass of C^{12} (see Table 2.4). Because of the natural variation in the relative abundance of their isotopes, the atomic weights of boron and sulphur have ranges of ± 0·003 about the values quoted.

(2) Melting and boiling points at atmospheric pressure are given to the nearest degree. All values, except those for At, Cm, Fr, Pm, Pa and Ra and the boiling points of Np and Tc are taken from *Selected Values of the Thermodynamic Properties of the Elements*, American Society for Metals, 1973.

(3) The structure given is that at room temperature, except for elements which are not solid at that temperature. Many elements have other structures at higher or lower temperatures. Structures are denoted as follows:

b.c.c.	body-centred cubic
b.c.t.	body-centred tetragonal
c.p.h.	close-packed hexagonal
cub.	cubic
d.	diamond structure, two interpenetrating f.c.c. lattices
d.c.p.h.	double close-packed hexagonal (ABAC stacking)
f.c.c.	face-centred cubic
f.c.orth.	face-centred orthogonal
f.c.t.	face-centred tetragonal
hex.	hexagonal
monoc.	monoclinic
orth.	orthogonal
r.	rhombohedral
t.	tetragonal

TABLE 2.1—(*Continued*)—SOME PHYSICAL PROPERTIES OF THE ELEMENTS

Element	Symbol	Atomic number	Atomic weight[1]	M.P. °C[2]	B.P. °C[2]	Crystal structure[3]	Lattice constants[4] nm *a*	*b*	*c*
Krypton	Kr	36	83·80	– 157	– 153	f.c.c.	0·568	(at – 191 °C)	
Lanthanum	La	57	138·91	920	3457	d.c.p.h.	0.3770		1·2159
Lawrencium	Lw	103	[257]						
Lead	Pb	82	207·19	327	1750	f.c.c.	0·49502		
Lithium	Li	3	6·939	181	1342	b.c.c.	0·35100		
Lutetium	Lu	71	174·97	1663	3395	c.p.h.	0·35031		0·55509
Magnesium	Mg	12	24·312	649	1090	c.p.h.	0·32094		0·52105
Manganese	Mn	25	54·9380	1244	2062	cub.	0·89139		
Mendelevium	Mv	101	[256]						
Mercury	Hg	80	200·59	– 39	357	r.	·3005	α = 70° 32′ (at – 46 °C)	
Molybdenum	Mo	42	95.94	2617	4639	b.c.c.	0·31468		
Neodymium	Nd	60	144·24	1016	3068	d.c.p.h.	0·36579		1·17992
Neon	Ne	10	20·183	– 249	– 246	f.c.c.	0·452	(at – 268 °C)	
Neptunium	Np	93	[237]	637	3902	orth.	0·6663	0·4723	0·48872
Nickel	Ni	28	58·71	1453	2914	f.c.c.	0·35238		
Niobium (Columbium)	Nb (Cb)	41	92·906	2467	4744	b.c.c.	0·33066		
Nitrogen	N	7	14·0067	– 210	– 196	hex.	0·403	(at – 234 °C)	0·659
Nobelium	No	102	[256]						
Osmium	Os	76	190·2	3027	5012	c.p.h.	0·27353		0·43191
Oxygen	O	8	15·9994	– 219	– 183	cub.	0·683	(at – 225 °C)	
Palladium	Pd	46	106·4	1552	2964	f.c.c.	0·38907		
Phosphorus	P	15	30·9738	44	277	cub.	0·718	(at – 35 °C)	
Platinum	Pt	78	195·09	1769	3827	f.c.c.	0·39239		
Plutonium	Pu	94	[242]	640	3230	monoc.	0·6183	**0·4822** **β = 101° 47′**	1·0963
Polonium	Po	84	210	254	960	r.	0·3359	**α = 98° 13′**	
Potassium	K	19	39·102	63	759	b.c.c.	0·5247	**(at – 196°C)**	
Praseodymium	Pr	59	140·907	931	3512	d.c.p.h.	0·36725		1·18354
Promethium	Pm	61	[147]	~ 1080					
Protactinium	Pa	91	231	3000		t.	0·3925		0·3238
Radium	Ra	88	226·05	700	1140				
Radon	Rn	86	222	– 71	– 62				
Rhenium	Re	75	186·2	3180	5596	c.p.h.	0·2760		0·4458
Rhodium	Rh	45	102·905	1963	3697	f.c.c.	0·38044		
Rubidium	Rb	37	85·47	39	1016	b.c.c.	0·570		
Ruthenium	Ru	44	101·07	2250	4150	c.p.h.	0·27058		0·42816
Rutherfordium (Kurchatovium)	Rf (Ku)	104	[260]						
Samarium	Sm	62	150·35	1072	1791	r.	0·8996	α = 23° 13′	
Scandium	Sc	21	44·956	1539	2831	c.p.h.	0·33090		0·52733
Selenium	Se	34	78·96	221	683	hex.	0·43656		0·49590
Silicon	Si	14	28·086	1412	3267	d.	0·54307		
Silver	Ag	47	107·870	961	2163	f.c.c.	0·40862		
Sodium	Na	11	22·9898	98	883	b.c.c.	0·42906		
Strontium	Sr	38	87·62	768	1377	f.c.c.	0·60849		
Sulphur	S	16	32·064	113	445	f.c.orth.	1·04646	1·28660	2·44860
Tantalum	Ta	73	180·948	3014	5458	b.c.c.	0·3298		
Technetium (Masurium)	Tc (Ma)	43	[99]	2200	4877	c.p.h.	0·2735		0·4388
Tellurium	Te	52	127·61	450	988	hex.	0·44566		0·59268
Terbium	Tb	65	158·924	1357	3223	c.p.h.	0·36010		0·56936
Thallium	Tl	81	204·37	304	1473	c.p.h.	0·34566		0·55248
Thorium	Th	90	232·038	1755	4788	f.c.c.	0·50845		
Thulium	Tm	69	168·934	1545	1947	c.p.h.	0·35375		0·55546
Tin	Sn	50	118.69	232	2603	b.c.t.	0·58315		0·31814
Titanium	Ti	22	47·90	1670	3289	c.p.h.	0·29511		0·46843
Tungsten	W	74	183·85	3380	5555	b.c.c.	0·31652		
Uranium	U	92	238·03	1132	4134	orth.	0·28537	0·58695	0·49548
Vanadium	V	23	50·942	1902	3509	b.c.c.	0·30231		
Xenon	Xe	54	131·30	– 112	– 108	f.c.c.	0·624	(at – 185 °C)	
Ytterbium	Yb	70	173·04	824	1194	f.c.c.	0·54862		
Yttrium	Y	39	88·905	1526	3338	c.p.h.	0·36474		0·57306
Zinc	Zn	30	65·37	420	907	c.p.h.	0·26649		0·49468
Zirconium	Zr	40	91·22	1852	4409	c.p.h.	0·32312		0·51477

(4) Lattice constants are given for room temperature except where otherwise stated. Values for metals are taken from *Handbook of Lattice Spacings and Structure of Metals*, Vol. 2, by W. B. Pearson, Pergamon, 1967.

(5) Value at 36 atmospheres pressure. Arsenic sublimes at 610 °C at a pressure of one atmosphere.

(6) Graphite sublimes without melting at atmospheric pressure.

(7) At 30 atmospheres pressure. Helium does not solidify at one atmosphere pressure.

Table 2.2—The Periodic Table of the Elements (after Mendelejeff)

Period	Group 0	Group I A	Group I B	Group II A	Group II B	Group III A	Group III B	Group IV A	Group IV B	Group V A	Group V B	Group VI A	Group VI B	Group VII A	Group VII B	Group VIII
1		H														
2	He	Li		Be		B		C		N		O		F		
3	Ne	Na		Mg		Al		Si		P		S		Cl		
4	A	K	Cu	Ca	Zn	Sc	Ga	Ti	Ge	V	As	Cr	Se	Mn	Br	Fe Co Ni
5	Kr	Rb	Ag	Sr	Cd	Y	In	Zr	Sn	Nb	Sb	Mo	Te	Ma	I	Ru Rh Pd
6	Xe	Cs	Au	Ba	Hg	La and rare earths	Tl	Hf	Pb	Ta	Bi	W	Po	Re	—	Os Ir Pt
7	Rn	—		Ra		Ac		Th		Pa		U				

neutral inert gas, and the remaining elements show a gradual transition from the strongly electropositive alkali metals (lithium, sodium, etc.) to the strongly electronegative halogens (fluorine, chlorine, etc.).

In each group, the elements show a similarity in properties. The elements of each group in the fourth and subsequent periods are divided into two sub-groups, the elements of one sub-group more closely resembling the elements of that group in the second and third periods than do those of the other sub-group. For example, in Group I, potassium (K), rubidium (Rb), and caesium (Cs) have a greater resemblance to lithium (Li) and sodium (Na) than do copper (Cu), silver (Ag), and gold (Au). The properties of the elements in Group VIII are intermediate between those of Group VII A elements and Group I B elements. Thus iron (Fe), cobalt (Co), and nickel (Ni) show a gradual transition of properties between manganese and copper.

Lothar Meyer (1869) pointed out that many of the physical properties of the elements are also periodic functions of atomic weight. A curve of atomic volume (atomic weight/specific gravity) plotted against atomic weight (fig. 2.1) shows a periodicity, similar elements occupying similar positions on the curve. It will be seen from fig. 2.2 that melting-point is also a periodic function of atomic weight and gives a similar sort of curve. Some other physical properties that show a periodicity when plotted against atomic weight are specific gravity, hardness, thermal and electrical conductivities. One property that is not periodic is specific heat, which is considered more fully in Chapter 9.

2.2 Valency

The valency of an element is the number of atoms of hydrogen with which it will combine or which it will replace.

Valencies, in general, rise from 0 in Group 0 to 4 in Group IV and either fall again to 1 or increase to 7 in Group VII. Thus elements of Groups V, VI, and VII may exhibit two valencies which total to 8. Table 2.3 shows examples of compounds formed by representative elements of each group which illustrate these valency values.

TABLE 2.3

Group	0	I	II	III	IV	V	VI	VII
Valency	0	1	2	3	4	5 3	6 2	7 1
Element	He	Li	Be	B	C	N	S	Cl
Compound	—	LiH Li_2O	BeO	B_2O_3	CH_4 CO_2	NH_3 N_2O_5	H_2S SO_3	HCl Cl_2O_7

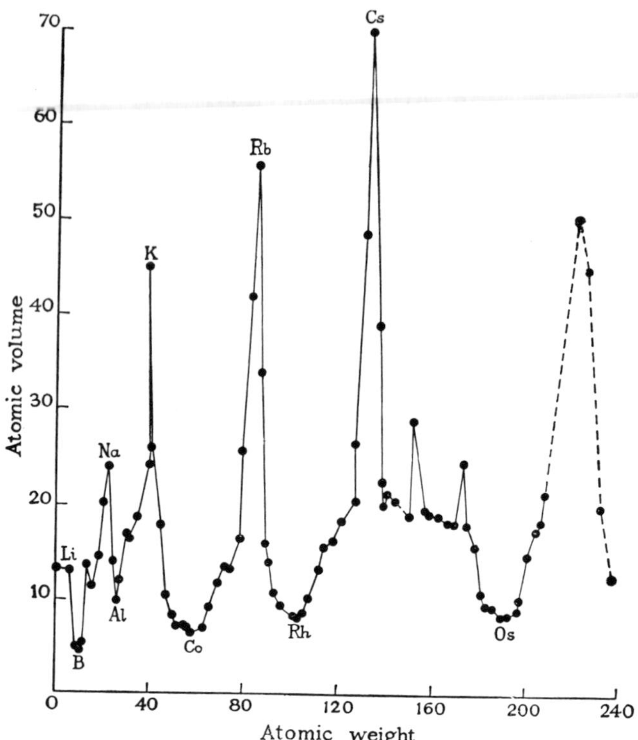

Fig. 2.1.—Variation of atomic volume (=atomic weight/specific gravity) of elements with atomic weight

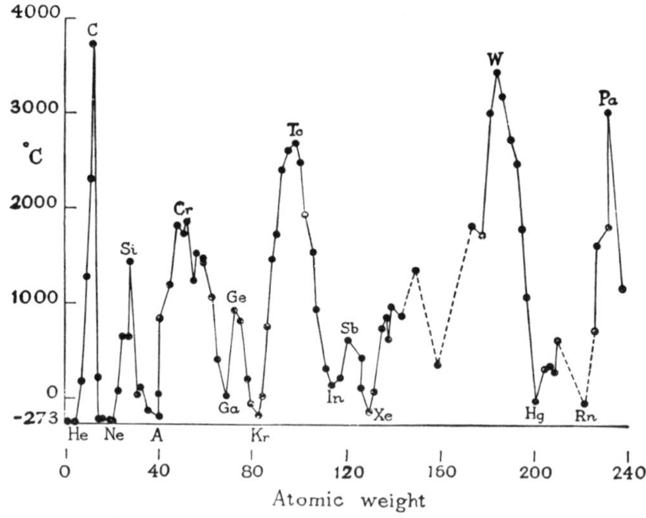

Fig. 2.2.—Variation of melting-points of the elements with atomic weight

When the periodic table was first put forward, it was necessary to leave gaps so that similar elements should fall in similar groups. It was possible to predict the properties of the elements that would fill these gaps, and the predictions were confirmed when the elements were actually discovered.

The similarities and dissimilarities of the elements must depend upon differences in the atoms of the various elements. An explanation is therefore dependent upon a prior knowledge of the structure of the atom.

2.3 Atomic number

The atom consists of a central positively-charged nucleus around which circulate a number of electrons, sufficient to give electrical neutrality to the atom. The nuclear charge is $+Ze$, where $-e$ is the charge on an electron and Z is an integer. Z is thus equal to the number of orbiting electrons and is known as the *atomic number*, each element having a different atomic number.

In general, the atomic weight increases as the atomic number increases. There are a few exceptions, for which it was found necessary in arranging the elements in the periodic table to reverse the order as given by the atomic weight to make the properties of the elements fit.

Thus the arrangement of elements in order of increasing atomic number also gives the periodic table, a more modern layout for which is shown in fig. 2.3. Each period is represented by a vertical line of elements, and elements in different periods with similar chemical and physical properties are linked by cross lines.

After the first short period comprising hydrogen (H) and helium (He) follow two periods of eight elements each. Each element in the third period falls alongside an element of the second period with very similar properties. In the fourth period, potassium (K) and calcium (Ca) bear a strong resemblance to sodium (Na) and magnesium (Mg) respectively, and then there are ten more elements before gallium (Ga), which resembles aluminium (Al), and five more elements corresponding to the last five elements of the third period. As was evident in Mendelejeff's periodic table, sodium also shows a weaker resemblance to copper (Cu), magnesium to zinc (Zn), etc. The weaker resemblances are shown by dashed lines in fig. 2.3. The ten elements, scandium (Sc) to zinc, are known as *transition elements.*

The fifth period is also a "long" period with all elements in it resembling the respective ones of the fourth period. The sixth period commences in a similar manner, but at the beginning of the transition elements fifteen *rare earth* elements, from lanthanum (La) to lutetium (Lu) appear. These resemble yttrium (Y) and each other so closely that they are placed in a single space of the table. After these rare earths or *lanthanides*, this period follows the previous one. The seventh period is incomplete, but as more elements of high atomic number are produced by nuclear reactions, it gradually grows. Fourteen trans-uranium elements with atomic numbers from 93 to 106 have been identified and all except 106 are shown in fig. 2.3.

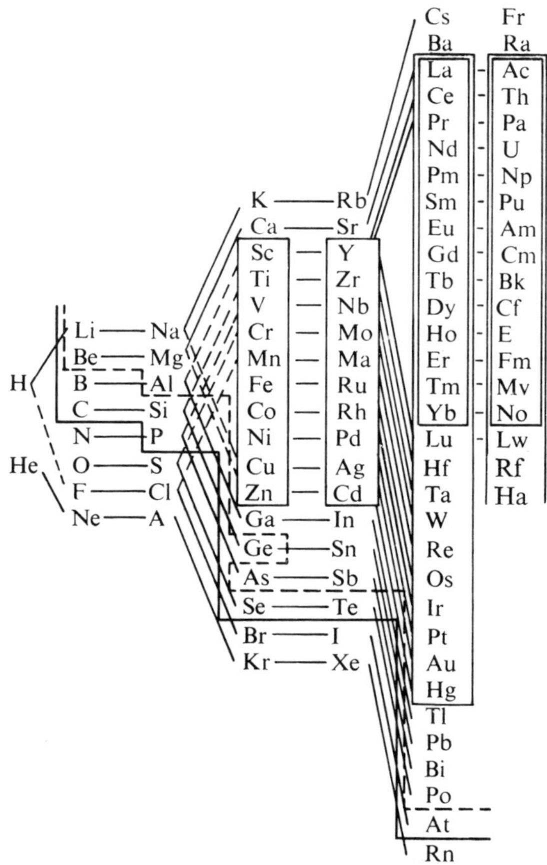

Fig. 2.3.—Periodic table. The bold dashed line separates metals from metalloids; the bold full line separates metalloids from non-metals

The elements following radium (Ra) belong to a group analogous to the rare earths and are known as the *actinides*.

It should be observed that the numbers in the periods are 2, 8, 8, 18, 18, 32, – –, which are 2×1^2, 2×2^2 twice, 2×3^2 twice, and 2×4^2.

Boundaries are marked on fig. 2.3 between metallic, metalloid, and non-metallic elements. Apart from the trans-uranium elements, the numbers in each group are:

metals	70
metalloids	8
non-metals	14

The picture of atomic structure will have to explain not only the cause of metallic properties, but also why there are so many metals.

2.4 The Bohr–Rutherford atom

The idea of an atom comprising a central positively-charged nucleus surrounded by negative charges was proposed by Rutherford in 1911, following the analysis of the results of experiments on the scattering of charged particles. The angular dependence of the scattering of α-particles (see p. 405) when beams of these particles were passed through different materials could be explained only by postulating this central nucleus whose diameter was of the order of 10^{-4} to 10^{-5} of that of an atom.

The negative charges surrounding the nucleus were identified with electrons which had been discovered by J. J. Thomson in 1897. The nucleus must exert an attraction on the electrons, so it is necessary to postulate that they would circulate in orbits in such a manner that the centrifugal force balances the electrostatic attraction. When electric charges move in an orbit, they emit electromagnetic waves, in the manner that oscillations of charge in an aerial can transmit wireless waves. Hence, on classical theory, the electron would be expected to emit electromagnetic radiation of the same frequency as the orbital frequency. This would involve loss of energy and an approach of the electron nearer the nucleus. This, in turn, would mean a change of frequency of the emitted radiation in a continuous manner and also the final collapse of the electron into the nucleus—neither of which is observed to happen. When atoms do emit radiation, it is in the form of fixed frequencies.

In 1913, Bohr proposed a quantum model for the atom, his assumptions being:

(1) The electrons exist only in stable circular orbits of fixed energy, the angular momentum of an electron in an orbit being an integral multiple of $h/2\pi$, where h is Planck's constant.

(2) An electron will emit or absorb energy only when making a transition from one to another possible orbit.

Some years earlier, Planck introduced the idea that light and all other forms of electromagnetic radiation possess energy, the smallest unit of which, a *quantum*, or *photon*, has energy $h\nu$, where ν is the frequency of the radiation, i.e., $\nu = c/\lambda$, where c is the velocity of light and λ is the wavelength. h is Planck's constant and has the value $6{\cdot}62 \times 10^{-34}$ J s.

From values of wavelengths observed in the spectra of the light emitted by hot gases, the values of energy of the photons can be calculated. Bohr's hypothesis was an attempt to find a basis for possible electron orbits, the energy differences between which would account for the observed spectral lines.

2.5 The single-electron system

Consider one electron of mass m moving with velocity v in a circular orbit of radius r around a fixed nucleus which has a charge $+Ze$ (fig. 2.4).

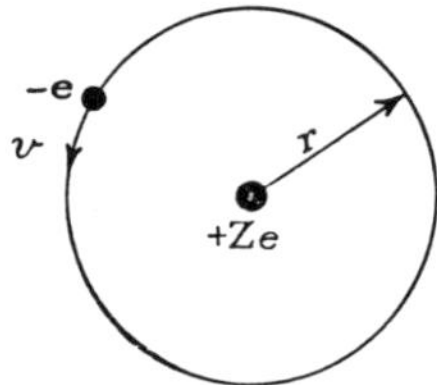

Fig. 2.4.—The Bohr-Rutherford atom with one electron

Assume that the Coulomb inverse-square law for attraction and repulsion between electrostatic charges is valid for atomic distances. Then the attractive force between the electron and the nucleus is

$$Ze^2/4\pi\varepsilon_0 r^2$$

where ε_0 is the permittivity of free space. Assuming that Newton's laws of motion apply, then the centrifugal force on the electron is mv^2/r. If the electron remains in its circular orbit, these forces are equal, that is

$$\frac{mv^2}{r} = \frac{Ze^2}{4\pi\varepsilon_0 r^2} \qquad (1)$$

or

$$v^2 = \frac{Ze^2}{4\pi\varepsilon_0 mr}$$

Thus for any value of r there is a particular value of v.

The energy of the electron is made up of two parts: its kinetic energy and its potential energy. The kinetic energy is

$$\text{K.E.} = \tfrac{1}{2}mv^2$$

The potential energy of the electron at any position is the work that has to be performed to move it to that position from a position at which the potential energy is taken to be zero. In this case, if the zero position is taken at infinity, the potential energy is the work done on the electron in moving it from infinity to a distance r from the nucleus, i.e.,

$$\text{P.E.} = -\frac{Ze^2}{4\pi\varepsilon_0 r} = -mv^2$$

Thus the total energy of the electron when in the orbit of radius r, relative to its energy when at rest at infinity, is

$$\begin{aligned}\text{K.E.}+\text{P.E.} &= \tfrac{1}{2}mv^2 - mv^2 \\ &= -\tfrac{1}{2}mv^2\end{aligned}$$

Introducing Bohr's condition that the angular momentum mvr can have certain values only,

$$mvr = \frac{nh}{2\pi} \qquad (2)$$

where n can take the integral values 1, 2, 3, etc. Combining equations (1) and (2),

$$v = \frac{Ze^2}{2nh\varepsilon_0}$$

and the total energy is

$$-\tfrac{1}{2}mv^2 = -\frac{Z^2me^4}{8n^2h^2\varepsilon_0^2}$$

By inserting the numerical values of the constants (Table 2.4), the total energy becomes

$$-2{\cdot}18\frac{Z^2}{n^2}\times 10^{-18}\ \text{J}$$

As n, called the *principal quantum number*, takes the values 1, 2, 3, 4, etc., the energy levels will be in the proportions 1, $\frac{1}{4}$, $\frac{1}{9}$, $\frac{1}{16}$, etc., as represented in fig. 2.5*a*.

TABLE 2.4—FUNDAMENTAL CONSTANTS AND CONVERSION CONSTANTS

Electronic charge	e	$1{\cdot}60202\times10^{-19}$ C
Rest mass of electron	m	$9{\cdot}1083\times10^{-31}$ kg
Rest mass of proton	M_p	$1{\cdot}67238\times10^{-27}$ kg
Rest mass of neutron	M_n	$1{\cdot}67470\times10^{-27}$ kg
Avogadro's number	N_0	$6{\cdot}0225\times10^{26}$ kmol^{-1}
Planck's constant	h	$6{\cdot}625\times10^{-34}$ J s
Boltzmann's constant	k	$1{\cdot}38048\times10^{-23}$ J K^{-1}
Velocity of light	c	$2{\cdot}99793\times10^{8}$ m s^{-1}
Permittivity of space	ε_0	$8{\cdot}854\times10^{-12}$ F m^{-1}
Unified atomic mass unit		$1{\cdot}6605\times10^{-27}$ kg
Electron volt		$1{\cdot}60202\times10^{-19}$ J

As an electron jumps from infinity or an outer orbit to an orbit nearer the nucleus, there will be a decrease in its total energy equal to the difference in its energies in the two orbits.

$$\delta E = 2{\cdot}18 \times 10^{-18} Z^2 \left(\frac{1}{n_1^2} - \frac{1}{n_2^2}\right) \text{J}$$

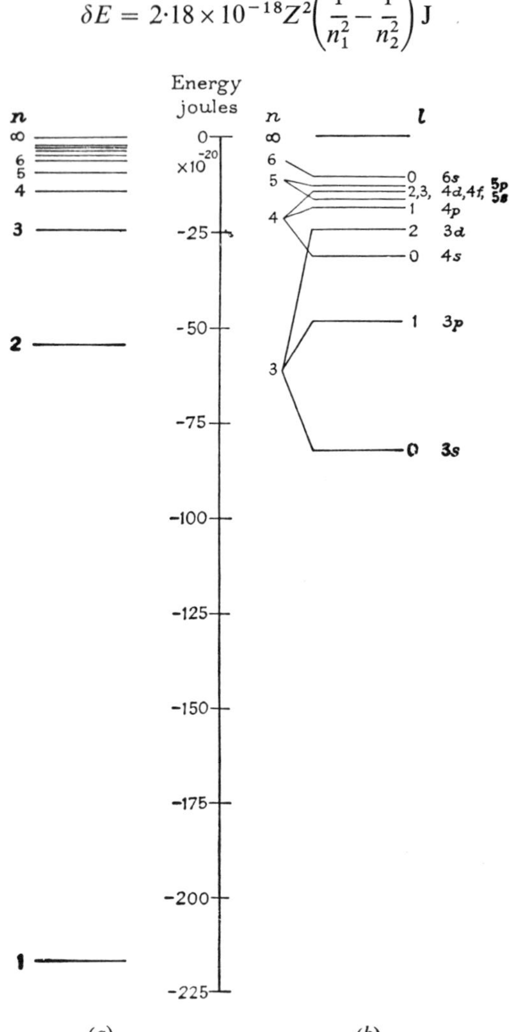

Fig. 2.5.

(*a*) Electron energy levels in hydrogen atom.

(*b*) Energy levels for valency electron in sodium atom. The energies of the $n = 1$ and $n = 2$ levels are out of the range of the diagram. The $4f$ level moves above the $5p$ level in heavier atoms

This energy is released as a quantum of electromagnetic radiation of frequency $\nu = \delta E/h$, or of wavelength $\lambda = c/\nu = ch/\delta E$.

It is possible to check the values of the wavelengths emitted in the spectra of atomic hydrogen ($Z = 1$) and ionized helium ($Z = 2$), (i.e., a helium nucleus with only one orbiting electron). If either of these is excited, i.e., the electron knocked out of the innermost possible orbit, as for example by moving electrons in a gas discharge tube, then an electron can return to the innermost orbit by a single jump or a series of jumps. The wavelengths of the emitted radiation (which are of such order of magnitude that they fall in the visible and near-visible portion of the spectrum) are found to confirm generally the above formula. Slight differences have led to refinements of the theory.

Firstly, Sommerfeld postulated elliptical as well as circular orbits, the possible orbits again being restricted. This involved the introduction of a second quantum number, commonly known by the symbol l.*

Then further work resulted in the introduction of two more quantum numbers, one m_l to allow for the restriction of possible directions of the axes of the electron orbits, and the other m_s to allow for possible values of the spin of the electron.

Bohr's assumption that the angular momentum of the electron should be an integral multiple of $h/2\pi$ gave the right results, but there seemed to be no logical basis for it. A new system of mechanics, called *quantum mechanics* or *wave mechanics*, was developed which dealt satisfactorily with the behaviour of electrons and which led naturally to the quantum numbers of the electrons around a nucleus.

2.6 Wave-particle duality

Classical mechanics assumes that at any definite instant of time, both the position and velocity of a particle can be determined simultaneously. While this is so for large particles, e.g., a golf ball, it is easily shown that it is far from true for small particles of the order of size of an electron. The more accurately one is determined, then the greater the error in any simultaneous determination of the other.

Thus the position of an electron may be determined by observing the reflection of a photon of electromagnetic radiation. A more precise determination requires a photon of shorter wavelength, but this would have a greater energy, and at the impact between photon and electron there would be a larger momentum change of the electron, making any determination of the original velocity less precise.

* Sommerfeld used a symbol k which is equal to $(l+1)$.

This is summed up in Heisenberg's *uncertainty principle*, which may be expressed mathematically as follows: the minimum uncertainties δx and δp in determining the simultaneous values of the position x and the momentum p of a body are related by

$$\delta x . \delta p = \frac{h}{2\pi}$$

or, alternatively, the minimum uncertainties δE and δt in the energy E and the time t for which it has that energy value are related by

$$\delta E . \delta t = \frac{h}{2\pi}$$

An electron in motion about an atom is in a stationary state, i.e., the energy remains constant for as long as the electron remains in that state, so that δt is large and δE will be very small. That is to say, electrons in stationary states would have sharply defined energy values.

Quantum mechanics deals with the probability of an electron being at any specified position at any time. From the curves of probability the observable properties of electrons in atoms and molecules can be calculated.

2.7. Wave mechanics

Wave mechanics deals with a wave function ψ which varies with position and whose meaning is that $|\psi|^2 \ \delta V$ is the probability of finding the electron in an element of volume δV. Alternatively, the value of $|\psi|^2$ at a point can be regarded as the average electrical charge density at that point. The value of ψ can be calculated from Schrödinger's wave equation, which is a standard equation of the kind that deals with the motion of waves in strings, etc.

De Broglie, in 1924, pointed out that just as electromagnetic radiation had both a wave and a particle nature, so might all matter in general. He derived an appropriate wavelength for particles of matter:

$$\lambda = h/p$$

where λ is the wavelength of the particle and p is its momentum. The wave nature of electrons has been demonstrated by electron diffraction phenomena.

2.8. Electron states in atoms

The various permitted stationary states of motion of an electron in an atom are represented by various patterns of the wave function, each pattern being a possible solution of the wave equation. The quantum numbers n, l, and m_l appear as possible roots of the equation.

The most conspicuous features of the patterns are the nodes, i.e., places where $\psi = 0$ or where the probability of finding an electron is nil. These nodes lie on certain surfaces in the atom which are called *nodal surfaces*. There will be a nodal surface at infinity, since ψ will be zero there. The total number of nodes for a given state of motion is given by n (= 1, 2, 3, etc.). In changing from one state of motion to another, the number of nodes must change by an integral number, so that the electrons can only take one or another of a fixed set of energy values.

The nodal surfaces are of two kinds: spherical surfaces concentric with the nucleus, and surfaces passing through the nucleus. The number of nodal surfaces passing through the nucleus is denoted by the second quantum number l, which can take any of the values 0, 1, 2, ..., $(n - 1)$. The possible orientations of these surfaces are given by the values of the third quantum number m_l, which can take any integral value from $-l$ to $+l$ including 0.

To define completely the state of an electron, values must be given for each of the four quantum numbers. n is the principal quantum number and is a measure of the energy of the particle. For any given value of n (which must be an integer) the other quantum numbers are restricted to certain values as follows:

Angular momentum quantum number

$$l = 0, 1, 2, \ldots, (n-1).$$

Magnetic quantum number $\quad m_l = -l, -(l-1), \ldots, -1, 0, 1, \ldots, l.$

Spin quantum number $\quad m_s = \pm\frac{1}{2}.$

Thus, for example, if $n = 1$, then $l = 0$, $m_l = 0$, $m_s = \pm\frac{1}{2}$, so that there are only two sets of quantum numbers for an electron of this energy level. If $n = 2$, then either

$$l = 0, \quad m_l = 0, \quad m_s = \pm\tfrac{1}{2},$$

or

$$l = 1, \quad m_l = -1, 0 \text{ or } +1, \quad m_s = \pm\tfrac{1}{2},$$

giving 8 sets of quantum numbers.

For $n = 3$ there will be 18, etc.; that is, for a particular value of n there are $2n^2$ possible sets of quantum numbers.

2.9. Many-electron atoms

According to Pauli's exclusion principle, which applies when more than one electron is present, not more than one electron may occupy a state described by any one set of values of the four quantum numbers. In the stable condition the electrons will occupy the states of lowest energy consistent with this principle.

Thus for helium, which has two electrons per atom, these electrons will be in the energy level corresponding to $n = 1$. The lithium atom has three electrons, of which two will occupy the $n = 1$ level, and the other one will go into the $n = 2$ level. Beryllium ($Z = 4$) will have two electrons in each of the first two levels, and so on. Each level of energy defined by a different value of n is termed a *shell.*

For the one-electron system, the energy of any possible electron state is governed exclusively by n. When more than one electron is present, the simple analysis previously given does not apply entirely, and the energy depends to a lesser extent upon l as well. Each shell thus has a series of sub-levels. Some of the sub-levels for the valency electron of a sodium atom are shown in fig. 2.5*b*. It will be seen that for the higher shells, where the difference in energy between successive values of n becomes less, the spread of the sub-levels is sufficient to give overlapping between the levels of different shells.

As it is often convenient to refer to the sub-level of a shell, these have been given the letters *s*, *p*, *d*, and *f*, corresponding to $l = 0, 1, 2,$ and 3, respectively. In any one shell, there could be two *s* electrons, each shell beyond the first can have six *p* electrons, etc. The letters may be preceded by a number denoting the value of n. Thus a boron atom ($Z = 5$) in the stable condition would have two 1*s* electrons, two 2*s* electrons, and one 2*p* electron.

Wave mechanics also leads to the result that, when all shells are full or contain 8 or 18 electrons, a completely symmetrical structure is formed which is extremely stable and will not react chemically.

Thus the inert gases have the electron configurations:

helium	2
neon	2, 8
argon	2, 8, 8, even though the $n = 3$ shell is not full
krypton	2, 8, 18, 8, etc.

2.10 Relationship of chemical behaviour to electron structure

The two 1*s* electrons of lithium are closely bound to the nucleus by the attraction of the $+3e$ charge, but the single 2*s* electron is much farther away from the nucleus and partly screened from its attraction by the negative charges of the two 1*s* electrons. Consequently the lithium atom readily loses this outer electron forming a positively charged *ion* (written as Li^+). This ion has the same electron structure as the inert gas helium and is stable, though unlike helium it has a net positive charge.

Fluorine ($Z = 9$) has two electrons in the first shell and seven in the second shell (two 2*s* and five 2*p* electrons). It is short of one electron to form the stable structure of the neon atom, which has eight electrons in the $n = 2$ shell, and will readily absorb any available electron to give this stable structure. Thus it acquires a resultant negative charge and so becomes a negative ion (F^-).

Hydrogen may either lose an electron to become H^+ or take one up to form H^-.

In the third period of the periodic table, sodium ($Z = 11$) and chlorine ($Z = 17$) are one electron over and one electron short, respectively, of stable structures, and so behave like lithium and fluorine. Elements near to lithium and sodium in these periods, which have two or three electrons in their outer shells, also lose electrons to form stable positive ions, which carry two or three units of charge, respectively. Elements near fluorine and chlorine similarly tend to form negative ions.

It will be seen from fig. 2.5*b* that the 4*s* sub-level has less energy than the 3*d* sub-level. The two elements of atomic number just above argon will thus have electron structures

potassium	2, 8, 8, 1
calcium	2, 8, 8, 2

Since chemical properties are dictated mainly by the number of electrons in the outermost shell, these will be similar to sodium and magnesium. The next ten elements in increasing order of atomic number will be formed by the filling of the 3*d* level with two electrons in the 4*s* shell in each case.* These ten elements are the transition elements to which reference was made on p. 9. Following zinc, the period is completed by the filling of the 4*p* level for the next six elements up to the inert gas krypton.

The next period follows a similar pattern, but the sixth period has added complications. In building up successive elements, the first two electrons go into the 6*s* level, then one goes into the 5*d* level, after which the as yet vacant 4*f* level receives electrons. These electrons are unable to exert much influence on the chemical behaviour, being screened by the

* This is not quite true, there being two exceptions—chromium and copper which have only one 4*s* electron.

$n = 5$ shell. Thus these elements from lanthanum to lutetium (known as the rare-earth metals or lanthanides) are so similar chemically that they are put into a single space in the periodic table. After lutetium, the period completes in the same manner as the previous period.

The seventh period, which is incomplete, appears to follow the same pattern as the sixth.

It will be noticed that the majority of elements (which includes the transition elements) have only a few electrons in the outermost shell. These are the elements that form positive ions, a property typical of metals.

2.11 X-ray spectra

Energies of a high order are required to remove the electrons in the inner shells of the heavier atoms from their orbits—energies such as would be acquired by an electron in traversing a potential difference of a few kilovolts. When an electron is removed in such a manner, the vacancy is filled by another electron jumping in from an outer orbit with a consequent evolution of energy. This is given out as a quantum of electromagnetic radiation, the wavelength being of the order of 1/1000 of that of visible light. This intensely energetic radiation is known as X-rays.

Each element has its characteristic X-ray spectrum, the lines of which are denoted partially by a letter indicating the shell into which the electron jumps. The letters K, L, M, N, O, P, and Q refer to the shells for which $n = 1, 2, 3 \ldots, 7$, respectively.

2.12. The bonding between atoms

Some of the physical properties, such as melting point and boiling point, and to a large extent the mechanical properties depend upon the strength and nature of the bonding between atoms.

The various types of bonds fall into two groups—primary and secondary bonds, the former being distinguished as considerably stronger. Primary bonds are described in the first four and secondary bonds in the last of the following sections.

1. *The electrovalent bond*

It has been shown that, in forming ions, atoms favour structures which have a stable outer shell of electrons. This ion formation is simply the losing or gaining per atom of a number of electrons, and the number will be equal to the valency. Thus sodium and chlorine form monovalent ions by reactions which may be written as:

$$\mathrm{Na} \rightarrow \mathrm{Na}^{+} + e^{-}$$

$$\mathrm{Cl} + e^{-} \rightarrow \mathrm{Cl}^{-}$$

where e^- represents an electron. The ions can combine by the attractive force between unlike charges, giving an *electrovalent* or *ionic* bond.

$$Na^+ + Cl^- \rightarrow Na^+Cl^-$$

An alternative form of presentation of these reactions is to show the electron structure of the outermost electron shell of each atom as follows:

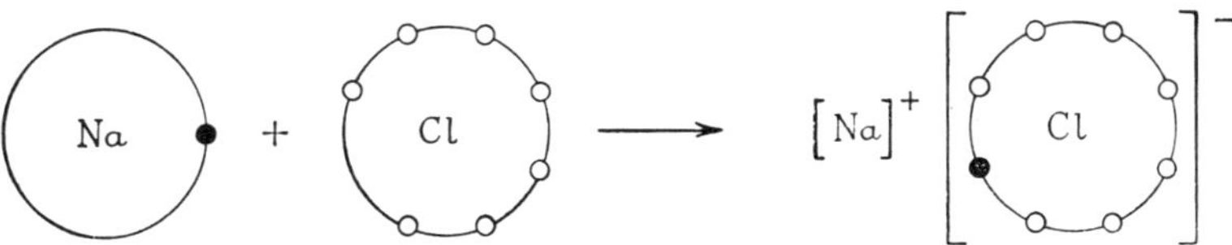

In calcium chloride, the $+2e$ charge on the divalent calcium ion is balanced by the charges on two monovalent chlorine ions:

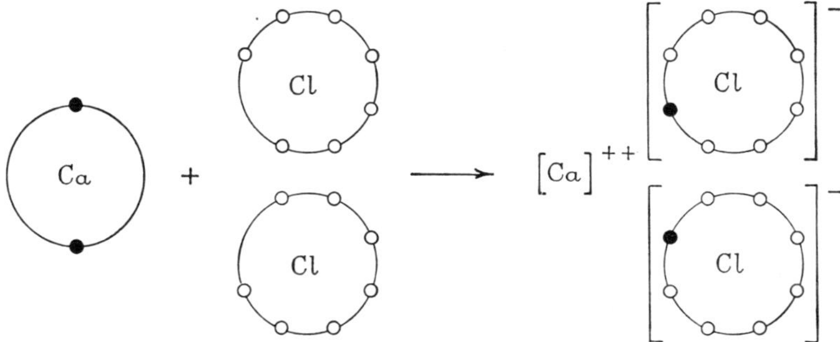

2. *The covalent bond*

Another way in which atoms can form the inert gas structure is exhibited by a form of bonding between atoms which have nearly eight electrons in their outer shells. Electrons are shared, so that some electrons orbit around both nuclei. An example is found in the chlorine molecule, where two electrons, one from each atom, are shared. The number of atoms in the outer shell of each atom is raised to the stable number of eight. Covalent bonds tend to be very stable.

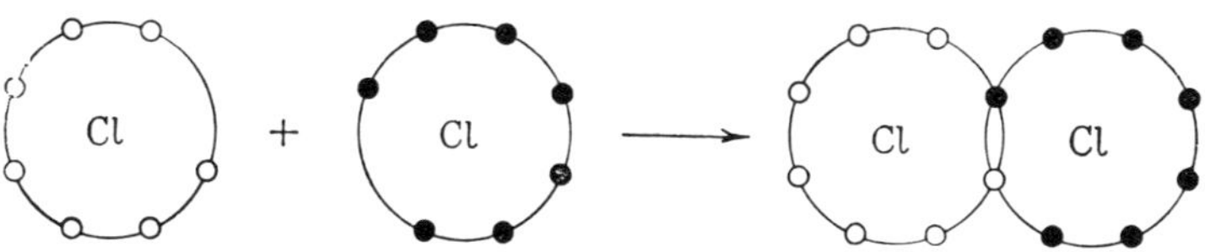

This form of bonding is favoured by hydrogen and by atoms which have four or more electrons in their outer shells. As many bonds are formed as will bring the number of electrons up to a stable, inert-gas structure, which in the case of hydrogen is only two.

Elements lying between the alkali metals (Li, Na, etc.) at one end of the periodic table and the halogens (F, Cl, etc.) at the other end are less definite in character than these extreme groups. The tendency is for the formation of covalent rather than ionic bonds.

In general, atoms forming positive ions are seldom able to lose more than three electrons, or those forming negative ions to gain more than two. This may be easily understood; for example, although aluminium loses three electrons to form Al^{+++}, this tendency is strongly opposed by the attraction of the triply charged ion for the electrons.

Fully ionic and fully covalent bonding are completely different, but even so only represent extremes, since many bonds are partly ionic and partly covalent in character. An example is found in the compound between hydrogen and chlorine. These elements each form diatomic molecules with covalent bonding. In the HCl molecule, the bonding is covalent, but the electrons are concentrated more around the chlorine atom than the hydrogen atom.

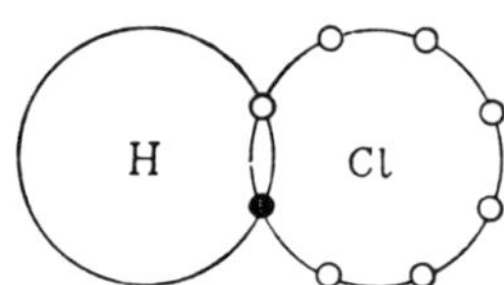

This means that the chlorine atom has a resultant small negative charge and the hydrogen atom has a small positive charge. This constitutes an electric dipole—the electrical equivalent of a bar magnet.

+  −

When dissolved in water, the bonding of most of the hydrogen chloride molecules changes to ionic, dissociation to H^+ and Cl^- ions occurring.

Whereas the electrostatic forces exerted by an atom in ionic bonds can act equally in any direction, the covalent bonds of an atom that combines with two or more other atoms tend to have a definite spatial relationship to one another. An example is found in the water molecule where two hydrogen atoms are bonded to an oxygen atom. Wave mechanical treatment shows that the two covalent bonds would act in directions which subtend an angle of 90° at the centre of the oxygen atom. The repulsion of the hydrogen atoms, however, causes the angle to be somewhat greater. Due to asymmetry, the molecule has a permanent dipole moment.

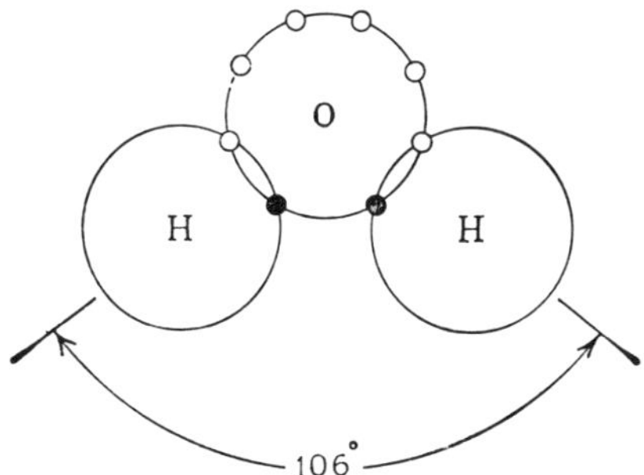

Further examples of the spatial arrangements of covalent bonds will be given in Chapters 20 and 21.

3. *The coordinate bond*

The coordinate bond is a covalent bond in which both of the shared electrons are provided by one of the atoms. An example is found in the bonding of oxygen to sulphur in the sulphate ion. The six electrons in the outer shell of the sulphur atom are shared in pairs with oxygen atoms, so that each oxygen atom acquires an outer shell of eight electrons. Two more electrons gained from metal or hydrogen atoms provide the bond for the fourth oxygen atom.

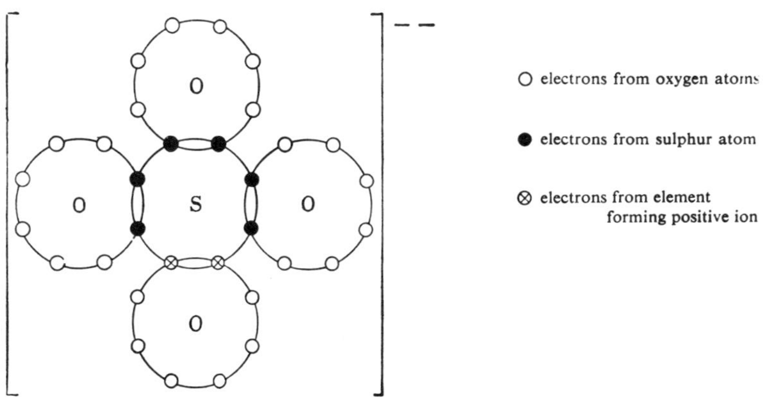

4. *The metallic bond*

Metals obviously cannot form covalent bonds, as in general each atom has one or two electrons to lose.

Even when uncombined with other elements, metals are ionized. A metal may be described as a *cloud* or *gas* of free electrons in which the positive ions are embedded at the lattice positions, and the mutual repulsion due to their positive charges is balanced by the negative charge of the electron cloud which holds them together. This picture of a metal is called the *free-electron theory* and gives an explanation of the electrical conductivity of metals, as the electrons would be free to move. The thermal conductivity and optical effects can also be explained by it. The free-electron picture is highly simplified and not entirely correct.

By consideration in terms of wave mechanics, the electron density is obtained from the solution of the wave equation for the case where the potential is a function which repeats periodically in three dimensions within the boundaries of the crystal, the potential reaching a maximum at each position occupied by a positive ion. The results show that there is a series of possible energy levels for these valency electrons. In accordance with Pauli's exclusion principle, each level can be occupied by only two electrons (these having opposite spins), and in the stable state those levels of lowest energy will be occupied. The energy distribution that results from this theory can explain several features, such as semiconductors, which the free-electron theory does not.

5. *Intermolecular forces*

Molecules are formed when all the atoms in a group are held together by covalent bonds and all the valencies are saturated. Atoms of the inert gases do not combine and so for these substances, single atoms may be considered to be molecules. At sufficiently low temperatures molecular compounds exist as liquids and at still lower temperatures as solids, so that intermolecular forces must exist. However, these forces must be considerably smaller than the interatomic forces within the molecules.

These attractive forces, which are known as van der Waals forces, are due to the forces between the electrical dipoles of the two molecules. Some molecules, like water, have permanent dipole moments and so will attract one another when the positive end of one is directed towards the negative end of the other. This is known as the *hydrogen bond* or *hydrogen bridge*; the hydrogen atom in addition to being covalently bonded to an electronegative atom within the molecule also has this electrostatic attraction, weaker than an ordinary ionic bond, to the electronegative atom of another molecule.

Other molecules, which have no permanent dipole moments may yet have temporary ones. The electrons are constantly in motion and there will be times when their distribution is not symmetrical. The dipole moment due to one molecule causes a surrounding electric field which will tend to induce a dipole moment in any other molecule which lies within the field. The magnitude of the induced dipole will be proportional to the field, which itself is, apart from geometrical factors, proportional to the inducing dipole moment. The sign of the induced dipole is such that there will be an attractive force between the two molecules. Hence over a period of time, the average attractive force is proportional to the mean-square dipole moment.

In general, the larger the number of electrons in a molecule, the larger is the dipole moment that can be formed. As an example, with increase of atomic number in the inert gases, corresponding to more electrons per atom, the higher are the melting and boiling points, which implies a stronger bond.

QUESTIONS

1. What are the valencies of the metal atoms in each of the following compounds: $MgCl_2$, $SnCl_4$, SnO, P_2O_3, PCl_5, SbH_3, $CrSO_4$, $Cr_2(SO_4)_3$, CrO_3, Mn_2O_7?

2. Describe the electron structure of the light atoms up to that of atomic number 19. Discuss the formation of ionic and covalent bonding between atoms in terms of the electron structure. [MST]

3. Describe the atomic structure of helium, aluminium, and bromine. What types of chemical bond can each form? [MST]

4. Derive an expression for the energy levels of a hydrogen atom according to the Bohr–Rutherford theory. Hence, calculate the wavelengths of the spectral lines corresponding to electrons jumping from the $n = 2$ to $n = 1$ and $n = 4$ to $n = 3$ levels.

5. Describe the arrangement of electrons in atoms of the alkali metals and the nature of the bond between atoms in crystals of these metals.

How are some of the physical properties typical of metals explained in terms of his structure? [MST]

6. A diffraction experiment shows that a certain beam of electrons exhibits a wavelength of 0·4 nm. Calculate the velocity of the elctrons and their energy in electron-volts.

7. Explain with examples what is meant by the wave-particle duality of matter and of electromagnetic radiation.

A beam of light falling on a metal surface produces photoelectrons with energies up to a maximum of 1·30 eV. If the threshold frequency for photoemission is $6{\cdot}03 \times 10^{14}$ s^{-1}, what is the smallest wavelength present in the incident radiation? What is the smallest wavelength associated with the photoelectrons? [E]

CHAPTER 3

Aggregations of Atoms—The Fluid States

3.1. Introduction

In general, the engineer does not need to consider atoms as individuals, but is concerned with the properties or behaviour of assemblages of atoms in one or more of the three states of aggregation: gas, liquid, and solid. In gases, the spacing of the atoms or molecules is large (except at very high pressures), but in liquids and solids each atom is in close proximity to its neighbours. The state of aggregation in which a particular group of atoms or molecules exists depends on the nature of the attractive forces present and may vary with changes in temperature and pressure. Under any particular set of conditions, the state or states will be such that the energy of the system is a minimum.

GASES

3.2. The behaviour of a gas

The more obvious qualitative features of gas behaviour are that it is homogeneous, that it has a large compressibility, that gases diffuse through one another, and that they never settle.

Quantitatively, it was found that at low pressures gases obeyed the laws of Boyle and Charles, which may be summarized by the equation

$$pV = R_m T$$

where p and V are the pressure and volume of a mass m of the gas at an absolute temperature T, and R_m is a constant appropriate to the mass m.

Clearly R_m is proportional to m, since at the same pressure and temperature V will be proportional to m. Also, Avogadro's hypothesis, which was deduced originally from chemical evidence and since verified in other ways, states that at the same temperature and pressure equal volumes of all gases contain the same number of molecules. This hypothesis is accurate to the extent to which gases obey the simple gas laws.

Gases deviate in practice from these laws, but at sufficiently low temperature and pressure the deviation is negligible. A gas in such a condition that it obeys these laws is known as an *ideal gas*.

3.3 Mole

The definition of the *mole* is given on p. 424. If M is the molecular weight of a substance, then a kilogramme mole (kmol) has a mass of M kg.

Also, it follows that one mole of any substance contains the same number of molecules. This number is Avogadro's number and its value is

$$N_0 = 6{\cdot}0225 \times 10^{26}\ \text{kmol}^{-1}$$

As a consequence of Avogadro's hypothesis, 1 mole of any ideal gas at a given temperature and pressure always occupies the same volume. At *standard temperature and pressure* (S.T.P.), i.e. 0 °C and a pressure of 1 standard atmosphere, the volume of a kilogramme mole is 22·4 m^3.

Thus 1 m^3 at S.T.P. will contain

$$6{\cdot}0225 \times 10^{26}/22{\cdot}4 = 2{\cdot}69 \times 10^{25}\ \text{molecules.}$$

Hence if m is taken as the kilogramme mole of any gas, R_m will always have the same value, which is

$$R = 8314\ \text{J kmol}^{-1}\ \text{K}^{-1}$$

Although the value of the *gas constant* R is of great importance, it has no fundamental significance, since the size of the kilogramme mole is an arbitrary choice depending upon the standard unit of mass and the convention that the mass of the common isotope of carbon be taken as 12.

Hence it is often more convenient to use a more fundamental constant: viz.

$$\frac{R}{N_0} = k = \text{Boltzmann's constant}$$

$$= 1{\cdot}38 \times 10^{-23}\ \text{J K}^{-1}.$$

This may be regarded as the gas constant for a single molecule.

3.4. The kinetic theory of gases

Any theoretical model of a gas must explain the qualitative features listed above, and obey the laws quoted.

We suppose that a gas consists of molecules, which in any one gas are all alike—hence the homogeneity of a gas. The actual volume occupied by the molecules is very small compared with the total volume of the gas—explaining the high compressibility. Since diffusion can occur, the molecules must be in motion. Also we assume that the molecules exert no force on each other except when actually in contact, and that the collisions between molecules and of molecules with the container walls are perfectly elastic. If this were not so, there would be a loss of energy at each collision and the molecules, losing velocity, would gradually settle.

The pressure exerted by a gas is due to the collisions of the molecules with the walls of the container.

3.5. Calculation of the pressure of an ideal gas

Consider a closed cubical vessel of side d containing n molecules of gas, each of mass m.

Let a particular molecule have a velocity c_1, the components of which in the three directions perpendicular to the cube faces are x_1, y_1, z_1. Then

$$c_1^2 = x_1^2 + y_1^2 + z_1^2$$

Consider the two faces perpendicular to the x-direction. If it were not for collisions with other molecules, the particular molecule would traverse the cube x_1/d times per unit time and hence make $x_1/2d$ impacts on each face per unit time. Also, at each impact of the molecule, its momentum perpendicular to the face would change from mx_1 to $-mx_1$. Then the rate of change of momentum at each face would be

$$2mx_1 \cdot \frac{x_1}{2d} = \frac{mx_1^2}{d}$$

This would be the force exerted on each face perpendicular to the x-direction by a single molecule. Owing to collisions, one molecule will not, in general, traverse the cube without changes in velocity. Momentum in each direction will, however, be conserved and the total change of momentum at a face will be unaltered whether there are inter-molecular collisions or not.

Let $\overline{C^2}$ be the *mean-square velocity* such that

$$c_1^2 + c_2^2 + c_3^2 + - - = n\overline{C^2}$$

Then the total force on all sides of the cube is

$$\frac{2nm\overline{C^2}}{d}$$

The area of each face is d^2, and assuming an equal force on each face, the pressure is

$$p = \frac{2nm\overline{C^2}}{d} \cdot \frac{1}{6d^2}$$

$$= \frac{nm\overline{C^2}}{3d^3}$$

But d^3 is the volume V occupied by the gas.
Therefore

$$pV = \tfrac{1}{3}nm\overline{C^2}$$

$$= \tfrac{2}{3} \times \tfrac{1}{2}nm\overline{C^2}$$

$= \tfrac{2}{3} \times$ kinetic energy of translational motion of all the molecules in the gas.

If the kinetic energy is proportional to absolute temperature, i.e.

$$\tfrac{2}{3} \times \tfrac{1}{2}nm\overline{C^2} \propto T \quad \text{or} \quad \tfrac{2}{3} \times \tfrac{1}{2}nm\overline{C^2} = RT$$

where R is a constant, then

$$pV = RT$$

Hence the kinetic theory so far developed explains the laws of an ideal gas.

If we have equal volumes of two gases at the same temperature and pressure, then

$$\tfrac{2}{3} \times \tfrac{1}{2}n_1m_1\overline{C_1^2} = \tfrac{2}{3} \times \tfrac{1}{2}n_2m_2\overline{C_2^2}$$

where n_1 and n_2 are the numbers of molecules, and m_1 and m_2 are their masses for the two gases respectively.

Then if we assume that the average kinetic energies of the molecules of these gases are the same at the same temperature,

$$\tfrac{1}{2}m_1\overline{C_1^2} = \tfrac{1}{2}m_2\overline{C_2^2}$$

Hence $n_1 = n_2$, or equal volumes of two gases at the same temperature and pressure contain equal numbers of molecules, which is Avogadro's hypothesis.

3.6. Energy and velocity of molecules

The molecules in a volume of gas will not all have the same velocity. Even if this were possible at any instant, the molecules would immediately get different velocities owing to mutual collisions. Hence $\overline{C^2}$ really does represent an average value of C^2.

Consider 1 kmol of gas at S.T.P. The pressure is 0·101325 MPa and the volume is 22·4 m³, so that the kinetic energy of the gas molecules will be

$$\begin{aligned}\tfrac{1}{2}nm\overline{C^2} &= \tfrac{3}{2}pV \\ &= \tfrac{3}{2}\times 0{\cdot}1013\times 10^6\times 22{\cdot}4\ \text{J kmol}^{-1} \\ &= 3{\cdot}40\times 10^6\ \text{J kmol}^{-1}\end{aligned}$$

This is called the molecular energy of translation of a gas at S.T.P. and from it we can calculate the root-mean-square speed of the molecules. Now for a kmol, $nm = M$ kg, so that

$$\overline{C^2} = \frac{2\times 3{\cdot}4\times 10^6}{M}\ \text{J kg}^{-1}$$

For hydrogen, $M = 2$ kg, giving

$$\begin{aligned}\sqrt{(\overline{C^2})} &= \sqrt{(3{\cdot}4\times 10^6)} \\ &= 1840\ \text{m s}^{-1}\ \text{at } 0\ ^\circ\text{C}\end{aligned}$$

For carbon dioxide, $M = 44$ kg, giving

$$\begin{aligned}\sqrt{(\overline{C^2})} &= \sqrt{\left(\frac{2\times 3{\cdot}4\times 10^6}{44}\right)} \\ &= 393\ \text{m s}^{-1}\ \text{at } 0\ ^\circ\text{C}\end{aligned}$$

Now $\overline{C^2}$ is proportional to absolute temperature. Therefore at 400 K for carbon dioxide

$$\begin{aligned}\sqrt{(\overline{C^2})} &= 393\times\sqrt{(400/273)}\ \text{m s}^{-1} \\ &= 476\ \text{m s}^{-1}\end{aligned}$$

3.7. Maxwellian distribution of velocities

The results of the foregoing theory give the value of $\overline{C^2}$, but do not give any information about the *distribution* of velocities among the molecules. The form of the distribution curve, which will not be derived here, is given by *Maxwell's distribution law.* Of the total number of molecules, the fraction which have velocities lying between c and $(c+\delta c)$ is proportional to

$$\frac{1}{T^{3/2}}\,c^2\,e^{-mc^2/2kT}\,\delta c$$

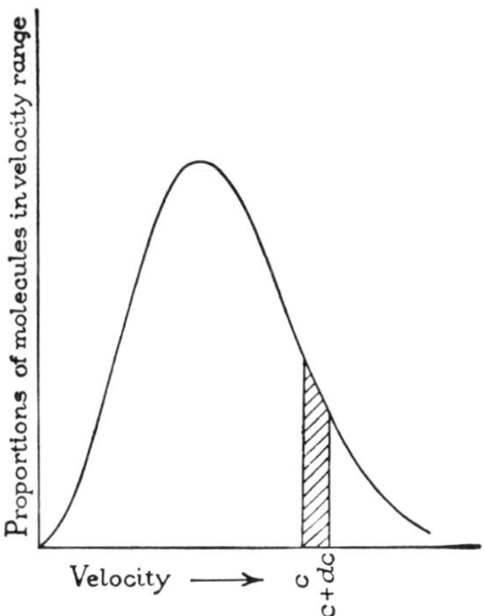

Fig. 3.1.—Distribution of velocities of molecules of a gas

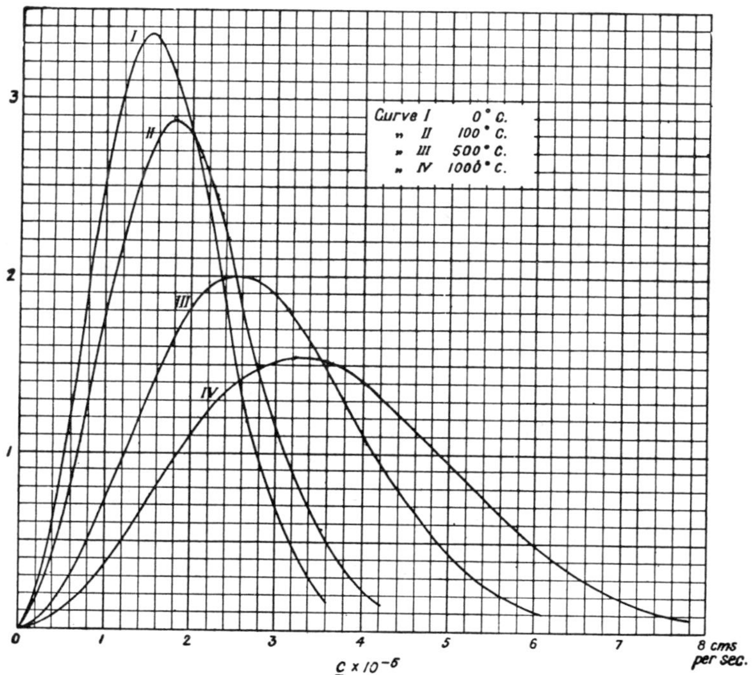

Fig. 3.2.—Effect of temperature on velocity distribution of hydrogen gas molecules [Roberts]

The curve of the distribution of velocities has the shape shown in fig. 3.1. The shaded area is proportional to the above expression. An increase of temperature alters the shape of the curve as may be seen from the distribution curves for hydrogen in fig. 3.2.

The mean velocity is found to bear a definite relation to the root-mean-square velocity.

$$\bar{c} = 0{\cdot}921\sqrt{(\overline{C^2})}$$

The most probable velocity α, i.e., the value of the velocity at the peak of the distribution curve, is given by

$$\alpha = 0{\cdot}816\sqrt{(\overline{C^2})}$$

3.8. Mean free path

The *mean free path* is the average distance that a molecule travels between successive collisions with other molecules.

A simple calculation of its value can be made by assuming all molecules but one to be at rest. Let

d = diameter of each molecule

n = number of molecules per unit volme.

The moving molecule will collide with any other molecule whose centre lies within a distance d of the path of the centre of the moving molecule. In travelling a distance l, it will collide with all molecules whose centres lie in a cylinder of length l and radius d (fig. 3.3). The number of molecules contained in the volume $\pi d^2 l$ is $\pi d^2 ln$, which is also the number of collisions.

Hence the mean free path λ is given by

$$\lambda = \frac{l}{\pi d^2 ln} = \frac{1}{\pi d^2 n}$$

If account is taken of the motion of all the molecules, and the velocities are assumed to be distributed according to Maxwell's distribution law, the mean free path is found to be

$$\lambda = \frac{1}{\sqrt{2}\pi d^2 n}$$

The term *diameter* as applied to a gas molecule does not have any precise significance, so that it is not possible to calculate λ from the above

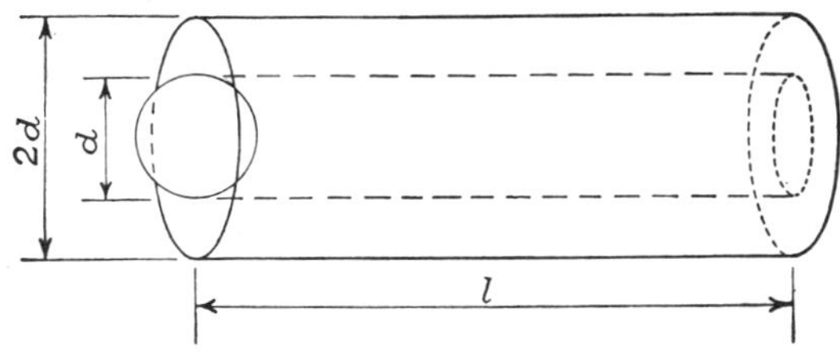

Fig. 3.3.

expression, but the mean free path is obviously of importance in connection with any property of a gas which involves transfer. An example is *thermal conductivity*. A simple approximate calculation leads to the expression

$$K = \tfrac{1}{3}nm\bar{c}\lambda C_v$$

where K = thermal conductivity and C_v = specific heat at constant volume (see p. 129). As in the previous calculation, a more refined treatment merely modifies the numerical constant.

K can be measured experimentally and hence λ and d calculated. Some approximate values at 15 °C and 1 atmosphere pressure are:

	λ	d
H_2	$16{\cdot}3 \times 10^{-8}$ m	$2{\cdot}3 \times 10^{-10}$ m
N_2	8·5	3·15
O_2	9·6	2·96

3.9. Variation of λ and K with pressure

Since $\lambda = 1/\sqrt{2}\pi d^2 n$ and because n is proportional to pressure at a given temperature,

$$\lambda \propto \frac{1}{p}$$

Therefore at low pressures λ may be quite large.

For example, in a radio valve, where the pressure is of the order of 10^{-9} atmosphere, the mean free path will be about 100 m. This is much greater than the dimensions of the valve, so that in general a molecule traversing the valve will not collide with any other molecules. (Even at this low pressure, there are still $2{\cdot}7 \times 10^{16}$ molecules m^{-3}.)

Since $K = \frac{1}{3}nm\bar{c}\lambda C_v$, and because n is proportional to pressure and λ to the reciprocal of pressure, K will be independent of pressure. This is found to be true at fairly high pressures, but not at low ones. Obviously the above expression is no longer valid when the mean free path becomes comparable with the size of the containing vessel.

3.10. Deviations from the gas laws

It was stated on p. 26 that at low pressures gases obeyed the laws summarized in the equation

$$pV = RT$$

At ordinary and high pressures gases depart from these laws to a measurable extent, the amount of departure varying from gas to gas.

Hydrogen, oxygen, and several others show extremely small departure at ordinary temperatures, while carbon dioxide shows considerable departure at ordinary temperatures and high pressures. With dihydrogen oxide (H_2O) the departure is considerable at ordinary temperatures and pressures.

The most famous experiments on the deviation from the ideal gas laws are those made by Andrews. He studied the behaviour of carbon dioxide using pressures up to 14 MPa and obtained the results shown in fig. 3.4. Each curve is an *isothermal*, showing the relationship between pressure and volume of a given mass of gas at a constant temperature.

Considering the isothermal for 21·5 °C as the pressure is increased from a low value, the volume decreases until the point q on the curve is

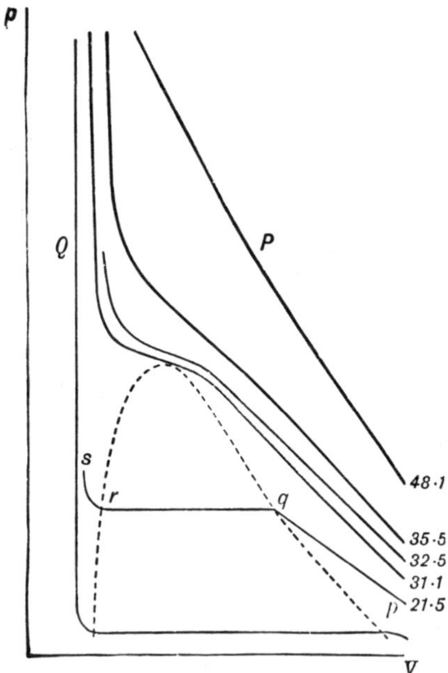

Fig. 3.4.—Andrews' isothermals for carbon dioxide [Roberts]

reached. On further compression, the pressure remains constant while the gas changes to liquid. At point r, the carbon dioxide is all liquid, and beyond this point the gradient is very steep, a large change in pressure producing only a small change in volume.

Liquefaction took place in a similar manner for all temperatures up to 31·1 °C, but at higher temperatures no liquefaction occurred, however much the pressure was increased. 31·1 °C is called the *critical temperature* for carbon dioxide, and the isothermal for this temperature is the critical isothermal.

Two dotted curves are shown, one drawn through all points such as q which represent the volume of gas when in equilibrium with liquid, and the other through points such as r which represent the volume of liquid when in equilibrium with gas. These conditions are known as saturated vapour and saturated liquid, respectively. These two curves meet at the critical isothermal at a point where the specific volumes of gas and liquid are equal. The pressure and specific volume at this point are known as the *critical pressure* and *critical volume*.

At P, the substance would be in the gaseous state, while at Q it would be in the liquid state. P and Q are single-phase regions, while between the two we have a two-phase region. (See p. 80 for the definition of a phase.) It is possible to pass from P to Q without any discontinuity occurring, by using a series of processes which do not pass through the two-phase region. This is referred to as the *continuity of the liquid and gaseous states*.

All gases show behaviour of a similar type, but the values of the critical temperatures, pressures, and volumes differ widely.

3.11. Van der Waals' equation

In deriving the ideal gas laws on the basis of the kinetic theory, two factors were neglected. These were the attractive forces between the molecules and the volume of the molecules. These factors were considered by van der Waals, who made the first attempt to modify the simple kinetic theory.

If an approximate value for the molecular " diameter " is taken as $3{\cdot}33 \times 10^{-10}$ m (see p. 75), then at least $2{\cdot}7 \times 10^{28}$ molecules could be packed into one cubic metre. At S.T.P. the actual number is $2{\cdot}7 \times 10^{25}$. Hence the molecules occupy only about 1/1000 of the total space at S.T.P. But when the gas is highly compressed, the volume of the molecules will be an appreciable part of the whole. A molecule thus has less far to go between collisions than would be apparent on simple theory, and hence will make more frequent collisions, giving a higher pressure. The effect is the same as if the molecules were of negligible size but contained in a smaller volume.

The equation is written as

$$p(V-b) = RT$$

where b is the correction for the volume of the molecules (but not equal to it).

The mutual attraction of the molecules has the effect of decreasing the volume, that is, it has the same effect as an external pressure would have on the molecules between which there is no mutual attraction. The pressure term of the simple theory therefore needs an additional term to allow for the effect of this *internal pressure.* The force on a single molecule depends upon the number of molecules within a distance such that the attractive force due to any of them is appreciable. Also the internal pressure depends upon the number of molecules in a given volume that are subject to the attractive forces. Each of these is proportional to the density, that is, to the reciprocal of the specific volume. Therefore the internal pressure term is proportional to $1/V^2$, or equal to a/V^2 where a is a constant. The equation of van der Waals is thus

$$\left(p+\frac{a}{V^2}\right)(V-b) = RT$$

for a kilogramme mole.

3.12. The properties of van der Waals' equation

Isothermal curves calculated from this equation are plotted in fig. 3.5. At low temperatures each curve has a maximum and a minimum, while at higher temperatures the curves merely have a point of inflexion and get closer to $pV = RT$ with increasing temperature.

One intermediate curve has a point of inflexion with a horizontal tangent at C. This corresponds to the critical isothermal as found by Andrews, and C is the critical point. The temperature of this critical isothermal, and the pressure and volume at C are denoted by T_c, p_c, and V_c, respectively

Now van der Waals' equation can be rewritten in the form

$$p = \frac{RT}{V-b} - \frac{a}{V^2}$$

This may be differentiated with respect to volume for an isothermal, giving

$$\frac{\partial p}{\partial V} = -\frac{RT}{(V-b)^2} + \frac{2a}{V^3}$$

which equals 0 when

$$\frac{RT}{(V-b)^2} = \frac{2a}{V^3}$$

or

$$p = \frac{2a}{V^3}(V-b) - \frac{a}{V^2}$$

$$= \frac{a(V-2b)}{V^3}$$

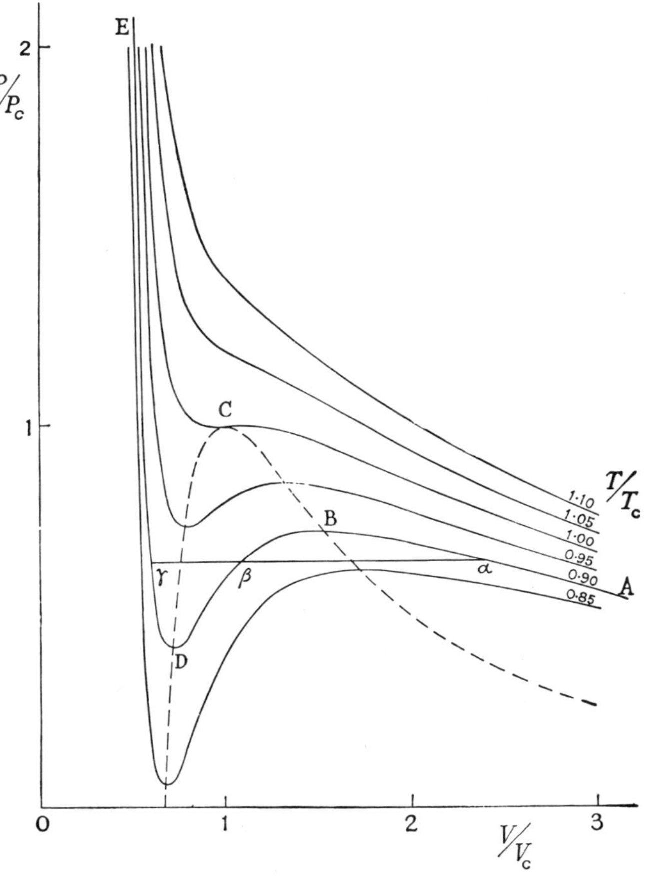

Fig. 3.5.—Isothermals calculated from van der Waals' equation. The pressure, volume, and temperature are expressed in terms of the values of these variables at the critical point *C*. In this form the curves are independent of the actual values of *a* and *b*

This is the equation of the dotted curve through the maxima and minima in fig. 3.5. For this curve

$$\frac{dp}{dV} = \frac{a}{V^3} - \frac{3a(V-2b)}{V^4}$$

$$= \frac{-2aV+6ab}{V^4}$$

The maximum of this curve is given by the point for which

$$\frac{dp}{dV} = 0$$

that is, when $V_c = 3b$ and $p_c = a/27b^2$.

It then follows that

$$\frac{RT_c}{p_c V_c} = \left(1+\frac{a}{p_c V_c^2}\right)\left(1-\frac{b}{V_c}\right)$$

$$= \left(1+a.\frac{27b^2}{a}.\frac{1}{9b^2}\right)\left(1-\frac{b}{3b}\right)$$

$$= \frac{8}{3}$$

Also

$$T_c = \frac{8}{3}.\frac{a}{27b^2}.\frac{3b}{R}$$

$$= \frac{8}{27}.\frac{a}{Rb}$$

3.13. Comparison of van der Waals' equation with experiment

Below the critical isothermal, van der Waals' equation gives an S-shaped curve *ABDE*, whereas experiment gives $A\alpha\beta\gamma E$, with discontinuities at α and γ. Here $\alpha\beta\gamma$ corresponds to the heterogeneous region of two phases in equilibrium.

The portions $\alpha\beta$ and γD, which correspond to a supercooled vapour and a superheated liquid respectively, can be obtained experimentally. *BD* represents an unstable region where increase of pressure causes an increase of volume, and could not be realized in practice.

The equation gives no indication as to where the point α corresponding to the start of liquefaction should be, but by thermodynamic reasoning it can be shown that the horizontal line $\alpha\beta\gamma$ should be drawn in such a position that areas $\alpha B\beta$ and $\beta D\gamma$ are equal.

The values of a and b deduced from the critical point give isothermal curves that differ somewhat from the experimental ones, so that van der Waals' equation is not completely correct. It is, however, a good general approximation and has the merit of simplicity. Other equations of state have been suggested, which are improvements on that of van der Waals, but none give complete agreement with experimental results.

LIQUIDS

3.14. The behaviour of liquids

A liquid, like a gas, is composed of molecules, but with much less free space. The molecules are still freely in motion, having kinetic energy which is dependent upon temperature, but the van der Waals attractive force has a significant value.

3.15. Surface tension

Whereas molecules in the bulk of the liquid will experience equal van der Waals attractive forces in all directions, those molecules nearer the surface will have an attractive force towards the bulk of the liquid which is not counterbalanced by any force from outside. This results in a definite force on the surface molecules which causes the liquid to behave as if enclosed in a skin. This force is known as the *surface tension.* The surface tension is measured in terms of the force necessary to separate an element of the surface layer of unit length. In increasing the surface area, work equal to the product of the surface tension and the increase of area has to be done. This is the *surface energy.*

The surface tension decreases with increasing temperature and becomes zero at the critical temperature (where vapour and liquid are indistinguishable). The surface tension of water is about 0·070 and that of alcohol about 0·025 N m^{-1} at room temperature. The surface tension of liquid metals is much higher, for example molten aluminium and iron have values of 0·5 and 1·5 N m^{-1}, respectively.

A drop of mercury resting on a horizontal surface will keep an almost spherical shape, whereas water would spread out to a thin layer because of the much higher surface tension of the metal, although the mercury has a much greater density.

3.16. Vapour pressure

A molecule moving towards the surface will have its velocity reduced as it passes through the region of the unbalanced force. The field of force will be sufficient to reduce to zero the translational energy of a molecule which has an initial energy equal to or less than the mean value for that temperature. Some molecules will, however, have a velocity sufficiently high to escape. These escaped molecules constitute the vapour of the liquid and exert a pressure known as the *vapour pressure* of the liquid. Owing to collisions some of these molecules will acquire velocities towards the surface, and when they strike it will be absorbed into the bulk of the liquid. If evaporation takes place inside a closed container, the number of vapour molecules will at first increase. As the density of vapour molecules increases, so also will the rate at which they return to the liquid. An equilibrium state will be reached when the rates at which molecules leave and re-enter the liquid are equal. The vapour is then saturated, and its vapour pressure is called the *saturation vapour pressure.*

When the saturation vapour pressure is equal to or greater than the total external pressure boiling can occur.

3.17. Viscosity

When a fluid is in a state of motion with different layers moving at different speeds, there is shear force between the layers which would bring all the fluid to the same average speed if external forces were not applied. This shear force is described as *viscosity*. In gases it is a transport phenomenon and has a relationship with mean free path. In that case molecules with the average velocity appropriate to their layers will be moving into adjacent layers carrying with them momentum, so that there will be a steady transfer of momentum tending to equalize the velocities of the two layers.

In liquids, the mean free paths are extremely short, so that the molecules will not in general pass from one layer to another. The intermolecular forces are, however, significant, so that each layer tends to drag the adjacent layer with it, thus reducing the relative motion.

3.18. Thermal conductivity

The conductivity of heat is also a transfer of molecular kinetic energy but, whereas with viscosity there was a definite transfer of momentum, the average momentum in this case will be zero.

QUESTIONS

1. 0·1015 g of an organic liquid, when vaporized, displaced a quantity of air whose volume was 27·96 cm^3 measured at 15 °C and 100 kPa. Assuming Avogadro's hypothesis to hold for the vapour, calculate the molecular weight of the liquid.

2. Derive an expression for the root-mean-square velocity of the molecules of a gas contained in a volume v at pressure p, if the molecular weight of the gas is M.
Hence calculate the root-mean-square velocity of nitrogen molecules at a pressure of 1 MPa and 150 °C. [P]

3. Define the *mean free path* of a molecule of a gas.
Derive an expression for the mean free path. Using this expression, calculate for argon at 15 °C the pressure at which the mean free path is 0·01 m, assuming that the diameter of the argon atom is $2{\cdot}88 \times 10^{-10}$ m. [P]

4. Explain the assumptions made by van der Waals in deriving his equation from the simple gas law $pv = RT$, where p, v, R, and T have their usual meanings.
State the van der Waals' equation. To what extent can it represent the experimental pressure, volume, and temperature relationship of a gas such as carbon dioxide? [P]

5. The critical point of a gas occurs at 10 ° C and a pressure of 5·17 MPa, the density then being 220 kg m^{-3}. Assuming van der Waals' equation to be correct in the region of the critical point, calculate the molecular weight of the gas.

CHAPTER 4

Aggregations of Atoms—Solids

4.1. Introduction

Continued abstraction of heat and lowering of the temperature of a liquid results in a continued decrease in the kinetic energy of the molecules until, at a certain temperature known as the freezing temperature, the molecules " fall " into relatively fixed positions with respect to each other. These positions might perhaps be better described as *centres of oscillation*, for, although the random molecular motion characteristic of the liquid and vapour states has disappeared, all molecular motion has not ceased. It is merely reduced to oscillation about a fixed point.

The change of state from liquid to solid is accompanied by a change of volume. The solid is usually of smaller volume than the liquid, but some substances, of which water is the best-known example, are exceptions to this. In ice, the H_2O molecules are arranged in a more regular and also more open structure than in the liquid state. Hence there is a volume increase on freezing.

4.2. The crystalline state

When a substance becomes solid, it does so in a crystalline form, that is, the atoms or molecules take up positions in a pattern that repeats periodically in three directions. One network of the pattern will continue so far and then meet another network which has a different orientation, the two networks being different crystals or grains of the substance. Where a crystal extends to a free boundary, regular faces may form naturally, giving the usual elementary idea of a crystal. Where crystals meet, the boundaries will not necessarily be regular plane surfaces, but usually are curved or irregular as shown in the micrograph in fig. 5.3. The size of the crystals or grains is a function of the past history of the specimen and will be discussed more fully in Chapter 5.

The " solids " which are non-crystalline are termed *amorphous solids*, but the modern view is that they are liquids of very high viscosity. An example is glass, which is a super-cooled liquid in which the molecules are not arranged in a regular pattern, but have become fixed in a more random arrangement. Such an amorphous solid has cooled from the liquid state by steady cooling with continuous increase of viscosity without an abrupt freezing-point. (See Section 20.5.)

4.3. Space lattices

The essential characteristics of a two-dimensional lattice, of which a wallpaper pattern is an example, is the regular repetition of a unit of pattern. The elementary units lie in equally-spaced rows, such that corresponding points of the units lie at the points of intersection of two sets of equally-spaced parallel lines as in fig. 4.1.

In three dimensions, there would be similar arrays parallel to this at equally-spaced intervals, such that lines can be drawn through each of the points 00, 01, 02, etc., which pass through the corresponding points in the parallel arrays. The whole set of points is formed by the intersection of three sets of parallel and equally-spaced planes. Such a collection of points is called a *space lattice.*

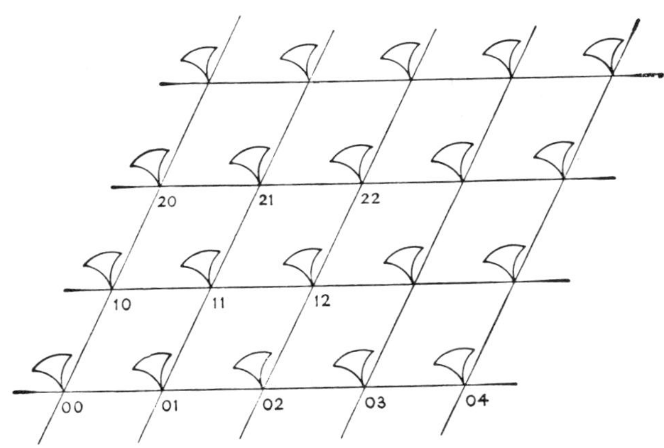

Fig. 4.1.—Regular repetition of a pattern in a two-dimensional lattice

The defining property of a space lattice is that each *lattice point* must have surroundings identical to every other lattice point. It can be shown that there are only fourteen ways of arranging points in space to meet this condition so that there are only fourteen space lattices, which are shown in fig. 4.2.

When atoms or clusters of atoms are positioned at lattice points, a *crystal structure* is obtained. Hence for any one space lattice there can be an unlimited number of crystal structures. Figure 4.3 shows some possible crystal structures, all of which have the same space lattice.

To specify the arrangement of atoms in a crystal structure, it is customary to give their coordinates with respect to a set of coordinate axes chosen with an origin at one of the lattice points. Each space lattice

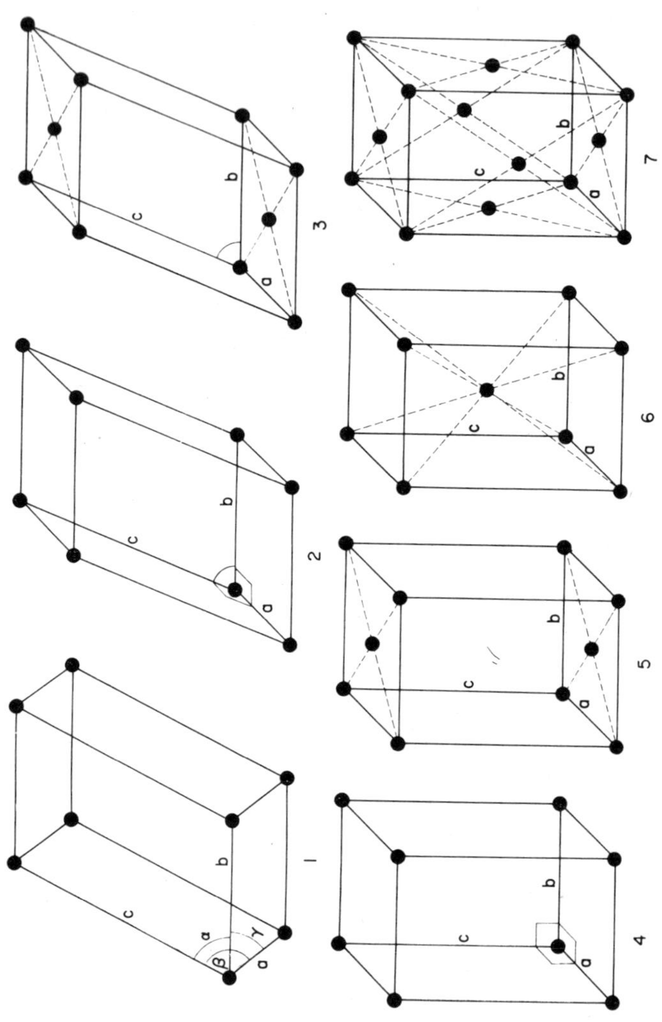

1
2
3
4
5
6
7
a
b
c
α
β
γ

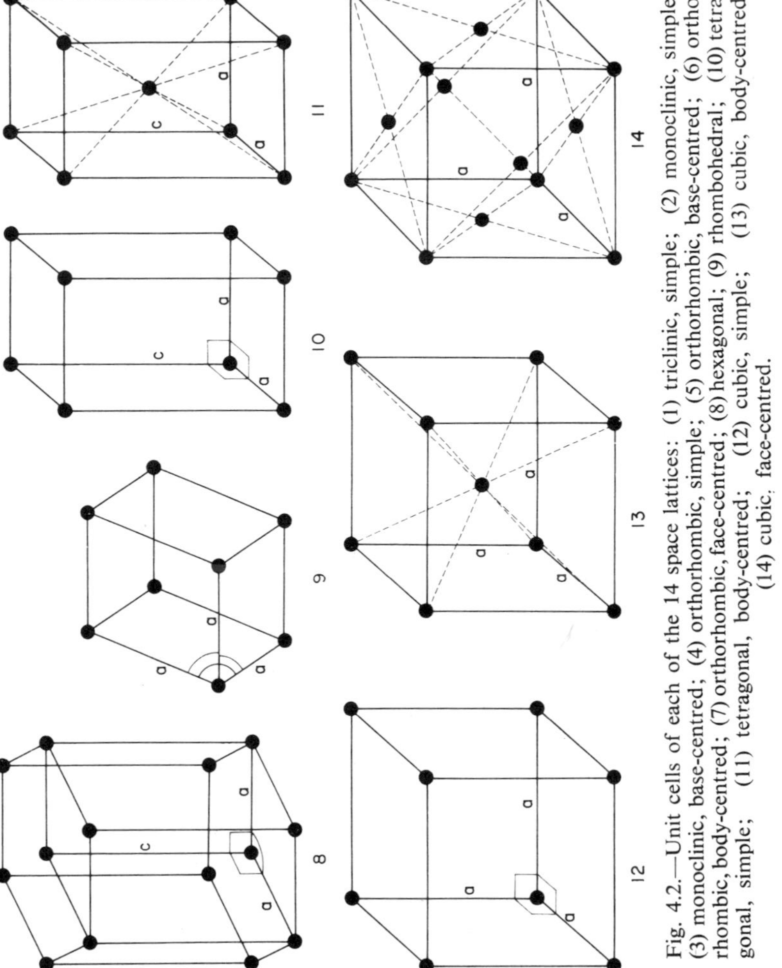

Fig. 4.2.—Unit cells of each of the 14 space lattices: (1) triclinic, simple; (2) monoclinic, simple; (3) monoclinic, base-centred; (4) orthorhombic, simple; (5) orthorhombic, base-centred; (6) orthorhombic, body-centred; (7) orthorhombic, face-centred; (8) hexagonal; (9) rhombohedral; (10) tetragonal, simple; (11) tetragonal, body-centred; (12) cubic, simple; (13) cubic, body-centred; (14) cubic, face-centred.

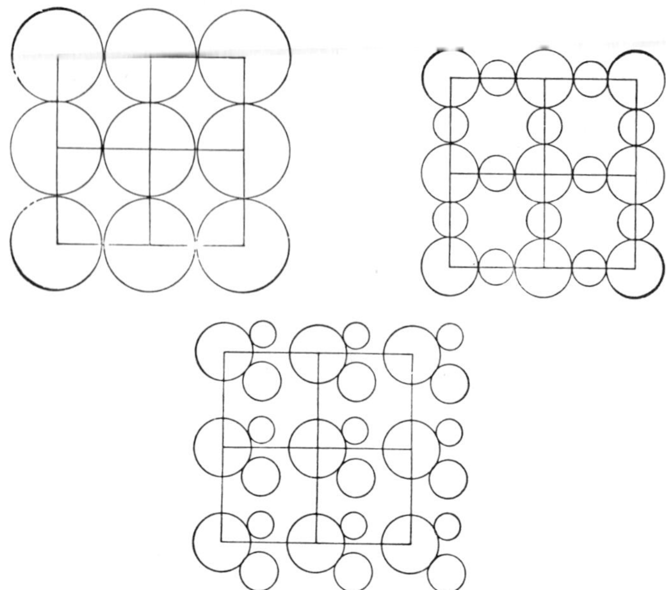

Fig. 4.3.—Three different two-dimensional " crystal structures " with the same space lattice

has some convenient set of axes that is conventionally used with it. These axes are not necessarily orthogonal, but in cases where there is sufficient symmetry, they are chosen so that two or all three are mutually perpendicular. The chosen units of length along the three axes are not necessarily equal, but again in cases where there is sufficient symmetry, the units along two or three axes are equal. The possible degrees of symmetry are considered in Section 4.4.

In the space lattice given by the intersections of the lines shown in fig. 4.4, the three axes are OX, OY, and OZ, and the units of length along them are the distances between the space lattice points, i.e. a, b, and c. The angles between the axes are α, β and γ, α being the angle between the OY- and OZ-axes, etc.

The space lattice is completely determined by a, b, c and α, β, γ.

The network of planes through the points of a space lattice divides the regions into prisms called *unit cells.* If the unit cell is the smallest possible, it is a *primitive* cell. It may now be obvious that there can be two or more perfectly valid descriptions of any particular structure. The choice is usually one of convenience, probably that which gives the highest degree of symmetry for the unit cell. In many cases the conventional choice of

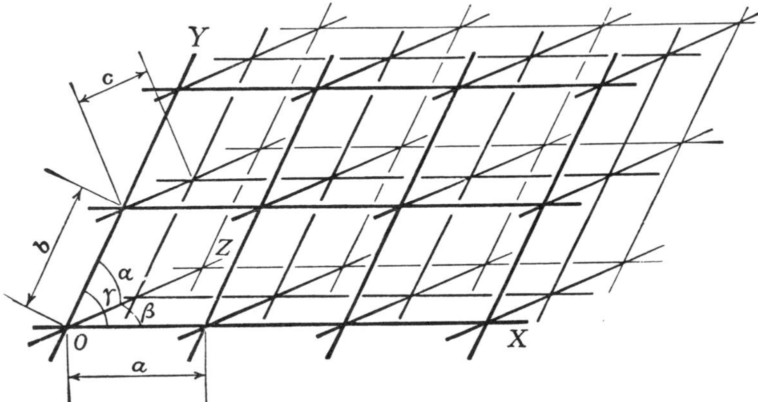

Fig. 4.4.—Lengths and angles specifying a space lattice

unit cell is a prism larger than the primitive cell, the convenience of its having a higher degree of symmetry or some other feature making it preferable to the primitive cell for discussion purposes.* Each unit cell in a space lattice is identical in size, shape, and orientation with every other in the same crystal. It is the building block from which the crystal is constructed by repetition in three dimensions.

By means of X-ray diffraction methods it is possible to determine the shape and size of the unit cell and also the distribution of atoms within it.

4.4. Crystal systems

As stated in the previous section, each space lattice has some convenient set of axes which are conventionally used with it. There are seven such sets which are known as the seven crystal systems. Table 4.1 lists these systems and the relationships that exist between the cell dimensions and between the angles in each case.

TABLE 4.1—THE SEVEN CRYSTAL SYSTEMS

Triclinic	$a \neq b \neq c$	$\alpha \neq \beta \neq \gamma \neq 90°$
Monoclinic	$a \neq b \neq c$	$\alpha = \gamma = 90° \neq \beta$
Orthorhombic	$a \neq b \neq c$	$\alpha = \beta = \gamma = 90°$
Tetragonal	$a = b \neq c$	$\alpha = \beta = \gamma = 90°$
Cubic	$a = b = c$	$\alpha = \beta = \gamma = 90°$
Hexagonal	$a = b \neq c$	$\alpha = \beta = 90°, \gamma = 120°$
Rhombohedral	$a = b = c$	$\alpha = \beta = \gamma \neq 90°$

*Some authors use the terms *structure cell* and unit cell to denote the unit cell and primitive cell respectively as defined here.

Within each system, there are one or more types of space lattice distinguished by differing degrees of symmetry within the unit cell. The cubic system is used for three space lattices: the simple cubic, the face-centred cubic, and the body-centred cubic, while the hexagonal system is used for only one lattice, the simple hexagonal. In each " simple " lattice, the primitive cell is also the unit cell.

Many physical properties are dependent more upon the shape of the unit cell and the arrangement of atoms within it, than upon exactly which atoms are present. Also, many properties vary with crystallographic direction. Hence, as well as specifying the form of a unit cell, it is also necessary to have a system for denoting the various planes of atoms and directions within a crystal. These are usually specified in terms of the *Miller index notation* which is described in the two following sections.

4.5. Indices of planes

A plane is defined by the length of its intercepts on the three crystal axes (the three edges of a unit cell), measured from the origin of co-ordinates. The intercepts are expressed in terms of the dimensions of the unit cell, which are the unit distances along the three axes. The reciprocals of these intercepts reduced to the smallest three integers having the same ratio are known as *Miller indices*.

As the origin may be taken at any lattice point, the lengths of the intercepts for a given plane are not specified, but their ratio is constant. Any other parallel plane would have the same indices, so that a particular set of Miller indices defines a set of parallel planes.

The shaded plane in fig. 4.5 has intercepts 1, 1, 1 and therefore indices (111). A plane with intercepts 2, ∞, 1 (i.e., parallel to the OY-axis) has reciprocal intercepts $\frac{1}{2}$, 0, 1 and Miller indices (102). If a plane cuts an

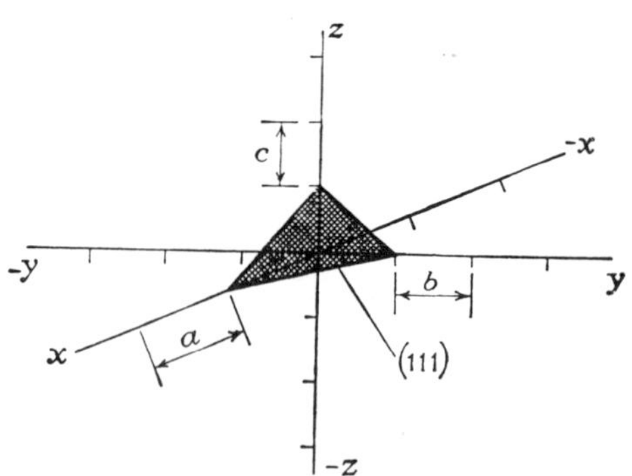

Fig. 4.5.—Intercepts of plane with crystal axis

axis on the negative side of the origin, the corresponding index will be negative and written with a line above the number. Some examples of indices of planes are shown in fig. 4.6.

Parentheses ( ) around a set of indices signify a single set of parallel planes.

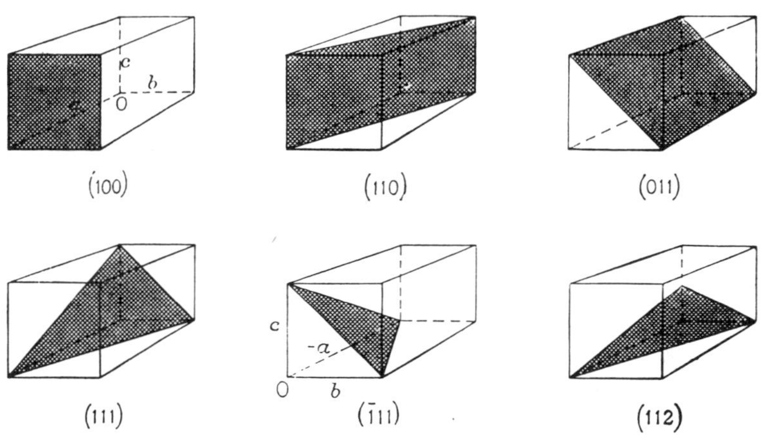

Fig. 4.6.—Examples of Miller indices of planes

Curly brackets or braces { } signify planes of a *form*, that is, sets of planes which are exactly equivalent to each other. Thus for a cubic crystal which has a high degree of symmetry {110} includes six sets of planes: (110), (101), (011), (1$\bar{1}$0), (10$\bar{1}$) and (01$\bar{1}$). It should be noted that reversal of the signs of all the indices merely denotes another parallel plane. Thus ($\bar{1}\bar{1}$0) is parallel to (110) and (10$\bar{1}$) to ($\bar{1}$01).

For the hexagonal system, an alternative indexing system is often used. Four numbers h, k, i, l are used in the *Miller–Bravais* indices referring to intercepts on four axes—a_1, a_2, a_3, and c as shown in fig. 4.7. The treatment is otherwise the same as for the simple Miller indices. The first three indices are not independent but must satisfy

$$h+k+i = 0$$

The relationship of planes to the symmetry of the hexagonal lattice is more obvious and planes of a form have the same sets of indices though in different orders. Conventional three-number Miller indices for a hexagonal lattice merely differ by omitting the third number and are used in preference to Miller–Bravais indices in the remainder of the book. Examples are shown in figs. 10.1 and 10.2.

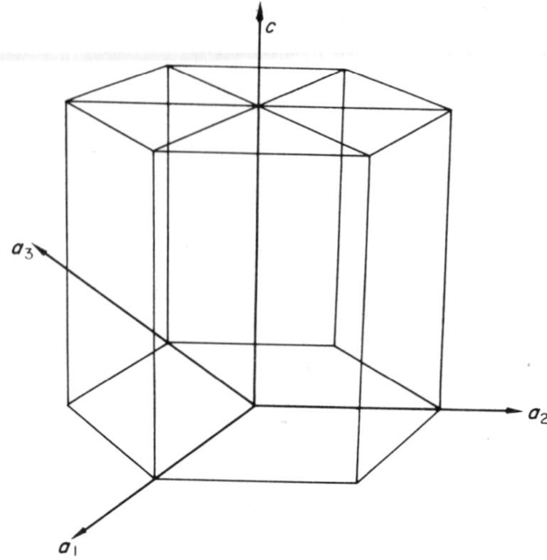

Fig. 4.7.—Axes in hexagonal lattice

4.6. Indices of direction

A direction is defined in terms of the successive motions parallel to each of the three axes necessary to move from the origin to another point which lies in the required direction. Suppose that the moves are a distance u times the unit distance a along the X-axis, v times b parallel to the Y-axis and w times c parallel to the Z-axis. If u, v, and w are the smallest set of integers that will perform the movement, then they are the indices of the direction and are written in square brackets thus: [uvw]. The X-axis will be [100], the negative Y-axis [0$\bar{1}$0], etc. Several directions are shown in fig. 4.8.

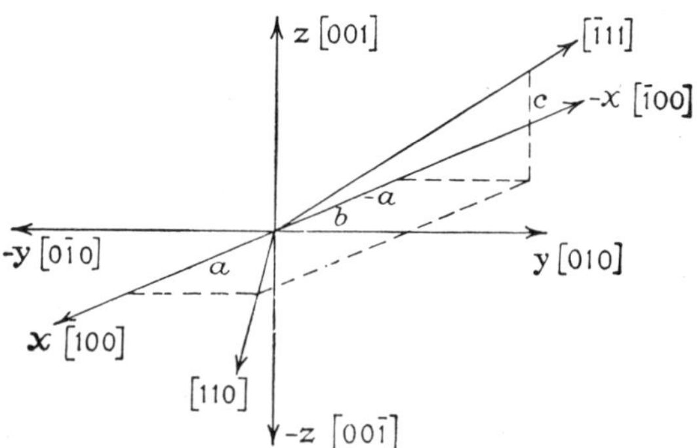

Fig. 4.8.—Examples of Miller indices of direction

A full set of equivalent directions (i.e., directions of a form) is indicated by carets $\langle\ \rangle$. Thus in the cubic system,

$$\langle 111\rangle = [111]+[11\bar{1}]+[1\bar{1}1]+[\bar{1}11]$$

It should be observed that reciprocals are *not* used in computing directions. In a cubic system *only*, the $[hkl]$-direction is perpendicular to the (hkl)-plane.

4.7. The common crystal lattices

Most metals crystallize in one of three structures, the face-centred cubic and body-centred cubic of the cubic system and the hexagonal close-packed structure, which is a special example of the simple hexagonal lattice. The features of each of these are described in the following sections.

4.8. Face-centred cubic system

The unit cell of the face-centred cubic system is shown in fig. 4.9. There is an atom at each corner of the cube and also one at the centre of each face. Figure 4.10 shows a model of a cluster of spheres packed in a face-centred cubic manner. By careful consideration of figs. 4.9 and 4.10, it can be seen that each of the atoms has surroundings identical with any other and so all of them are lattice points. The cubic unit cell has a volume which is four times that of a primitive cell. One possible primitive cell (which is rhombohedral) and its relationship to the unit cell are shown in fig. 4.11. Various other primitive cells of lower symmetry could also be chosen. Each atom at a corner of the unit cell is shared by eight unit cells,

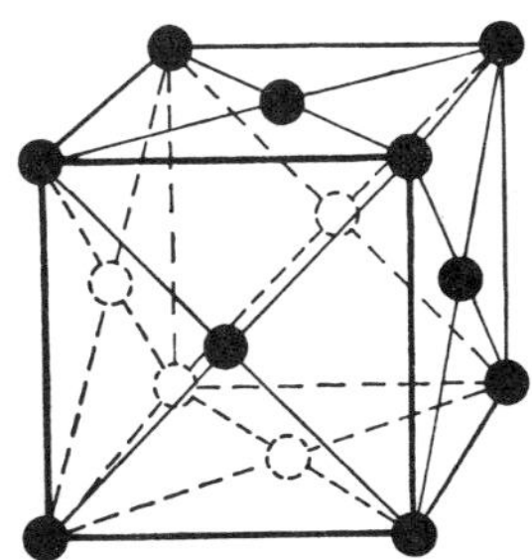

Fig. 4.9.—Face-centred cubic unit cell. The circles represent the nuclei of atoms

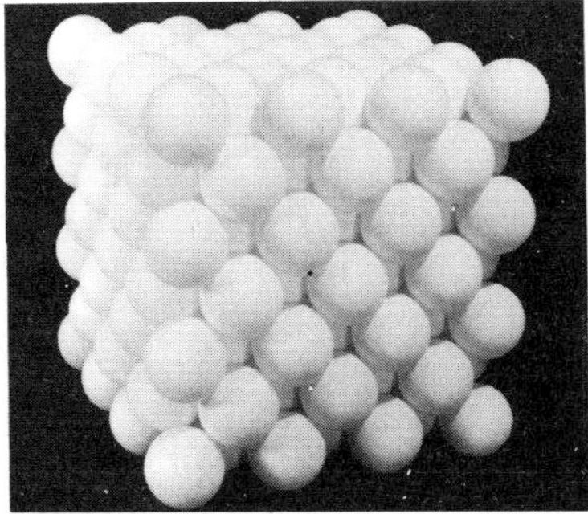

Fig. 4.10.—Face-centred cubic structure formed by spheres in contact

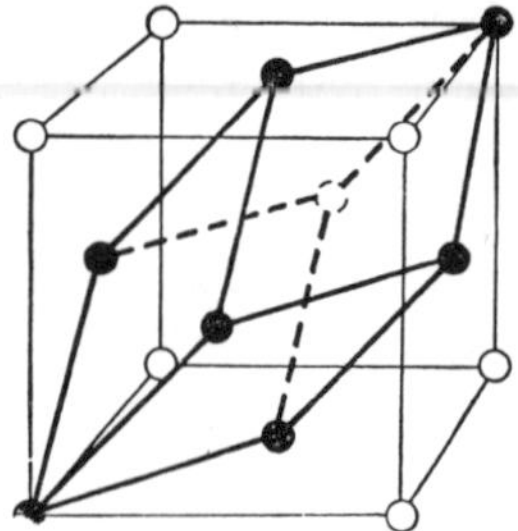

Fig. 4.11.—Relationship of face-centred cubic primitive cell to unit cell

and each face atom is shared by two unit cells. Hence the number of atoms per unit cell is $8 \times \frac{1}{8} + 6 \times \frac{1}{2} = 4$.

Each atom has twelve equidistant atoms near it, which is the closest possible packing. Six of these in one plane form a hexagon around the centre atom, and three are in parallel planes on each side, these being (111) planes. All the {111} planes in this system are planes of closest packing. The spheres touch along the ⟨110⟩ directions, which are therefore the directions of closest packing.

4.9. Body-centred cubic system

The unit cell of the body-centred cubic system has an atom at each corner and also one at the centre, as shown in fig. 4.12. The number of atoms per unit cell is $8 \times \frac{1}{8} + 1 = 2$. Each atom has eight equidistant near neighbours, so that this system is *not* one of closest packing. As may be seen from fig. 4.13, the spheres touch along the [111] direction, so that all ⟨111⟩ directions are close-packed. There is no close-packed plane.

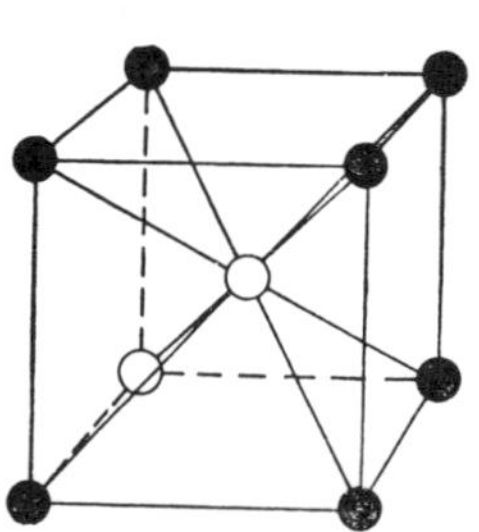

Fig. 4.12.—Body-centred cubic unit cell

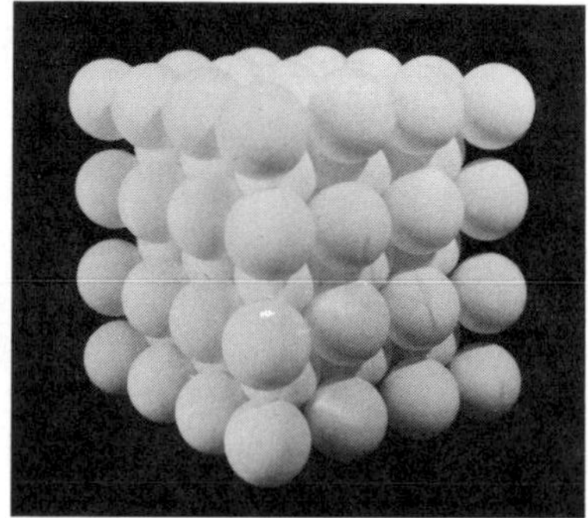

Fig. 4.13.—Body-centred cubic structure formed by spheres in contact

As for the face-centred cubic lattice, each atom site of a body-centred cubic lattice is a lattice point.

4.10. Close-packed hexagonal system

The close-packed hexagonal system is another structure in which close-packed layers are packed on top of one another, giving the structure shown in fig. 4.14. The unit cell, which is also a primitive cell of the simple hexagonal system, is referred to axes as specified in Table 4.1 and is related to the hexagonal structure as shown in fig. 4.15. Each unit cell contains two atoms, one at a corner, which may be taken as the origin of co-

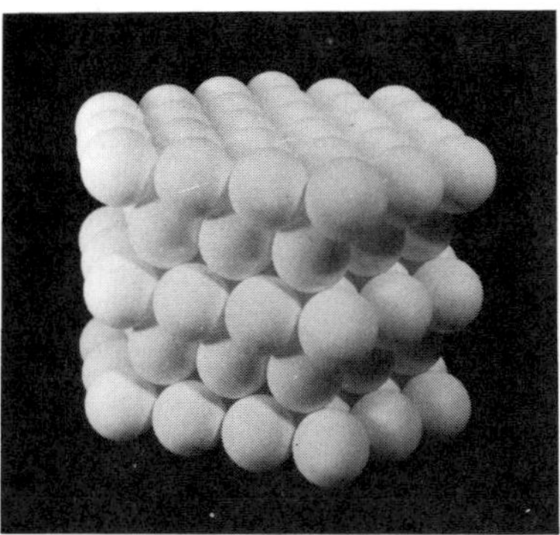

Fig. 4.14.—Hexagonal close-packed structure formed by spheres in contact

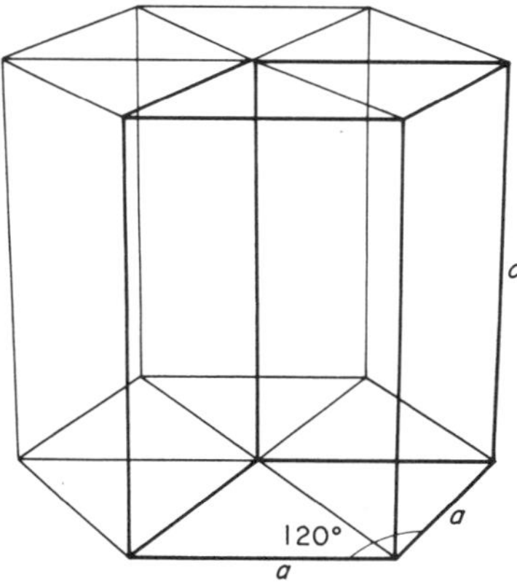

Fig. 4.15.—Relationship of unit cell to hexagon in hexagonal close-packed structure

ordinates, and the other with coordinates $\frac{2}{3}, \frac{1}{3}, \frac{1}{2}$ relative to it. The corner atoms (fig. 4.16) are at lattice points, but the atoms in the intermediate layer, shown in fig, 4.17, are not because the pattern of atoms surrounding

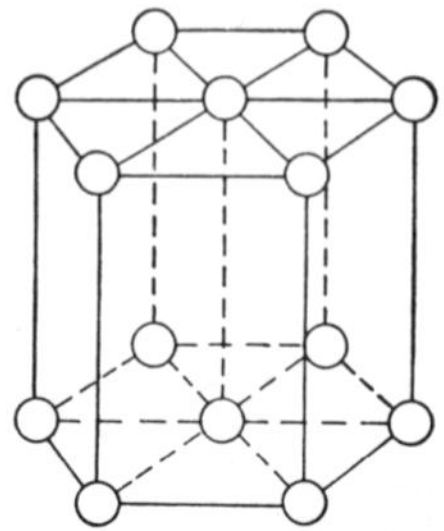

Fig. 4.16.—Lattice points of hexagonal system

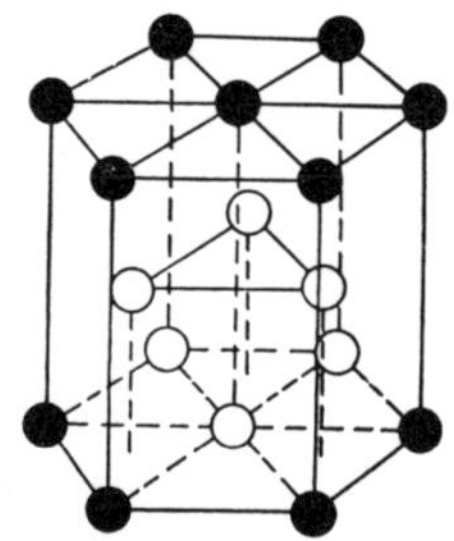

Fig. 4.17.—Atom sites of close-packed hexagonal system

them is different in orientation. A corner atom is common to eight unit cells, so that there are $8 \times \frac{1}{8} + 1 = 2$ atoms per unit cell. The volume of the unit cell is $a_1 . a_2 \sin 120° . c = \frac{1}{2} \sqrt{3}\, a^2 c$.

4.11. Stacking of close-packed layers

The face-centred cubic and the hexagonal close-packed systems are both equally dense and made by the stacking of close-packed layers, the difference being in the stacking pattern.

The atoms of the second layer can be placed on either of two positions on the first layer. In fig. 4.18, the positions of the centres of atoms of the first layer are denoted by A, and the possible positions for the centres of atoms of the second layer by B and C. If the second layer is placed, as shown by the dashed circles, with its centres at B, then the third layer can

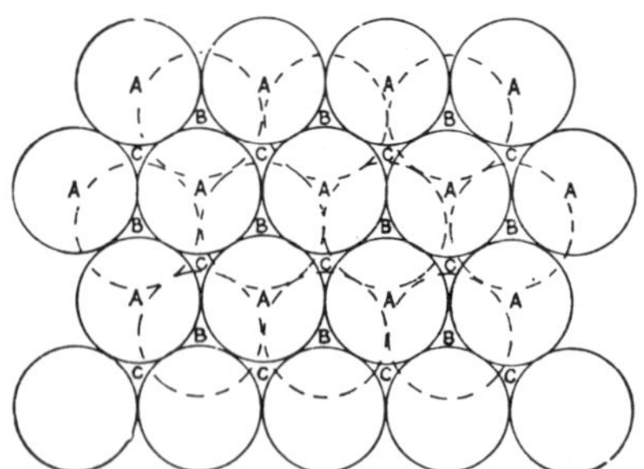

Fig. 4.18.—Stacking of close-packed layers

be placed with its centres at *A* or *C*. Any sequence of packing that does not have two adjacent layers with centres in identical positions will therefore give a close-packed structure. A crystal structure is formed only when a regular sequence is followed. If alternate layers are above one another giving the sequence *ABABAB* . . . (or *BCBCBC* . . . or *CACACA* . . .), then the structure is close-packed hexagonal, whereas if the sequence repeats at every third layer, i.e., *ABCABCABC* . . ., the structure is face-centred cubic.

The ideal axial ratio, c/a, of the close-packed hexagonal crystal should be 1·633 if the crystal were made of close-packed and equal spheres. In practice, it is found that metals which form crystals of this type have an axial ratio somewhat different from this value.

The values of lattice constants of the elements are given in Table 2.1.

4.12. Types of crystal and relation to mechanical properties

Crystals may be classified into types depending upon the method of bonding between the units of the structure. The mechanical properties are closely related to the nature of the bonding.

1. *Covalent*

The atoms are linked throughout by covalent bonds. The best example of this class is the diamond, which is a crystal of carbon. As may be seen from the model in fig. 4.19, each carbon atom is linked to

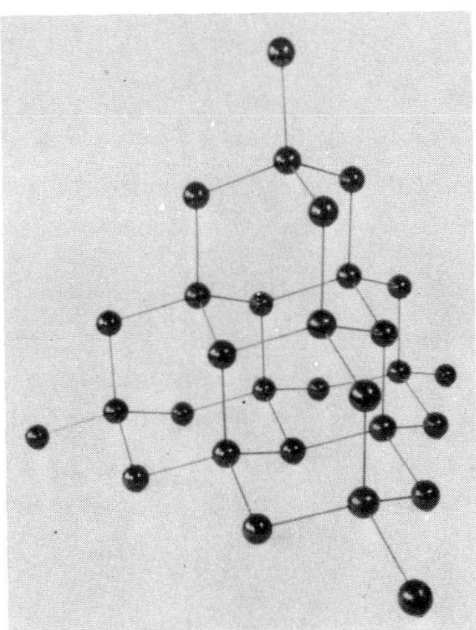

Fig. 4.19.—Structure of typical covalent solid—diamond

four others, which are symmetrically spaced around it so that their centres lie at the corners of a regular tetrahedron, at the centre of which the first atom is situated. Diamond has a face-centred cubic structure with two atoms per lattice point as shown in the diagram of the unit cell in fig. 4.20. When an atom forms more than one covalent bond, the directions of these bonds have definite angular relations with each other. This geometrical limitation restricts the number of possible crystal formations, so that this class is very small. Such crystals are hard and have sharply defined melting-points. Since a covalent bond is between two particular atoms, the attractive forces extend only to these two atoms. Hence mechanical deformation will produce irreversible rupture of covalent bonds, so that these crystals are completely brittle.

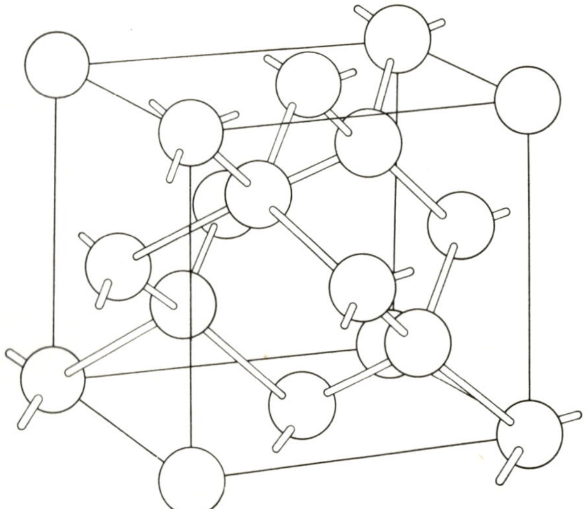

Fig. 4.20.—Unit cell of diamond structure

2. *Molecular*

Covalent bonds generally give rise to molecules. In a molecular crystal, the atoms within the molecules are strongly bonded, but the intermolecular van der Waals forces are much smaller. The strength, hardness, and melting-point, which depend upon these forces are all small. As the attractive force between molecules is due to their proximity only and not due to any covalent bonding, the crystals may be deformed permanently without rupture; that is to say, they possess *ductility* and this may give them a soapy feeling, particularly in the cases where the molecules tend to be flat in shape and take up parallel positions in the crystal.

3. *Ionic*

The structure adopted by any particular ionic crystal depends upon the relative valencies and relative sizes of the various ions. One of the simplest patterns is that exhibited by common salt (fig. 4.21). In a compound containing divalent ions, for example magnesium oxide, the bonding is approximately four times as strong, so that crystals are much stronger with much higher melting points. Since the attractive forces can extend over long distances and are non-directional, ductility is possible. One layer of atoms can slide over the next until it occupies a similar position, when the bonding will be as strong as before (fig. 4.22).

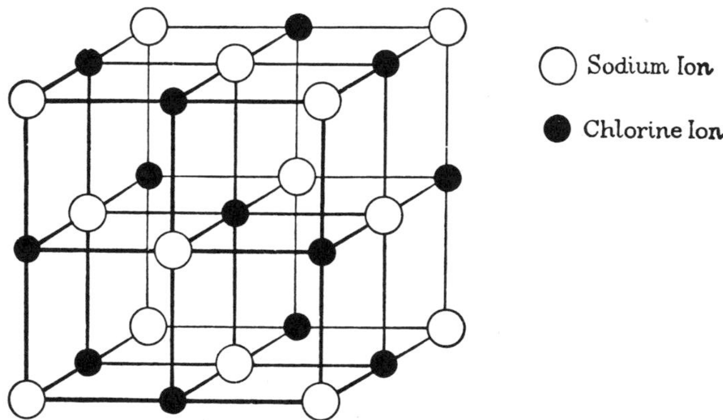

Fig. 4.21.—Structure of a typical ionic crystal—sodium chloride

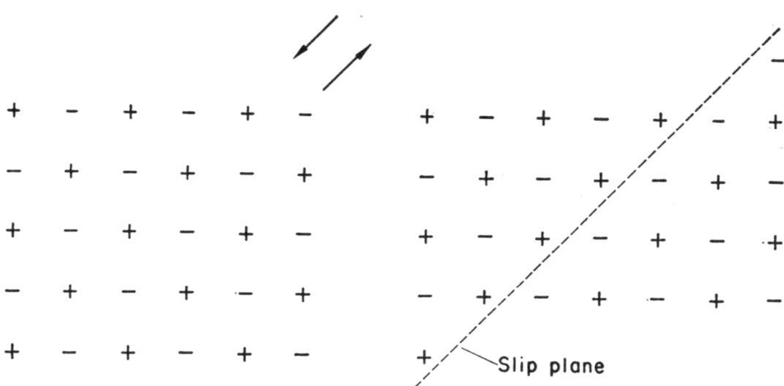

Fig. 4.22.—Slip plane in simple ionic crystal

4. *Metallic*

As described on p. 24 the atoms become positive ions held by the attractive force of the electron cloud. Though somewhat softer than classes (1) and (3), metallic crystals are at least comparable in strength and infusibility. They have high ductility because one atom is not tied to any other particular atom.

4.13. Interatomic forces in solids

In Section 2.12, the types of bonding between atoms were discussed and in Section 4.12 they were considered in relation to crystal structure. In addition to the attractive forces, there is also, when two atoms are sufficiently close, a repulsive force between them due to the interaction of the electron shells. This repulsive force is of shorter range than the attractive force and increases more rapidly with decreasing interatomic distance so that there is an equilibrium position when the two forces balance.

In considering the forces between atoms in a crystal it is convenient to consider also the total energy of the system or that total energy divided by the number of atoms in the crystal.

As the atoms are brought together from infinity, the attractive forces do work so that if the atoms are at rest at a finite value of the interatomic distance, r, the energy is less than if they were at an infinite distance apart.

Conversely, work is done against the repulsive forces causing an energy increase.

For an atom in a crystalline solid (e.g., rock salt (NaCl) or iron), its total energy U will be due to its interaction with all the surrounding atoms and will vary in some way with the lattice spacing as shown in fig. 4.23, having a minimum value at the stable spacing r_0. The relationship between U and r can usually be expressed in the form

$$U = -\frac{A}{r^m} + \frac{B}{r^n}$$

where A, B, m, and n are constants with $n > m$.

If a force is applied to the crystal so that the atomic spacing alters, the product of the force and displacement equals the work done or the change in energy. Hence the force f per atom will be given by

$$f = -\frac{dU}{dr}$$

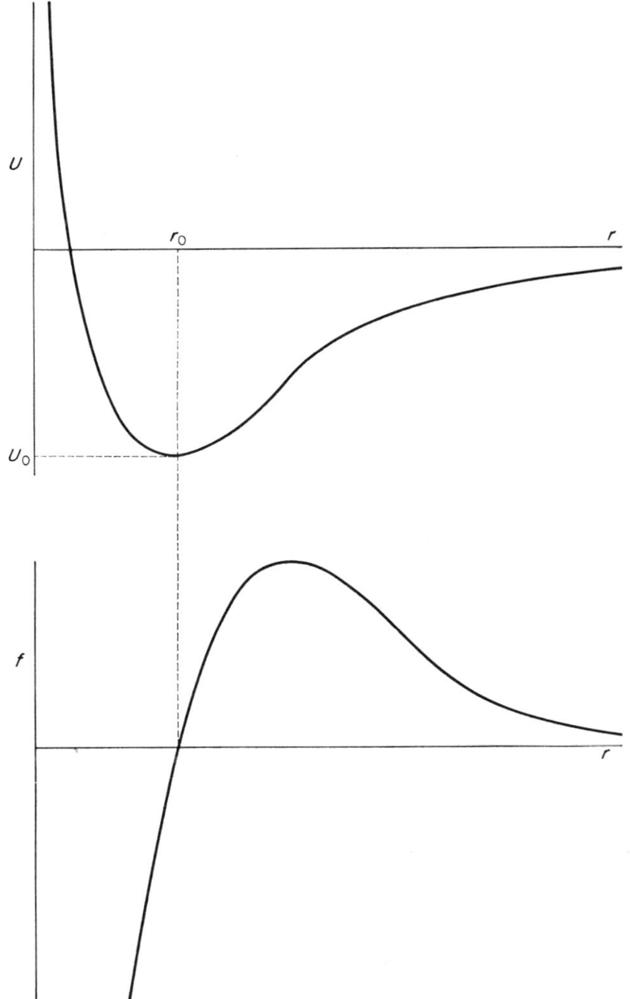

Fig. 4.23.—Variation of energy of atom and force on atom with interatomic spacing

As r is increased from r_0, the force is attractive, increases to a maximum and then decreases to zero as r tends to an infinite value.

If $-U_0$ is the energy of an atom at its equilibrium position, then energy U_0 would have to be added to remove one atom to infinity. This is the latent heat of sublimation (direct passage from the solid to the vapour phase).

Suppose the lattice is strained so that the interatomic distance is increased to

$$r = r_0 + \delta r$$

Then, since the $U-r$ curve is smooth, we can apply Taylor's theorem to give

$$U(r) = U_0 + \left(\frac{dU}{dr}\right)_{r_0} \delta r + \frac{1}{2}\left(\frac{d^2U}{dr^2}\right)_{r_0} \delta r^2 + \frac{1}{6}\left(\frac{d^3U}{dr^3}\right)_{r_0} \delta r^3 + \ldots$$

The condition for equilibrium spacing is that

$$\frac{dU}{dr} = 0 \quad \text{at} \quad r_0$$

so that for small displacements

$$U(r) = U_0 + \frac{1}{2}\left(\frac{d^2U}{dr^2}\right)_{r_0} \delta r^2$$

neglecting terms in δr^3 and higher powers.
Hence

$$f = \frac{dU}{dr} = \left(\frac{d^2U}{dr^2}\right)_{r_0} \delta r$$

or, force is proportional to displacement (which is Hooke's Law). Now on the crystal face there will be $1/r_0^2$ atoms per unit area for a simple cubic lattice (and a similar quantity differing by a numerical factor near 1 in other lattices) so that the stress becomes

$$\sigma = f/r_0^2$$

and the strain will be

$$\varepsilon = \delta r/r_0$$

Therefore

$$\sigma = \left(\frac{d^2U}{dr^2}\right)_{r_0} \frac{\delta r}{r_0^2}$$

$$= \left(\frac{d^2U}{dr^2}\right)_{r_0} \frac{\varepsilon}{r_0}$$

The definition of Young's modulus is

$$\sigma = E\varepsilon$$

so that

$$E = \frac{1}{r_0}\left(\frac{d^2U}{dr^2}\right)_{r_0}$$

Hence the elastic constant is a function of the curvature of the energy curve at its minimum, a sharper curvature corresponding to a higher elastic constant.

In an ionic crystal, the attractive force is electrostatic and varies as $1/r^2$. Hence the attractive energy per ion is $-e^2\alpha_M/4\pi\varepsilon_0 r$, where r is the shortest distance between ions of unlike sign and α_M is the *Madelung constant*, the value of which depends upon the crystal geometry and allows for the contribution to the energy of all the other ions, positive for ions of unlike sign and negative for ions of like sign. For the rock salt (NaCl) lattice, α_M has the value 1·7476.... The repulsive energy is due to interaction forces and usually only that due to the immediate neighbours is significant. In ionic crystals, this interaction energy of repulsion is usually taken to be proportional to r^{-9}.

For forms of bonding other than ionic, the energy terms are functions of different powers of r.

Example

From the relevant information in Tables 2.1 and 2.2, calculate for molybdenum

(*a*) the nearest distance between centres of two atoms,

(*b*) the density,

(*c*) the distance between successive (110) planes, and

(*d*) the angles between the [111] and [112] directions.

Answer

(*a*) Nearest distance between atoms is half length of cube diagonal (see fig. 4.12), i.e. $\frac{1}{2} \times \sqrt{(3)} \times 0{\cdot}31468$ nm

$$= \underline{0{\cdot}2725 \text{ nm}}$$

(*b*) Volume of unit cell $= (0{\cdot}31468 \times 10^{-9})$ m³.

Unit cell contains 2 atoms.

Mass of one Mo atom $= 95{\cdot}95 \times 1{\cdot}6605 \times 10^{-27}$ kg.

$$\text{Hence density} = \frac{2 \times 95{\cdot}95 \times 1{\cdot}6605 \times 10^{-27}}{(0{\cdot}31468 \times 10^{-9})^3} \text{ kg m}^{-3}$$

$$= \underline{10200 \text{ kg m}^{-3}}$$

(*c*) Distance between successive (110) planes (see fig. 4.24)

$$= a/\sqrt{2}$$

$$= \underline{0{\cdot}2225 \text{ nm}}$$

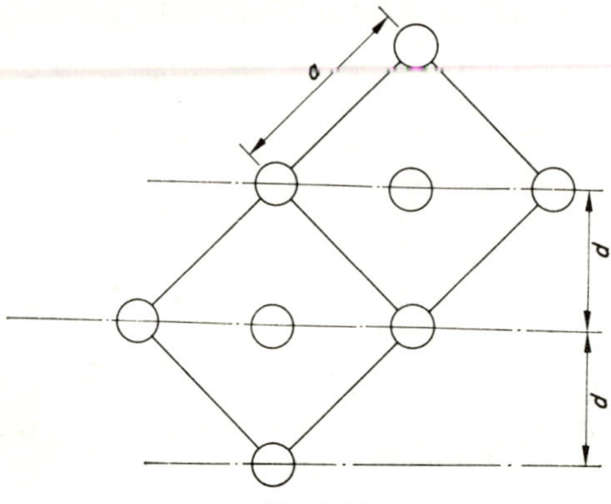

Fig. 4.24

(*d*) $(1\bar{1}0)$ plane which contains the directions is shown in fig. 4.25 as section of two unit cells.

$$\theta = \tan^{-1} 1/\sqrt{2}$$
$$= 35° \, 15'$$
$$\phi = \tan^{-1} 2/\sqrt{2}$$
$$= 54° \, 45'$$

Required angle $= \phi - \theta$

$$= \underline{19° \, 30'}$$

Fig. 4.25

QUESTIONS

1. By considering a crystalline array explain the difference between a primitive cell and a unit cell.

Sketch the unit cells of copper, sodium, and zinc. Indicate the close-packed directions and the positions of any close-packed layers of atoms in these structures, and explain their significance in the plastic deformation of crystals. [P]

2. Show by considering a two-dimensional array the difference between a space lattice point and an atomic site.

Sketch the unit cells of aluminium and titanium (at 20 °C), marking lattice points by +, atomic sites by ○ and a combination of the two by ⊕.

Indicate the (100), (010), and (112) planes and the [211] direction in α-iron. [P]

3. Calculate the nearest distance between the centres of two atoms and the density of nickel from the relevant information in Table 2.1.

4. Calculate the distances between successive (100), (110), and (111) planes in gold.

5. Describe the influence of the type of inter-atomic bonding upon the mechanical properties of crystals.

6. Calculate the angle between the (110) and (111) planes and between the [110] and [123] directions in a cubic lattice.

7. Describe, with sketches, three of the simpler crystal structures adopted by metals.

From the information in Table 2.1, calculate the density of rhenium.

In a cubic structure, what are the angles between (*a*) [112] and $[12\bar{3}]$, (*b*) (111) and $[\bar{2}34]$? [P]

8. The total potential energy of a molecule composed of two univalent ions is given by

$$V = -\frac{e^2}{4\pi\varepsilon_0 r} + \frac{B}{r^m},$$

where e is the magnitude of the charge of each ion, r is the distance of separation of the ions, B is a constant and $m > 1$. Explain the physical basis of each term in the expression and sketch a graph of the variation of V with r.

Replacing B by $Ce^2/4\pi\varepsilon_0$, where C is another constant, determine the equilibrium separation of the ions in terms of m and C only. [MST]

9. The potential energy of a pair of ions of unlike charge, each of magnitude q, and a distance r apart consists of a term $-q^2/4\pi\varepsilon_0 r$ due to attractive forces and a term $+B/r^m$ due to close-range repulsive forces.

Show that the total potential energy of a pair of ions, each of valency Z, at their equilibrium spacing r_0 is given by

$$V_0 = -\frac{Z^2e^2}{4\pi\varepsilon_0 r_0}\left(1 - \frac{1}{m}\right).$$

In an ionic crystalline solid, each ion experiences attractive and repulsive forces from many near neighbours so that the bond energy of a single ion is M times that of an isolated ion pair. Calculate the bond energy per mole for $Mg^{++}O^{--}$ using $m = 9$, $M = 1{\cdot}75$ and $r_0 = 0{\cdot}21$ nm. ($\varepsilon_0 = 8{\cdot}854 \times 10^{-12}$ Fm^{-1}). [P]

CHAPTER 5

Crystal Growth and Size

5.1. Micrographic examination

To examine the microstructure of a metal visually, it must be suitably sectioned and prepared in a manner that will show up the grain structure. After the surface to be examined has been exposed, either by removing any unwanted surface layer or by sectioning, and flattening with an emery wheel or file, it is polished with successively finer grades of abrasive material. Emery paper may be used for this purpose, or alternatively, diamond dust on a cloth-covered rotating disc. The final lapping is by rouge or magnesia or alumina powder on a cloth-covered disc, after which the surface should be free from scratches that would be visible under a microscope. It is then attacked by a suitable etchant, which may have one of two effects depending upon the metal and the etchant used: either, atoms will be removed at the grain boundaries, or, the surface of each grain will be attacked to expose facets of certain crystallographic planes as in fig. 5.1.

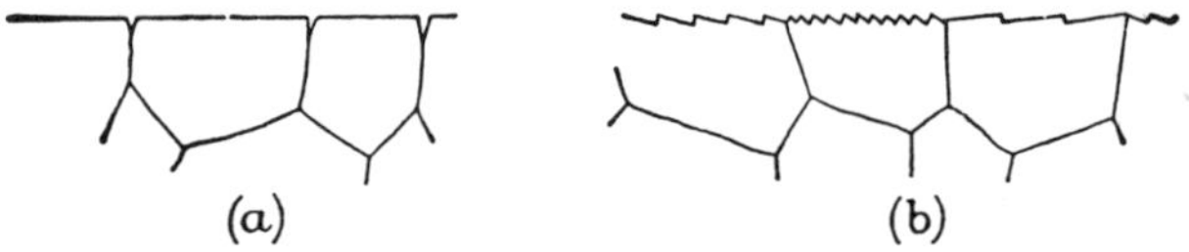

(a) (b)

Fig. 5.1.—Action of etchant upon metal surface
(*a*) grain boundary attack
(*b*) grain surface etch, producing facets of differing orientation in neighbouring grains

The metal is then examined under a metallurgical microscope. As the specimens are opaque, a vertical illuminator which will illuminate the specimens via the microscope eyepiece must be used. The optical system of a typical metallurgical microscope is shown in fig. 5.2. Suitable magnifications for metallurgical purposes are from $\times 50$ to $\times 2000$. The appearance of single-phase material after each of the two types of etchant acting are shown in figs. 5.3 and 5.4.

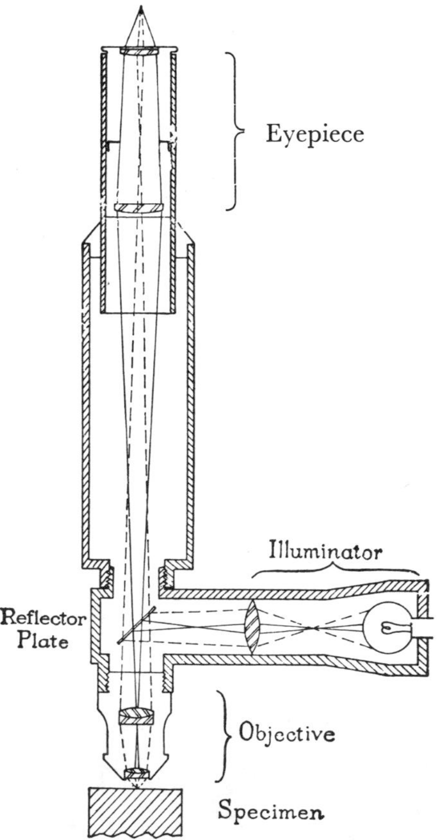

Fig. 5.2.—Metallurgical microscope with vertical illuminator

5.2. Formation of liquid drop from vapour

The growth of metal crystals from molten liquid is by a process of nucleation and grain growth, which is very similar to the formation of liquid drops from a vapour. In a vapour, due to the random motion of the atoms by variation of their thermal energy, they will come together in groups of varying numbers for an instant and then disperse. Above a certain size the group may be stable and continue to grow. The condition for stability depends upon the energy change as the group is formed. This consists of two parts: firstly, the energy of the interface which in the case of a liquid drop is the surface tension, and secondly, the energy change due to the difference in the free energy of the atoms in the two states of aggregation.

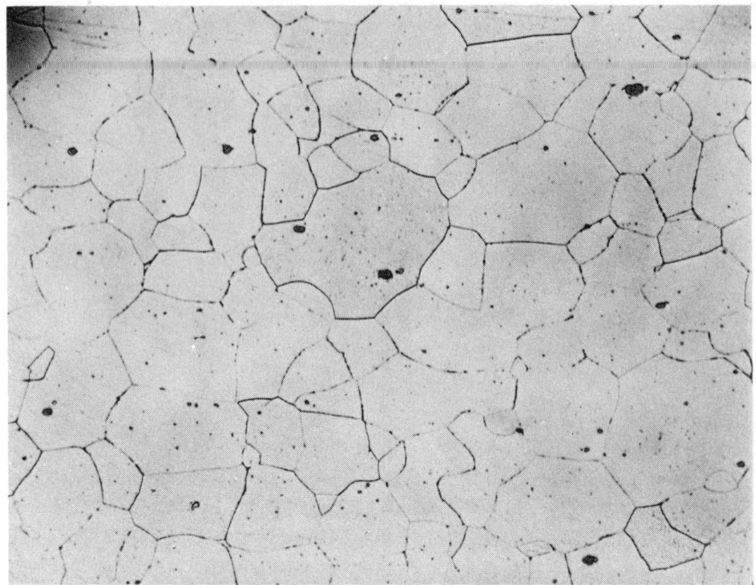

Fig. 5.3.—Armco iron (× 200) showing grain boundaries which were attacked preferentially by etchant

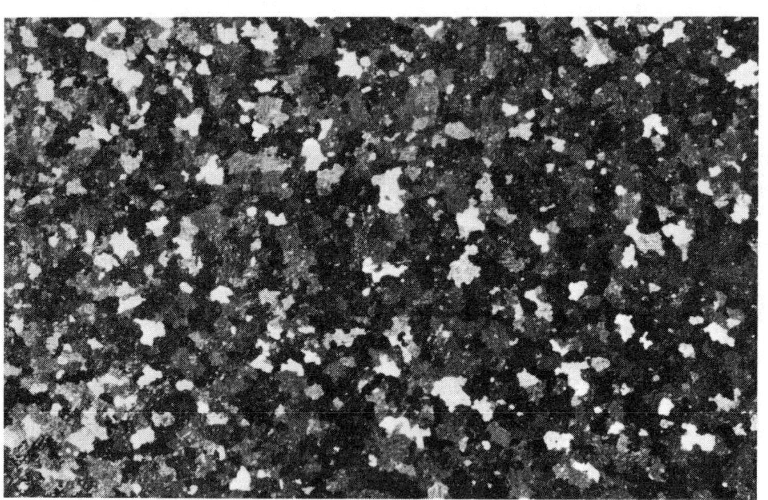

Fig. 5.4.—Aluminium (× 5). Grains reflect light to differing extents depending upon the orientation of the facets exposed in etching

If ΔF is the change of free energy* per unit volume of liquid in passing from the vapour to the liquid phase, then ΔF is zero at that temperature at which the saturation vapour pressure equals the pressure of the vapour and becomes negative as the temperature falls below that value.

Suppose a spherical group of atoms or molecules of radius r has formed due to the random motion. The energy used to create the interface is the product of the surface area $4\pi r^2$ and the surface tension τ. Then the total energy change is

$$\Delta f = 4\pi r^2\tau + \tfrac{4}{3}\pi r^3 \Delta F$$

If the group is increased in size by addition of more atoms or molecules

$$\frac{d\Delta f}{dr} = 8\pi r\tau + 4\pi r^2 \Delta F$$

Only if this is negative will the drop be stable and continue to grow, which requires that ΔF is negative and

$$r > -\frac{2\tau}{\Delta F}$$

Smaller drops will generally disperse.

At low temperatures $-\Delta F$ is larger and so the critical size of drop is smaller. Hence more of the drops formed by random motion will be stable, that is, stable nuclei form at a faster rate. Also the rate of growth of stable drops will be greater.

5.3. Crystal growth from molten metal

Similarly in the solidification of a metal from the melt, there is a minimum size of nucleus for stability. Once stable nuclei are formed, they grow by accretion of atoms on certain crystallographic planes, hence producing definite crystal faces, or frequently tend to grow in the form of dendritic shapes as in fig. 5.5. As growth continues, different crystals will meet along boundaries that do not follow a crystallographic pattern.

At the melting-point, both the rate of nucleation N and the rate of grain growth G are zero. Below the melting-point, as the temperature falls, G and N at first increase at different rates, the relative importance of which varies from material to material. If G has predominance, then few crystals will nucleate and will grow big, whereas if N predominates many crystals will nucleate before much growth occurs and the grain size will be small. At lower temperatures still, when the atom mobilities decrease, N may fall off faster than G.

* See Appendix II.

Fig. 5.5.—A dendrite

In glass, N is negligible until temperatures much below the melting-point are reached. Hence glass does not crystallize (except very slowly over a period of centuries), but forms an amorphous solid. Metals cannot be retained in an amorphous state.

5.4. Crystal shapes

As the crystal boundaries are regions where the atom pattern is distorted, they have a high energy, and hence there will be a tendency for the boundaries to assume the smallest possible surface area. Just as a single soap bubble assumes a spherical shape, this being the shape of minimum surface area for a given volume, so a collection of soap bubbles assumes shapes that make the total surface area a minimum.

Along an edge where three boundaries meet, the surface tension of each being the same, the boundaries will have angles of 120° between them. In a similar way, the boundaries between metal grains will tend to give angles of 120°, although this may not be apparent if the section examined is not perpendicular to the line of intersection of the grain boundary surfaces.

5.5. Crystal pattern in castings

In casting, the metal near the mould surface is cooled quickly and many small crystals tend to form. Then grain growth continues towards the centre of the mould, the rate of solidification being governed mainly by the rate at which the latent heat of fusion can be extracted. As the grain growth rate varies with crystallographic direction, some of the crystals in the surface layer are more favourably oriented for growth perpendicular to the surface of the casting than others. Hence only some of the grains continue to grow and columnar crystals are formed as in fig. 5.6. If a mould material is used which does not chill the metal so rapidly, or if the mould is heated before the metal is poured, the grain size in the surface layers is large.

Fig. 5.6.—Section from centre of cast ingot of copper showing columnar crystals ($\times 3$)

5.6. Grain growth in strain-free solid metal

When the strain-free metal is heated to a sufficiently high temperature for migration of atoms to occur, the grain boundaries can migrate, firstly to produce the minimum surface area discussed in Section 5.4, and then to produce a uniform increase in grain size up to a specific size for each temperature. This is known as *normal grain growth*. It is a process of energy decrease by decreasing the grain surface area and occurs extremely slowly.

5.7. Grain growth in strained solid metal

If the metal has been deformed plastically, then there are highly distorted regions left which possess high energy. The number and amount of stored energy of these regions is dependent upon the amount of deformation. On heating, these regions readily form nuclei of new grains; a process known as *recrystallization*. On heating to higher temperatures, grain growth will occur around the new nuclei. Hence, the final crystal size will depend upon the amount of straining and the temperature reached. There is a critical amount of straining that gives the largest crystal size. Further straining would give more nuclei and hence smaller grains. The variation of grain size with amount of deformation and annealing temperature for copper is shown in fig. 5.7.

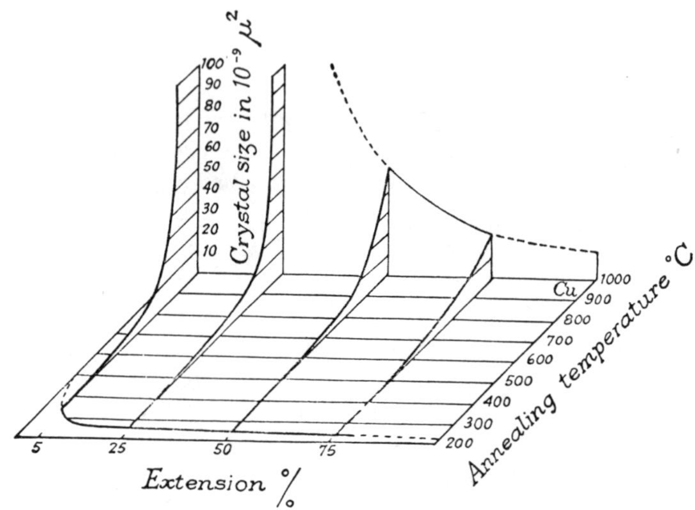

Fig. 5.7.—Variation of grain size in copper with amount of deformation and annealing temperature [Elam]

5.8. Specification of grain size

The most satisfactory way of indicating grain size would be the number of grains per unit volume. This, however, would be very difficult to determine. The method used is to count the number of grains that appear in unit length or in unit area of a prepared section of the metal. In the latter case, the grain size may then be quoted as the number of grains that appear per square inch or square millimetre of cross-section, or may be expressed in a logarithmic manner as the A.S.T.M. (American Society for Testing Materials) grain size.

If the number of grains per square millimetre of the actual section, be m, then the A.S.T.M. grain size N is given by

$$m = 8 \times 2^N$$

Since small variations in grain size are insignificant in their effect upon mechanical properties, N is quoted to the nearest whole number.

For example, a grain size of 9800 grains per mm^2 would give N = 10 ($\log_2 9800/8 = 10{\cdot}3$).

When the metal is sectioned, some grains are sectioned centrally and some are sectioned near their corners, so that even if the grains are of uniform volume the section will show a scatter in grain size.

5.9. Growth of single crystals

For investigations of the fundamental properties of metal crystals it is necessary to prepare single crystals for experimental work. The following methods are used for this purpose:

(i). A " seed " in the form of a small piece of single crystal is dipped into the surface of molten metal and withdrawn extremely slowly, so that grain growth can occur without nucleation of any new grains.

(ii). A specimen in the form of a wire is heated by a travelling furnace, so that each portion in succession melts and then cools slowly. In general, a single nucleus will form, and the wire finally becomes a single crystal. A modification of this method is to place a seed crystal at one end which does not get melted. By this means the orientation of the crystallographic planes in the final crystal can be predetermined.

(iii). A specimen of the metal is given a critical amount of strain, discussed in Section 5.7, and is then annealed. If the critical strain is exactly right, the number of nuclei formed per unit volume will be very small and large crystals are obtained.

QUESTION

1. The free energy change ΔG_{sl} per unit volume when a liquid metal freezes to the solid state at a temperature T which is below the equilibrium melting point T_E is

$$\Delta G_{sl} = \frac{\Delta H_{sl}(T_E - T)}{T_E}$$

where ΔH_{sl} is the enthalpy of melting per unit volume.

If freezing commences by atoms of the liquid clustering to form solid spherical nuclei, show that the critical radius r^* of such nuclei is

$$r^* = \frac{2\gamma_{sl} T_E}{\Delta H_{sl}\,(T_E - T)}$$

where γ_{sl} is the free energy of the solid–liquid interface. Evaluate in terms of the same quantities the critical amount of energy ΔG^* necessary to form such stable nuclei.

Calculate the values of r^* and ΔG^* at 900 °C for a copper alloy for which γ_{sl} = 0·2 J m^{-2}, the enthalpy of melting is 212 kJ kg^{-1}, the density is 8·9 Mg m^{-3} and the equilibrium melting point is 1050 ° C. [E]

CHAPTER 6

Aggregations of Two Sorts of Atoms—Binary Alloys

6.1. Introduction

A mixture of atoms of two or more elements, of which the principal one at least is metallic and which exhibits metallic properties, is known as an *alloy*. The mechanical and physical properties of the mixture may be appreciably different from those of any one of the constituent pure elements. Because of the great variation of properties obtainable by alloying, alloys are of the greatest importance in engineering; pure metals have a much more restricted use.

A study of the atomic arrangements possible in alloy systems must precede any discussion of the resulting mechanical and physical properties.

6.2. Mixtures of two liquids

When two liquids, including liquid metals, are mixed, the resulting mixture may fall into one of three classes:

(i). One liquid dissolves completely in the other in all proportions so that one homogeneous solution is formed. A common example is that of water and ethyl alcohol.

(ii). Each liquid is partially soluble in the other, so that if a little of one liquid is added to a lot of the other it all dissolves, forming a single homogeneous solution, but if more is added so that the limit of solubility is reached, then two solutions form and, on standing, separate into two layers, the denser at the bottom. Each layer will have one of the constituents as the solvent with a limited quantity of the other dissolved in it. An example is phenol and water, to which further consideration is given on p. 82.

(iii). Each liquid is completely insoluble in the other, so that a mixture will always separate into two layers, each being the pure substance. It is the limit of case (ii) with the solubilities tending to zero. It is difficult to say whether an example of this case actually exists. A mixture of liquid lead and aluminium is a case where the solubilities are extremely small.

The majority of liquid metal mixtures fall into the first category. It is not possible to make homogeneous metal castings from those metal mixtures which do not form a single homogeneous solution in the liquid state. In such cases, alloys can be made only by the method of powder metallurgy.

6.3. Mixtures in the solid state

In the solid state, a mixture of two elements may again crystallize separately or may be soluble in one another giving solid solutions. Certain compositions of mixture behave as crystalline compounds of fixed composition, possessing a unique crystal structure and having sharp melting-points. These are known as *intermetallic* or *intermediate compounds.*

6.4. Solid solutions

In a solid solution, the solute atoms are distributed throughout the crystal grains, the crystal structure being the same as that of the pure metal, which is the solvent.

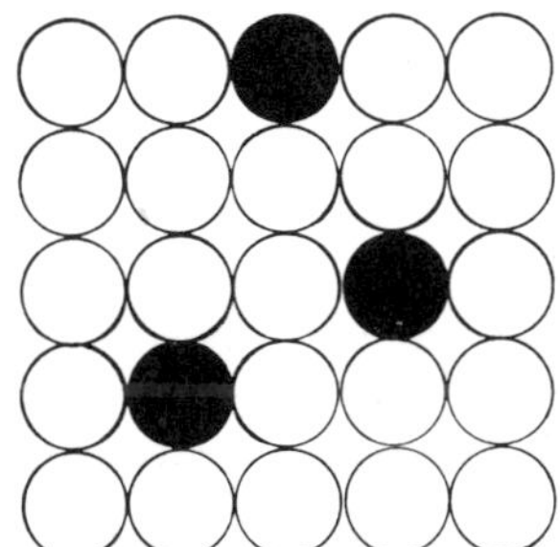

Fig. 6.1.—Substitutional solid solution

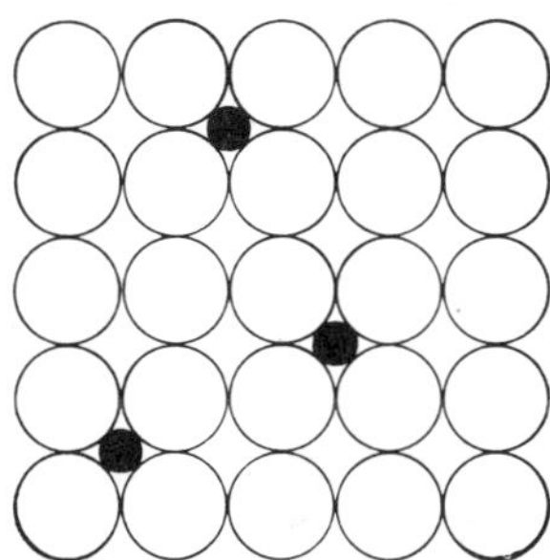

Fig. 6.2.—Interstitial solid solution

A solid solution may be formed in two ways:

(i). A *substitutional* solid solution, in which solute atoms replace some solvent atoms so that they lie at normal atom sites of the crystal structure (fig. 6.1).

(ii). An *interstitial* solid solution, in which the solute atoms are located in the interstices of the solvent lattice (fig. 6.2).

In the former case, the solute atoms may replace solvent atoms at random, or may take up a more *ordered* structure (fig. 6.3*a*). Such an ordered structure can exist at a particular composition only. For near-by compositions the structure would be as ordered as possible, as in fig. 6.3*b*. Where ordered structures (also known as *superlattices*) exist, they usually

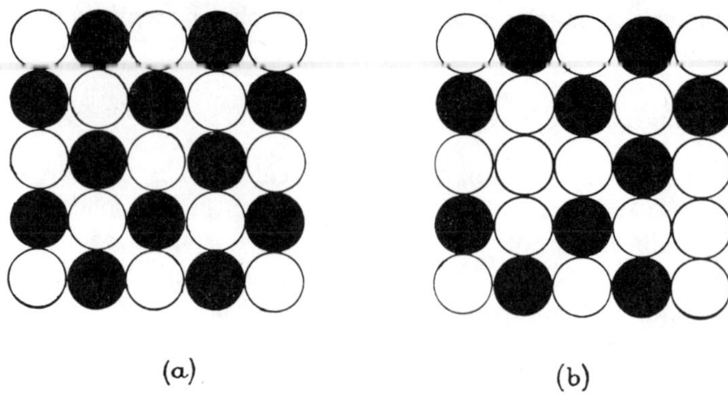

Fig. 6.3.—Ordered structure
(*a*) complete ordering with 1:1 ratio of atoms
(*b*) incomplete ordering when atom ratio is not quite 1:1

do so at lower temperatures, becoming *disordered* at higher temperatures. For an example, see the β-brass in fig. A.11.

6.5. Effect of atomic size

The type of solid solution formed and the limits of solubility are governed partly by the relative sizes of the solute and solvent atoms.

The distances between atoms in metals and between ions in ionic crystals approximately obey an additive law, each atom or ion being packed in a structure as if it were a sphere of definite size. The radius of the sphere is not constant for any given ion or atom, but depends on the number of equally-spaced near neighbours—the *coordination number*. In a close-packed lattice (p. 52) the number is 12, in a body-centred cubic lattice it is 8, in a diamond it is 4, etc. In passing from 12-fold to 8-fold coordination, there is a contraction of 3% in the radius and a greater contraction when passing to lower coordination numbers. For purposes of comparison, atomic radii are best expressed as the value appropriate to one particular coordination number. The atomic radii for 12-fold coordination of the majority of the elements are given in Table 6.1. In some cases where the element has a crystal structure with a low degree of, or no coordination, the values quoted are one-half of the smallest inter-atomic distance.

Interstitial solid solutions can form only when the solute atom is small enough to fit into a space between solvent atoms without causing excessive distortion. In a close-packed lattice, there are two kinds of space, one with four atoms and the other with six atoms arranged symmetrically around the void. The centres of these surrounding atoms lie at the corners of regular tetrahedra and octahedra, respectively, so that the voids are known as *tetrahedral* and *octahedral spaces* (see fig. 6.4). The position of an interstitial atom in a body-centred cubic lattice is discussed in Section 8.2.

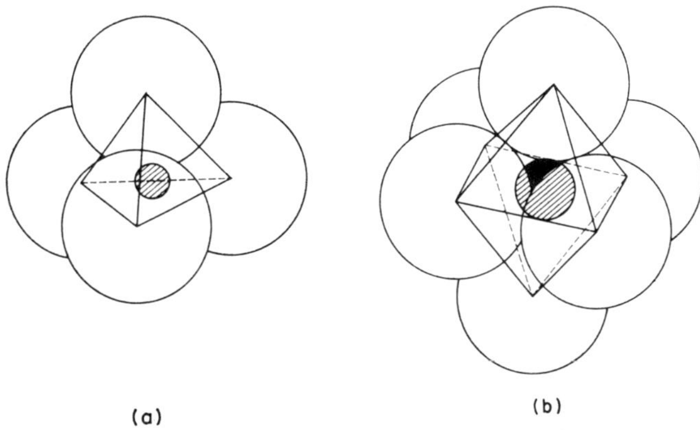

Fig. 6.4.—(*a*) Tetrahedral and (*b*) octahedral spaces in close-packed lattice, showing interstitial atoms

There can be appreciable solid solubility only if the diameter of the solute atom is about 0·6 or less of the diameter of the solvent atom. Apart from the inert gases, the alkali and alkaline-earth and some of the rare-earth metals, nearly all atoms have diameters which lie in the range 0·21 nm to 0·36 nm, and the metals which form the basis of commercially important alloys lie in the range 0·25 nm to 0·32 nm. The only atoms small enough to form interstitial solid solutions are the first five in Table 6.1.

A more detailed treatment of the formation of interstitial solid solutions of carbon in the different allotropic forms of iron is given on pp. 108–111.

6.6. Factors controlling substitutional solid solubility

Substitutional solid solutions are favoured when the atoms are more nearly of the same size. When the difference between the diameters is more than 14%, solid solubility is likely to be restricted. Complete solid solubility can, of course, exist only when both pure metals have the same crystal structure.

Two factors which also influence the extent of solid solubility are the relative electronegative valence and the relative valency. When one of

TABLE 6.1—THE RADII OF THE ATOMS IN ORDER OF MAGNITUDE

The values quoted are those appropriate to 12-fold coordination, except those given in parentheses which are half the smallest inter-atomic distance in the pure material (values taken from *Metals Reference Book*, second edition edited by Smithells, Butterworth).

	nm		nm		nm
H	0·046	Ir	0·135	Mg	0·160
O	0·060	V	0·136	Ne	0·160
N	0·071	I	(0·136)	Sc	0·160
C	0·077	Zn	0·137	Zr	0·160
B	0·097	Pd	0·137	Sb	0·161
S	(0·104)	Re	0·138	Tl	0·171
Cl	(0·107)	Pt	0·138	Pb	0·175
P	(0·109)	Mo	0·140	He	0·179
Mn	(0·112)	W	0·141	Y	0·181
Be	0·113	Al	0·143	Bi	0·182
Se	(0·116)	Te	(0·143)	Na	0·192
Si	(0·117)	Ag	0·144	A	0·192
Br	(0·119)	Au	0·144	Ca	0·197
Co	0·125	Ti	0·147	Kr	0·197
Ni	0·125	Nb	0·147	Sr	0·215
As	(0·125)	Ta	0·147	Xe	0·218
Cr	0·128	Cd	0·152	Ba	0·224
Fe	0·128	Hg	0·155	K	0·238
Cu	0·128	Li	0·157	Rb	0·251
Ru	0·134	In	0·157	Cs	0·270
Rh	0·134	Sn	0·158	Rare earths	
Os	0·135	Hf	0·159	0·173–0·204	

the elements is more electropositive and the other is more electronegative, i.e., appear in lower and higher groups, respectively, of the periodic table, the formation of intermediate compounds is probable, and this restricts the possible range of solid solutions.

Secondly, adding a metal with more valency electrons to one with fewer valency electrons increases the ratio of the total number of valency electrons to the total number of atoms. Such increase is limited to some more or less fixed value for any particular crystal structure. The reciprocal effect of decreasing the electron concentration by adding a metal of lower valency to a metal of higher valency is even more restricted.

Some examples chosen to illustrate the effects of these factors are shown in Table 6.2. The first three examples are all metals of Group I with a valency of 1 and have the same crystal structure. Hence size is the factor that will control the extent of solid solubility. In the three cases Ag-Au, Au-Cu, and Ag-Cu the differences in size are 0·2%, 12·8%, and 13%, respectively, of the size of the smaller atom. The first case should obviously show complete solid solubility while the others will be border-line cases. The Ag-Cu system shows limited solubility, while the Au-Cu system does not, but the equilibrium diagram (fig. A.8) shows it to be near the limit with a tendency to form a eutectic (see p. 86).

In the Ag-Cd and Cu-Cd systems, the crystal structure and valency of cadmium differ from those of silver and copper, factors that would restrict the solid solubilities. The size differences are 5% and 16% respectively, so that the solubilities are much smaller in the second case.

The Cu-As system is an example of an alloy system between two elements that have different crystal structures and are well separated in the periodic table, arsenic being almost a non-metal. The size difference is only 2%, but solubilities are small and an intermediate compound is formed.

6.7. Intermediate compounds

These compounds have crystal structures different from either of the pure elements. They are variously referred to as intermetallic, intermediate, and chemical compounds. The term intermetallic is applicable only when both elements are metallic. Many of the intermediate compounds of engineering importance contain a non-metallic element, especially carbon and nitrogen in carbides and nitrides, respectively. Three types of intermediate compounds are discussed.

1. *Valency compounds*

Valency compounds are compounds of electropositive metals with electronegative elements of Groups IVb, Vb, and VIb of the periodic table (Table 2.2). The bonding may be either ionic or covalent, and the normal valency rules are obeyed. They are usually hard and brittle, and poor conductors. They have sharp melting-points which may be above the melting-point for either pure element.

2. *Electron compounds*

Metals not greatly different in electrochemical properties and having a fairly favourable size factor but having different numbers of valency

TABLE 6.2—EXAMPLES OF SUBSTITUTIONAL SOLID SOLUTIONS

Metals	Atom diameter nm	Crystal structure	Group of Periodic Table	Valency	Relative solubility	Compounds formed
Ag Au	0·2889 0·2884	F.C.C. F.C.C.	I I	1 1	Continuous	—
Au Cu	0·2884 0·2556	F.C.C. F.C.C.	I I	1 1	Continuous (but liquidus shows a minimum)	—
Ag Cu	0·2889 0·2556	F.C.C. F.C.C.	I I	1 1	Restricted 8·8% Cu, 8% Ag	—
Ag Cd	0·2889 0·304	F.C.C. H.C.P.	I II	1 2	Restricted 6% Ag, 44% Cd	Electron compounds
Cu Cd	0·2556 0·304	F.C.C. H.C.P.	I II	1 2	Slight 0·12% Cu, 1·7% Cd	Electron compounds
Cu As	0·2556 0·250	F.C.C. Rhombohedral	I V	1 3	Restricted 8% As, no data on Cu solubility	Valency compound Cu_3As

electrons may show intermediate phases which are essentially metallic in properties and are not of exactly fixed composition.

Hume-Rothery has pointed out that the crystal structure is often related to the ratio of the total number of valency electrons to the total number of atoms. Thus monovalent copper will dissolve bivalent zinc in the face-centred cubic structure until the electron/atom ratio is about 1·4 (corresponding to 40% zinc by numbers of atoms). The body-centred structure is more stable when the electron/atom ratio is about 1·5, and further structures appear at higher ratios. When trivalent aluminium is alloyed with copper, the limit of solubility in the face-centred cubic copper lattice is about 20% by numbers of atoms, again corresponding to a valency electron/atom ratio of 1·4, and a body-centred cubic phase exists in the vicinity of an electron/atom ratio of 1·5.

In the electron compounds, the bonding is metallic and the range of composition corresponds to a solid solution. The more complex structures tend to be brittle.

3. *Interstitial compounds*

Interstitial compounds form between the elements which can go into interstitial solid solution and metals in which they have little solid solubility. If the size factor is favourable, the metal takes up a simple crystal structure with the non-metallic atoms in interstitial positions. In some cases, such as cementite, which is discussed in the following section, the size conditions are borderline, and the structure is complex.

Interstitial compounds have high melting-points and are very hard and brittle. Carbides are of great importance in cutting tools, and nitrides are formed in one process of surface-hardening steel.

6.8. Formulae of intermediate compounds

In the iron-carbon system, an intermediate compound, *cementite*, is found at a composition of 6·67% carbon by weight. The ratio of numbers of atoms can be found by dividing the relative weights of elements by the atomic weights. Thus:

$$\frac{\text{atoms of carbon}}{\text{atoms of iron}} = \frac{6{\cdot}67/12}{93{\cdot}33/55{\cdot}85} = \frac{0{\cdot}556}{1{\cdot}671} = \frac{1}{3}$$

Hence the empirical formula is Fe_3C.

This should not be taken to mean that cementite is composed of molecules, each of which consists of one carbon and three iron atoms.

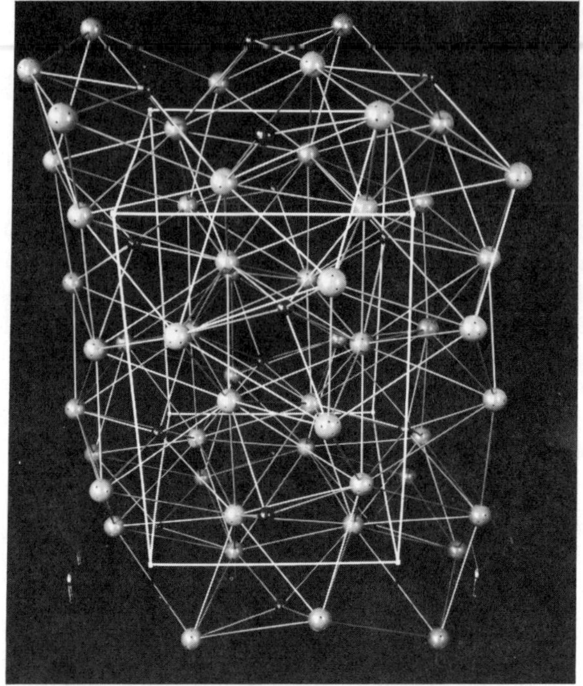

Fig. 6.5.—Model of cementite structure. White rods outline unit cell

The crystal merely contains three times as many iron as carbon atoms. The structure, of which fig. 6.5 shows a model, is orthorhombic, the unit cell sides being

$$a = 0{\cdot}452 \text{ nm}$$
$$b = 0{\cdot}508 \text{ nm}$$
$$c = 0{\cdot}637 \text{ nm}$$

and each unit cell contains twelve iron and four carbon atoms.

6.9. Phase

A phase is defined as a portion of matter, homogeneous in the sense that the smallest parts into which it can be divided mechanically are indistinguishable from one another. If two portions of matter co-existing in equilibrium are different in composition, crystal structure, or state, they are different phases of the matter considered.

A mixture of phenol and water in approximately equal portions at room temperature will separate into two layers, the upper being water with phenol dissolved, and the lower being phenol with water dissolved in it. The two layers are of different composition and hence are two phases of the phenol-water system.

Allotropic forms of the same metal, such as the α and γ forms of iron, and different types of crystal structures occurring at different compositions of an alloy system are different phases.

QUESTIONS

1. Distinguish between an interstitial and substitutional solid solution. What part does the ratio of the atomic radii play in determining the type and range of solid solution? Are any other factors involved?

2. Copper and nickel form a continuous solid solution at all compositions, whereas copper and silver form two solid solutions of limited solubility. How can this be explained in terms of atomic radii?

3. Discuss the factors that control the extent of solid solubility in alloy systems.
For each of the following alloy systems state, with reasons, whether one might expect complete, partial, or negligible solid solubility: (*a*) copper-silver, (*b*) silver-gold, (*c*) silver-cadmium, (*d*) copper-cadmium. [MST]

4. Tin and lanthanum form intermediate compounds containing 28%, 44%, and 70% of lanthanum. Determine the empirical formulae of the three compounds.

5. From the information given in Section 6.8, calculate the specific gravity of cementite.

6. In terms of the radius of the solvent atom, calculate the radius of the largest atom which can be fitted interstitially into a face-centred cubic lattice. [P]

7. Treating atoms as hard spheres, calculate the radius of the largest atom that can be accommodated in the smaller kind of void present in the hexagonal system of closest packing of atoms of radius R. [P]

CHAPTER 7

Equilibrium Diagrams

7.1. Introduction

A description of the internal structure of each of the alloys possible between two or more metals, giving the compositions and temperatures over which the various phases exist, is most conveniently presented as an equilibrium or constitution diagram.

The base line or abscissa of the diagram for an alloy system between two elements A and B shows the composition of all possible alloys from 100% of one component A and 0% of the other component B at the left-hand end, to 0% of A and 100% of B at the right-hand end, as in fig. 7.1.

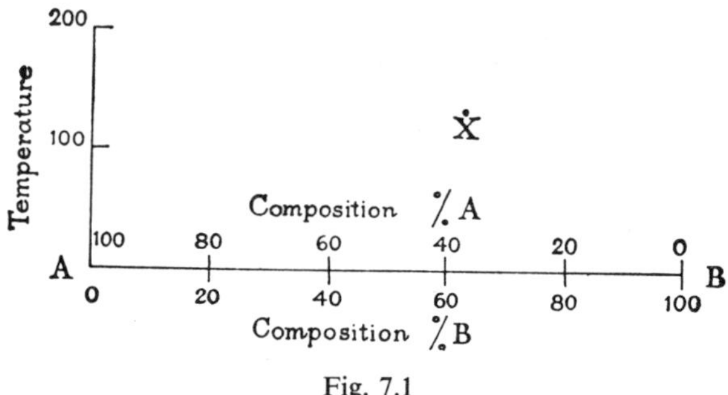

Fig. 7.1

The ends of the abscissa are labelled with the names or symbols of the components. Thus A appears at the left-hand end and B at the right-hand end. It is not necessary, though often convenient, to show the graduations of the scale. The composition is usually expressed by weight, but may be shown in terms of numbers of atoms. The ordinate scale represents the temperature range. Thus in fig. 7.1, the point X indicates an alloy containing 37% of A and 63% of B at a temperature of 130 °C.

Lines can be drawn on this diagram to show the boundaries of the composition-temperature ranges over which the various phases exist.

7.2. Equilibrium diagram for two liquids

A simple example for consideration is the phenol-water mixture previously mentioned. At 10 °C the relative solubilities are 7·5 kg of phenol in 100 kg of aqueous solution and 25 kg of water in 100 kg of phenol solution. Thus if 7 kg of phenol is mixed with 93 kg of water, it can all dissolve forming an unsaturated solution—a single phase. If more than 7·5% of phenol is used, then two phases will form, being two saturated solutions, the relative amounts of the two solutions being such as to give the correct overall composition. On increasing the temperature, the solubilities increase. The variations of solubility are shown in fig. 7.2.

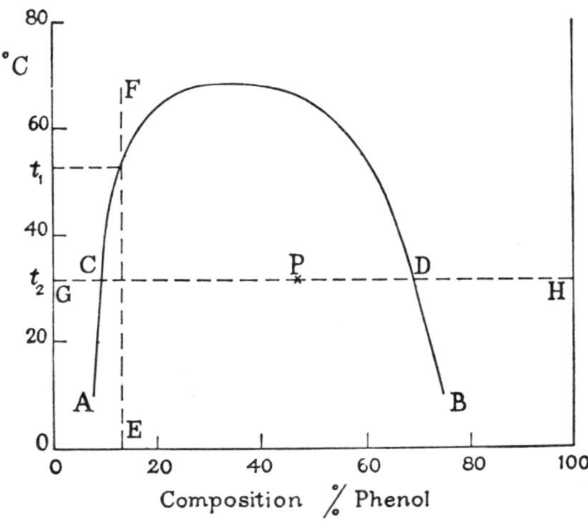

Fig. 7.2.—Solubility curve for water-phenol mixture

The lines AC and BD show the boundaries of the single-phase regions. At 68·3 °C it is seen that the two regions have come together so that at this and higher temperatures, one phase only will exist for any composition.

A mixture containing 13% of phenol, represented by EF, would have two phases at 10 °C, but on heating to a temperature t_1 the solubility of the aqueous layer would have increased sufficiently for there to be only one solution, and so it would be single-phase above that temperature.

7.3. Rules for interpreting binary equilibrium diagrams

At a temperature t_2, for a composition given by point P there would be two phases whose compositions would be the saturated solutions that can co-exist at that temperature, i.e., C and D. The relative amounts of the two phases can be determined as follows:

Let a, b, and p be the fractions of phenol in the mixtures of composition C, D, and P, respectively, i.e.

$$a = \frac{\text{GC}}{\text{GH}}, \quad b = \frac{\text{GD}}{\text{GH}}, \quad p = \frac{\text{GP}}{\text{GH}}$$

Let the mixture contain a fraction x of the phase represented by C and a fraction $(1-x)$ of D.

Then the proportion of phenol in the total is

$$p = ax + b(1-x)$$

whence

$$x = \frac{b-p}{b-a} = \frac{\text{PD}}{\text{CD}} \quad \text{and} \quad 1-x = \frac{\text{CP}}{\text{CD}}$$

These considerations lead to the two rules by which we can determine the compositions and proportions of phases present for any point in a two-phase region.

Rule 1.—Draw a horizontal line at the chosen temperature. The intersections of this line with the two boundaries of the two-phase field give the compositions of each of the phases existing at that temperature.

Rule 2 (The lever rule).—Let the point representing the composition and temperature be the fulcrum of a horizontal lever. The lengths of the lever arms from the fulcrum to the boundaries of the two-phase field multiplied by the weights of the phases present must balance.

While these two rules give the composition and structure of the phases, it is also possible to determine from the equilibrium diagram the way in which the phases are distributed. This will become clear in the following sections which deal with the various types of equilibrium diagrams. Discussion is restricted to those cases in which there is complete solid solubility in the liquid state.

7.4. Equilibrium diagram for the case of complete solid insolubility

When two substances are soluble in the liquid state but insoluble in the solid state, it is found that the freezing-point of one substance is lowered by the addition of some of the other substance to the liquid. Raoult's law states that *the amount by which the freezing-point is lowered is proportional to the concentration of the second substance and to its molecular weight.*

Thus the freezing-point curve would be of the form shown in fig. 7.3, beginning at the freezing-point of the pure substance A and dropping steadily as more B is added. Such a line which indicates the temperature at which freezing begins for any particular alloy is called a *liquidus* line.

Above the line only the liquid phase exists. When freezing commences on cooling, crystals of pure A form, so that the composition of the remaining liquid will change, becoming richer in B. This new liquid will have a lower freezing-point than the original liquid, and no more crystallization will occur until the temperature has fallen sufficiently. Hence crystallization of A occurs steadily over a range of temperature. Under the liquidus is a two-phase region, which is shaded in fig. 7.3.

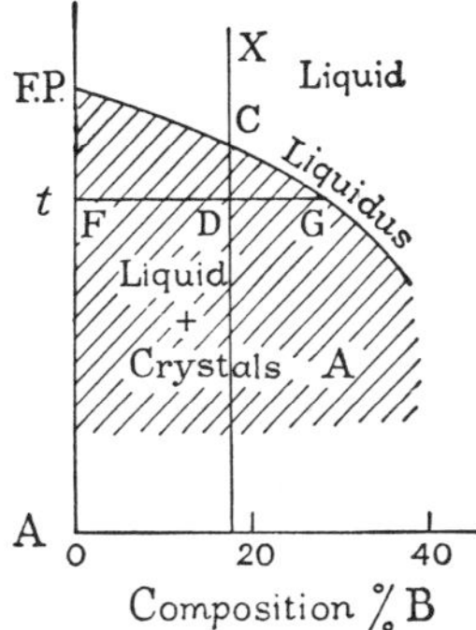

Fig. 7.3.—Depression of freezing-point of an element A by addition of an element B which is soluble in liquid A and insoluble in solid A

For an alloy of composition X cooled from the liquid phase, crystallization of A begins at a temperature given by C. At a lower temperature *t*, this alloy is in a two-phase region. Applying the two rules, it follows that (i) the compositions of the two phases are given by F and G, (ii) the relative amounts are:

$$100 \frac{DG}{FG} \% \text{ solid (pure A)}$$

$$100 \frac{FD}{FG} \% \text{ liquid (composition G)}$$

The liquidus curve for the freezing of pure B from liquids at the B-rich end of the system would be a similar curve starting from the freezing-point of pure B and falling as a higher proportion of A is present.

When the two liquidus curves are put on to one diagram as in fig. 7.4, they will intersect at E. A liquid of any composition will on cooling

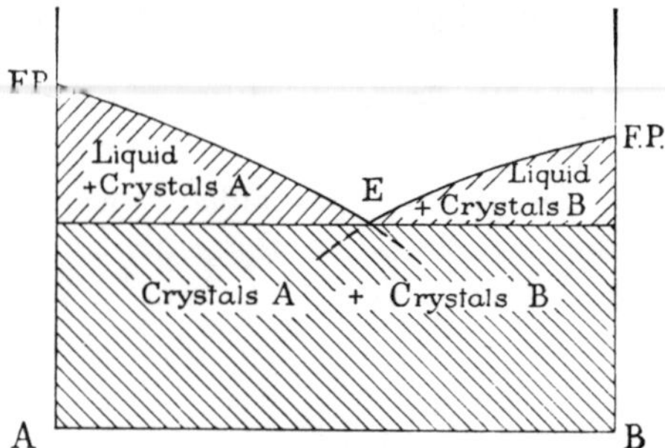

Fig. 7.4.—Eutectic formation between two elements insoluble in solid state

deposit either pure A or pure B, its composition thereby changing until point E is reached. At this composition and temperature, it is saturated with both A and B and can precipitate crystals of both elements. The result is that precipitation of both A and B proceeds as rapidly as removal of latent heat will permit until solidification is completed. During this process, the composition and temperature of the remaining liquid do not change.

The process can be regarded in more detail as follows. If a crystal of pure A nucleates in the liquid of composition E and continues to grow, the concentration of B will increase in the immediately adjacent liquid so that the liquid is supersaturated in B and nucleation of B crystals will occur. These will grow causing local supersaturation in A, leading to nucleation of more A, and so on.

Point E is known as the *eutectic point* and the change occurring there is the *eutectic reaction*. It may be written as

$$\text{liquid E} \underset{\text{heating}}{\overset{\text{cooling}}{\rightleftharpoons}} \text{crystals A} + \text{crystals B}$$

For any composition, solidification is completed at the temperature of E. A horizontal line on the equilibrium diagram at this temperature will divide the liquid+solid two-phase regions from the solid+solid two-phase region. This line below which everything is solid is called the *solidus*.

The arrangement of the solid phases, known as the microstructure, can be foretold. On cooling a liquid, firstly pure A or pure B will separate until the temperature reaches the eutectic temperature. This will be the primary or pro-eutectic crystals. Then the remainder undergoes the eutectic reaction forming a finely divided mixture of the two pure elements.

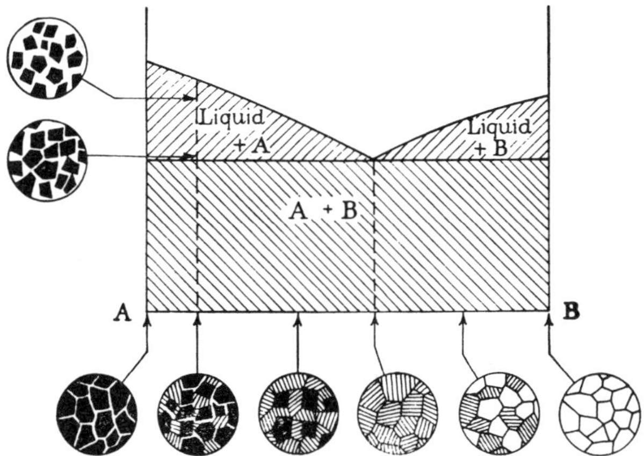

Fig. 7.5.—Microstructures of eutectic-forming alloy

Diagrammatic sketches of the microstructure found at various temperatures for a particular composition and for various compositions at a low temperature are shown in fig. 7.5. Here crystals of A are shown black and crystals of B white with typical grain boundary shapes. The liquid is shown white, while the eutectic is shown shaded, representing the mixture of A and B.

In practice, complete solid insolubility is unrealistic. Several metals do, however, have very small mutual solid solubilities and approximate to the equilibrium diagram shown in fig. 7.4, which is the limiting case of the form of diagram considered in Section 7.7.

7.5. Cooling curves

When a specimen is cooled over a temperature range for which its specific heat is constant and there is no phase change, the time-temperature curve is like (*a*) in fig. 7.6. Since the cooling will approximately follow Newton's law of cooling, which states that the rate of loss of heat is proportional to the temperature difference between a body and its surroundings, the curve will be roughly of logarithmic form.

A pure metal being cooled from the liquid state will have a similar curve down to the freezing-point. Then during solidification, the latent heat will be lost at constant temperature, after which the temperature will again fall in a logarithmic curve. A plateau will thus appear in the cooling curve at the freezing-point temperature (curve *b* in fig. 7.6).

For an alloy intermediate in composition between a pure metal and the eutectic composition, the cooling curve of the liquid phase will con-

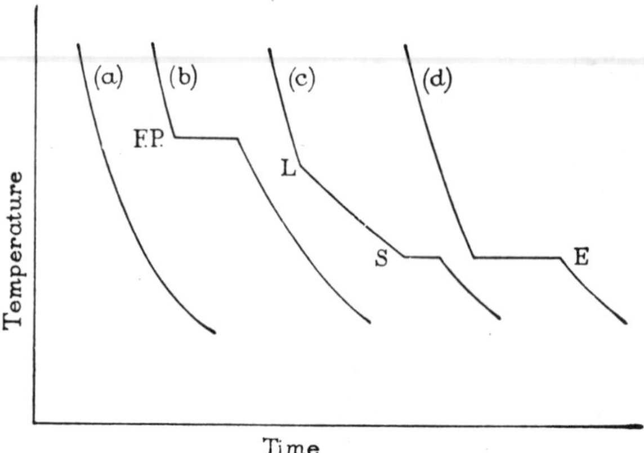

Fig. 7.6.—Cooling curves for (*a*) pure substance, (*b*) pure substance with change of state, (*c*) eutectic forming alloy with pro-eutectic component, (*d*) alloy of eutectic composition

tinue down to the temperature of the liquidus. Between the liquidus and solidus temperatures there is continuous solidification, and hence a steady loss of latent heat as well as specific heat while the temperature is falling. More heat has to be lost per degree fall of temperature than in the liquid state, and hence the rate of fall of temperature is less. There will be an abrupt change of slope at the liquidus temperature (curve *c* in fig. 7.6). At the eutectic temperature, the remaining liquid, now of eutectic composition, solidifies and the cooling curve shows a plateau at the solidus temperature. For a liquid, initially of eutectic composition, there will be a plateau at the eutectic temperature and no abrupt changes of slope elsewhere (curve *d* in fig. 7.6).

Hence by obtaining cooling curves for samples of such an alloy system, the points on the liquidus and solidus curves may be derived enabling the equilibrium diagram to be constructed.

7.6. Equilibrium diagram for complete solid solubility

Where the two elements of a binary alloy system show complete solid solubility in the solid state as well as in the liquid state, then only one phase exists for any point on the equilibrium diagram in the entirely solid region. It is found, however, in most cases that the crystals deposited from the liquid do not have the same composition as the liquid. A typical equilibrium diagram of this system is shown in fig. 7.7*a*. Raoult's law does not apply in this case and the freezing-point of a pure metal

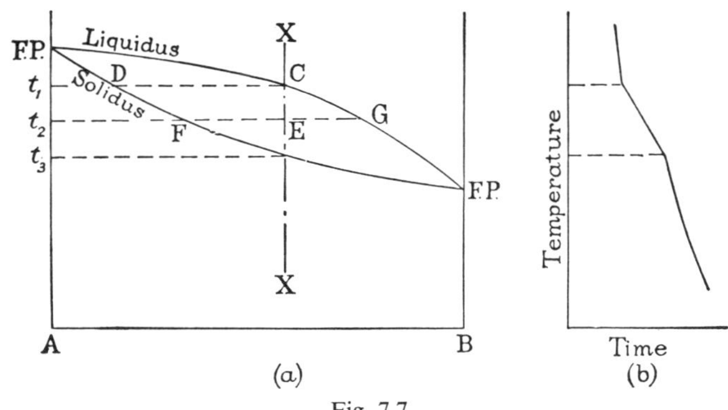

Fig. 7.7.
(*a*) Equilibrium diagram for two elements showing complete solid solubility
(*b*) Cooling curve for composition X

may be either raised or lowered by addition of the other metal. For an alloy whose composition is represented by the line XX in fig. 7.7*a*, the liquidus temperature is given by t_1, at which point crystals separate whose composition is given by D. The liquid is thus impoverished in A, and the temperature falls so that the composition of the remaining liquid follows the liquidus line CG.

At a temperature given by t_2, the application of Rule 1 (p. 84) shows that the liquid phase is of composition given by G and the solid phase of composition given by F. The crystals first formed were richer in A, but for equilibrium should now at temperature t_2 have a composition given by F. As long as cooling is slow enough, then solid-state diffusion will occur, atoms of B diffusing into the crystals already formed to bring their composition to F.

At temperature t_3, crystallization will be complete if cooling has been sufficiently slow for all crystals to homogenize. The cooling curve (fig. 7.7*b*) will show an abrupt decrease of slope at the liquidus and then an abrupt increase at the solidus.

When cooling is rapid, as may occur in practice when alloys are cast into cold moulds, this homogenization will not have had time to occur. The crystals then vary in composition, being richer in A at their centres, and the proportion of liquid remaining at any temperature will not be exactly that given by the equilibrium diagram. When a micro-structure is prepared it will be possible to see that there is such coring if the difference in composition in different parts of the crystal is sufficient to give a difference in appearance when etched with a suitable etchant. The cored structure may be homogenized by an annealing process, i.e., heating at a sufficiently high temperature for sufficient time to allow diffusion to occur.

7.7. Equilibrium diagram for partial solid solubility

In cases where the solid solubility is limited, no new principles, other than those considered in the two previous cases, are involved. The form of the diagram is shown in fig. 7.8. The outer boundaries of the (liquid + crystal) two-phase fields are not vertical lines, as in the case of complete solid insolubility, but curve away from these verticals so that the upper portion of each two-phase field has a similar appearance to the upper end of the two-phase region in the case of complete solid solubility. The crystals deposited are not pure metals but solid solutions, which, to preserve equilibrium, must change composition by solid-state diffusion as the temperature falls.

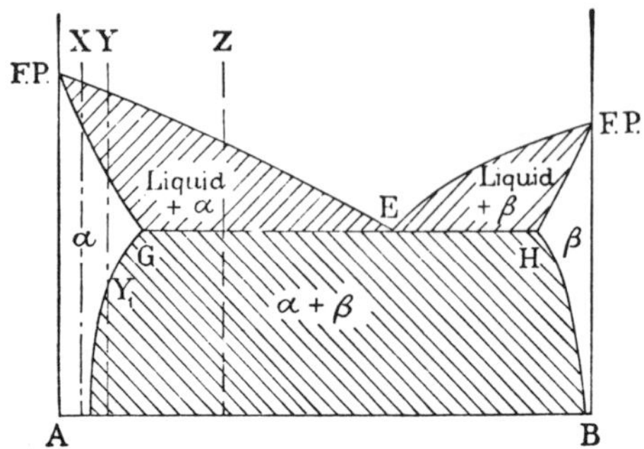

Fig. 7.8.—Equilibrium diagram for two elements showing partial solid solubility

The liquidi intersect at the eutectic point E, and the eutectic will be a mixture of crystals of composition G and of composition H.

It is usual to denote phases in alloy systems—both solid solutions and intermediate compounds—by Greek letters, usually in sequence from left to right. Here the solid solution of B in A has been termed the α-phase and that of A in B the β-phase.

Below the eutectic line GEH the solubilities decrease with falling temperature and the single-phase regions become narrower.

The compositions and proportions of phases for any alloy at any temperature can be found by direct application of the rules (p. 84). The arrangement of the phases can also be determined by a study of the equilibrium diagram.

Consider alloys of compositions denoted by the vertical lines X, Y, and Z in fig. 7.8. The structure of alloy X would be cored in cases of rapid cooling as for cases of complete solid solubility, but homogeneous after slow cooling.

Alloy Y when slowly cooled would show a homogeneous structure down to point Y_1. Below that temperature, it would become supersaturated with B-atoms and would precipitate them. The equilibrium B-rich structure is however the β-phase and so the precipitated B-atoms would take sufficient A-atoms to give β-crystals. Thus β-crystals would separate at the grain boundaries and within the grains of α-crystals. Precipitation within the parent grains will take place along certain crystallographic planes and show up as a distinct pattern in the microstructure. Such a pattern is known as a Widmanstätten structure. The name is due to the first discovery of the pattern in iron meteorites by Alois von Widmanstätten in 1808.

As grain growth occurs around the nuclei of the second phase, solute atoms have to migrate towards these nuclei so that the rate of precipitation of a second phase is limited by the rate of diffusion of the solute atoms.

Alloy Z will give, just below the eutectic temperature, primary or pro-eutectic α-crystals with a eutectic mixture of α and β of compositions G and H. Below this temperature, both α- and β-phases become supersaturated and precipitate β- and α-crystals, respectively.

The way in which the precipitation is carried out, i.e., by different rates of cooling, can cause differences in size and distribution of the second phase which can result in changes of properties which may be of engineering importance. This is dealt with later in Chapter 11.

Limiting cases of this diagram are those where G and H move towards E and just meet, so that only one solid phase is formed, but the liquidus shows a minimum (fig. 7.9); and where E moves to one boundary so that the eutectic liquid is a pure metal (fig. 7.10). Examples of these systems are the copper-gold (fig. A.8) and copper-bismuth systems, respectively.

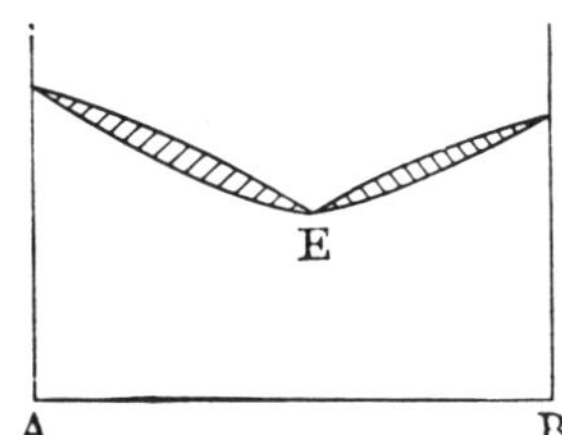

Fig. 7.9.—Equilibrium diagram for complete solid solubility with a minimum in the liquidus

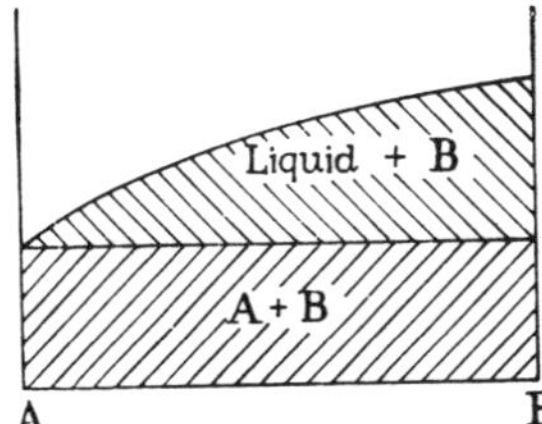

Fig. 7.10.—Equilibrium diagram showing eutectic at pure A

7.8. Intermediate compounds

The nature of intermediate compounds has been discussed in Section 6.7. Valency compounds and interstitial compounds have fixed composition and sharp melting-points, so the occurrence of one denoted by the general formula A_xB_y in a binary alloy system can be represented as in fig. 7.11.

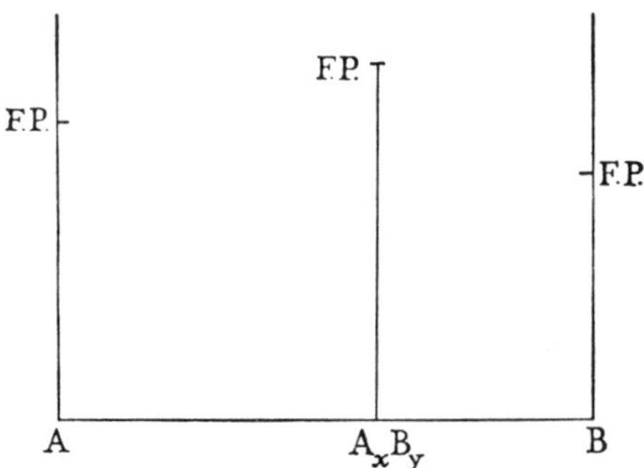

Fig. 7.11.—Freezing-points of pure element and intermediate compound

The equilibrium diagram is now divided into two parts, one showing the phases that occur between A and A_xB_y, and the other showing phases that occur between A_xB_y and B. Each of these parts may be one of the forms already described, i.e., there may be complete or partial solubility, or complete insolubility between either pure metal and the intermediate compound.

An example is shown by the lead-magnesium system in fig. A.13. The intermetallic compound occurs at 18% by weight of magnesium and can be shown to have the empirical formula Mg_2Pb. The pure lead is designated α, the intermetallic compound β, and the solid solution of lead in magnesium γ. The compound does not in this case have any range of solubility for either lead or magnesium.

In the copper-aluminium system (fig. A.3) there is an intermetallic compound at 54% by weight of copper, giving the formula $CuAl_2$. This compound, known as the θ-phase, has a limited range of solubility for both aluminium and copper, so that the θ single-phase region on the equilibrium diagram is of finite width.

Whereas valency and interstitial compounds have fixed compositions, an electron compound exists over a range of compositions, as for example, the β phase in brass (fig. A.11).

7.9. Peritectic reaction

This may be considered as an inverse of the eutectic reaction. During freezing of certain alloys, an interaction takes place between solid crystals already formed and the residual liquid to form another solid structure. This reaction occurs at a constant temperature and is known as a *peritectic* reaction. The solid phase formed may be either an intermediate compound or a solid solution as in figs. 7.12 and 7.13, respectively.

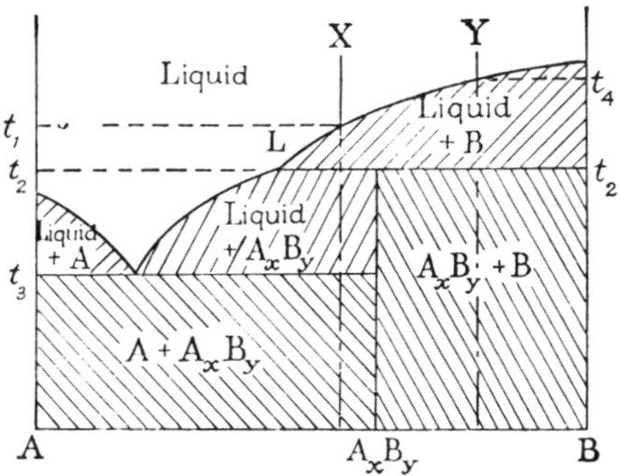

Fig. 7.12.—Peritectic reaction of intermediate compound

In fig. 7.12, the intermediate compound A_xB_y when heated to the peritectic temperature decomposes at a constant temperature to give liquid of composition represented by L and solid consisting of pure B. The peritectic reaction may be written

$$\text{liquid L} + \text{solid} \underset{\text{heating}}{\overset{\text{cooling}}{\rightleftharpoons}} \text{solid } A_xB_y$$

For an alloy of composition X cooled from the liquid state, crystals of B start to separate at t_1 and continue to do so down to t_2 which is the peritectic temperature. The peritectic reaction then occurs, all the B-crystals disappearing and an excess of liquid remaining. On further cooling, more A_xB_y is formed from the liquid, which gets richer in A until the eutectic reaction occurs at t_3.

For an alloy of composition Y, precipitation of B occurs from the liquidus temperature t_4 to the peritectic temperature t_2. There is now a higher proportion of B than was the case in alloy X and when the peritectic reaction occurs, all the liquid reacts with part of the solid B to give A_xB_y. Below the peritectic temperature we therefore have crystals of A_xB_y and B only, and their proportions do not change on further cooling.

Figure 7.13 shows a form of diagram that may occur when a solid solution is formed by a peritectic reaction. On cooling alloy W, β-crystals would separate between t_1 and t_2, the composition of the remaining liquid following the liquidus CD, and at t_2 all the remaining liquid would

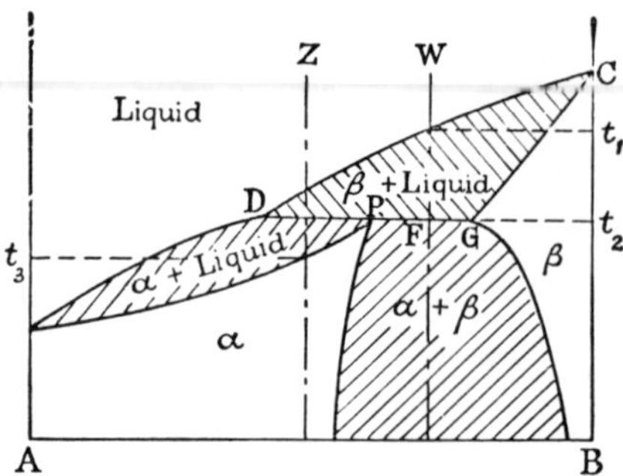

Fig. 7.13.—Peritectic reaction of solid solution

react with some of these β-crystals to give crystals of the α solid solution. Just above the peritectic temperature, the proportions of phases would be

$$100\,\frac{FG}{DG}\%\text{ of liquid (composition D)}$$

$$100\,\frac{DF}{DG}\%\text{ of }\beta\text{ (composition G)}$$

while just below this temperature the proportions would be

$$100\,\frac{FG}{PG}\%\text{ of }\alpha\text{ (composition P)}$$

$$100\,\frac{PF}{PG}\%\text{ of }\beta\text{ (composition G)}$$

Alloy Z commences to freeze in a similar manner, but at the peritectic temperature all the β-crystals react with some of the liquid to form α-crystals. On further cooling, the remaining liquid is absorbed into the α-phase, solidification being complete at the solidus temperature t_3.

Examples of the peritectic reactions are to be found in the β, γ, δ, ε, and η-phases of the copper-zinc system (fig. A.11) and in the γ-phase of the iron-carbon system (fig. 8.9).

7.10. Comparison of eutectic and peritectic reactions

It will be observed that a eutectic reaction involves one phase changing to two on cooling, whereas a peritectic reaction consists of two phases changing to one on cooling. The form of the equilibrium diagram around a eutectic and a peritectic temperature can be represented diagrammatically as in fig. 7.14.

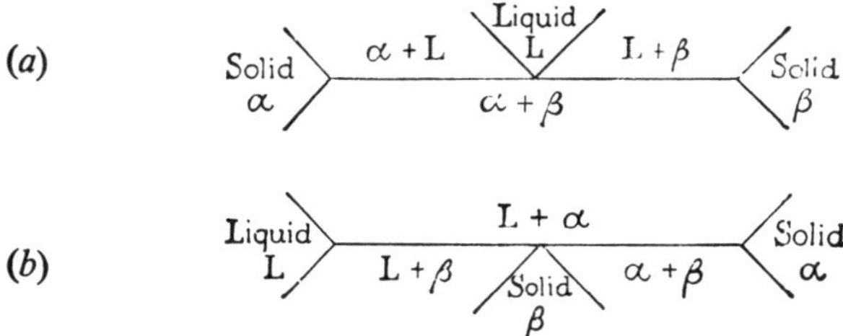

Fig. 7.14.—Diagrammatic representation of (*a*) eutectic and (*b*) peritectic reactions

7.11. Changes occurring below the solidus

The condition of an alloy when it has just solidified is not necessarily that which it will have at lower temperatures. The phase changes that occur are very similar to the phase changes occurring during the formation of the solid from the liquid state. However, diffusion in solids is extremely sluggish at temperatures that are low compared with the melting temperature range, and so equilibrium conditions are not readily obtained. The lower portions of equilibrium diagrams are therefore often uncertain. The changes that can occur are:

(i) Allotropy.
(ii) Precipitation of a second phase due to a decrease in solubility (see p. 91).
(iii) Eutectoid reaction.
(iv) Peritectoid reaction.
(v) Order-disorder changes (see p. 74).

7.12. Allotropy

Pure substances may exist in more than one crystalline form, each form being stable over more or less well-defined limits of temperature and pressure. This is known as *allotropy* or *polymorphism*, and is shown by many metals. Iron, one of the commonest examples, is considered in the next chapter. Two of the elements which show this form of behaviour and which are of importance in modern nuclear energy work are uranium and plutonium. Uranium has three and plutonium has six allotropic forms. Each of these forms, being different crystal structure, may have a different density of packing, so that there will be a change of volume at each change of temperature. The coefficient of thermal expansion of each allotropic modification may also differ, so that the changes of length with temperature can be far from regular. The variation of volume with temperature for plutonium is shown in fig. 7.15.

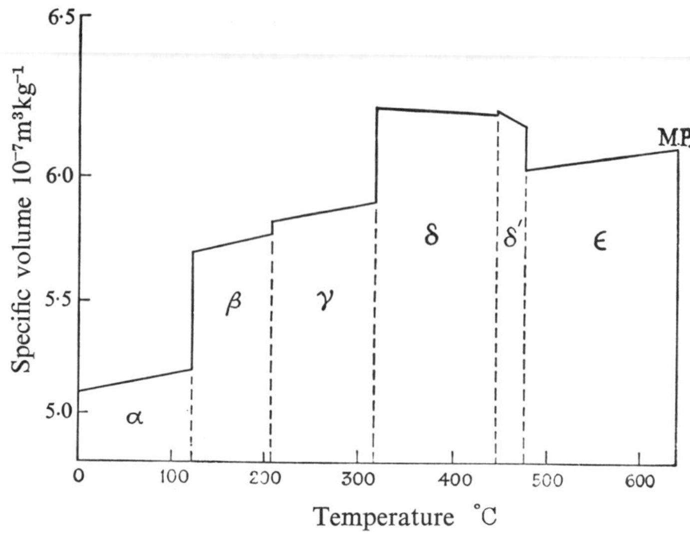

Fig. 7.15.—Variation of specific volume of plutonium with temperature

When the different allotropic forms give solid solutions in a binary system, then the transition temperature will vary with composition, the manner of variation depending upon the relative solubilities of the second element in the various allotropic phases. Some examples of the way in which this can vary are shown in fig. 7.16.

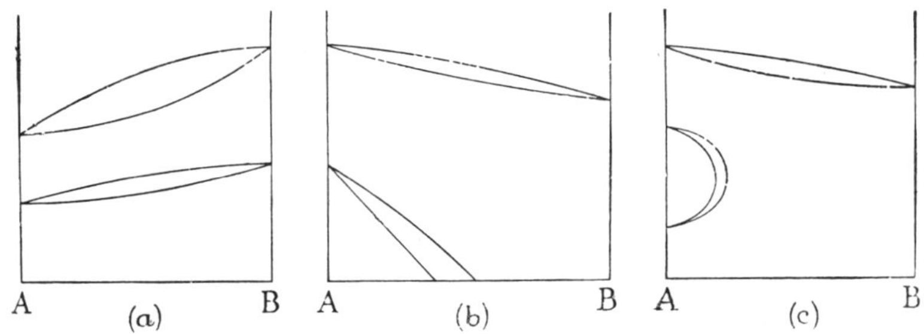

Fig. 7.16.—Examples of phase changes when polymorphism exists in binary systems (*a*) high-temperature and low-temperature forms of A and B isomorphous (i.e. of same crystal form); (*b*) A undergoes polymorphic change, high-temperature modification being isomorphous with B; (*c*) A has two polymorphic changes, high and low-temperature modifications are isomorphous with B

7.13. Eutectoid reaction

Just as a liquid solution can simultaneously precipitate two solid phases in a eutectic reaction, so a solid solution can exhibit exactly similar behaviour in what is termed a *eutectoid reaction.* The commonest example is that of pearlite formation in the iron-carbon system (p. 113).

7.14. Peritectoid reaction

This reaction is exactly similar to a peritectic reaction except that all three phases involved are solid phases. These changes are not common in alloy systems of engineering importance.

7.15. Ternary equilibrium diagrams

For plotting the equilibrium diagrams of ternary alloys, that is, alloys made of three components, three independent variables have to be considered. Two are necessary to specify the composition, and the third is temperature. A three-dimensional space model is therefore necessary to show the phases present for any composition and temperature.

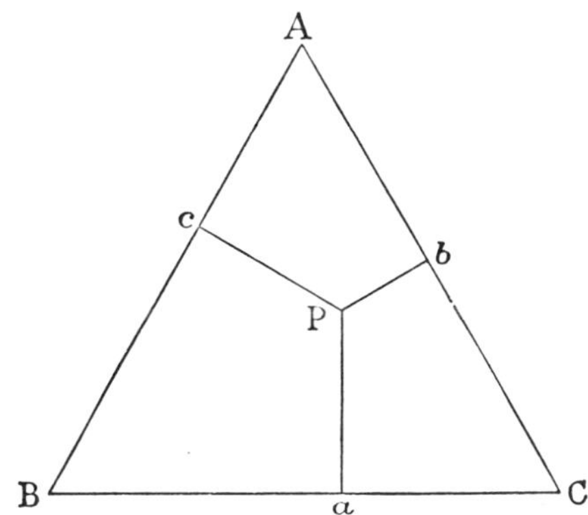

Fig. 7.17.—$Pa+Pb+Pc$ = constant

The most convenient form of diagram is that with an equilateral triangle as base and a temperature axis perpendicular to this base. The three perpendiculars to the sides of an equilateral triangle from any point P within the triangle have a constant sum whatever the position of P. If this sum be taken as 100%, then the three perpendicular distances can be used to represent the proportions of the three components. Thus in fig. 7.17, the distances P*a*, P*b*, and P*c* can represent the proportions of elements A, B, and C, respectively. The equilateral triangle can be divided up as shown in fig. 7.18, and the composition corresponding to any point can be read directly.

The side AB will represent binary alloys of elements A and B. The first line parallel to this side will represent all alloys containing 10% C.

By analogy with binary equilibrium diagrams, it is possible to develop the space models for various types of ternary alloys.

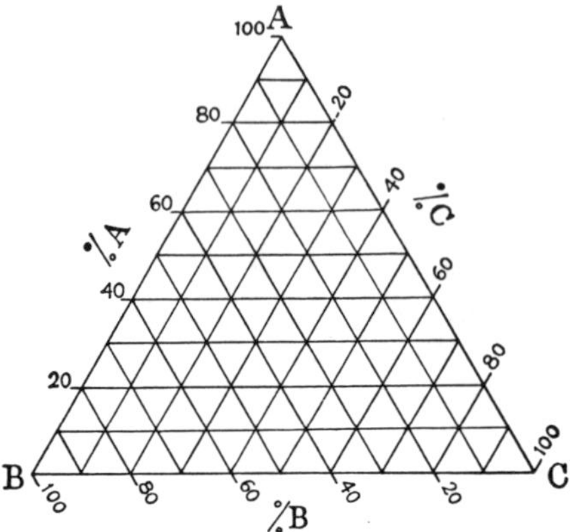

Fig. 7.18.—Scales for plotting composition on ternary equilibrium diagrams

7.16. Ternary system with complete solid solubility

For a ternary system of components completely miscible in the solid and in the liquid states, the sides of the model, i.e., the binary equilibrium diagrams, would be of the form shown in fig. 7.7. The liquids and solidus surfaces would be bounded by the liquidi and solidi of the binary diagrams and enclose a two-phase region in the form of a convex lens (fig. 7.19). While a simple diagram can be represented as a two-dimensional view, it does not lend itself to quantitative measurement, and for such purposes two-dimensional sections are used. These are usually isothermal sections or vertical sections.

The isothermal section will be an equilateral triangle on which phase boundaries are shown. Thus an isothermal section of fig. 7.19 which cuts the liquidus and solidus would look like fig. 7.20. Note that at X there exist a liquid and a solid phase, but further evidence is needed to give the actual composition of each phase. This information can be found only by experiment and is shown on the isothermal section by *tie-lines* which join compositions that co-exist in equilibrium. In fig. 7.20 tie-line PQ which passes through X shows that for the alloy of composition represented by X and for any other alloy whose composition lies on PQ, the two phases are liquid of composition P and solid of composition Q.

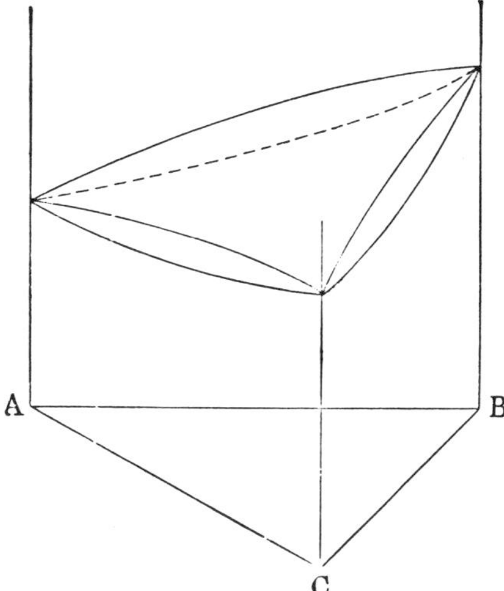

Fig. 7.19.—Ternary equilibrium diagram for complete solid solubility

A series of isothermal sections for different temperatures can give a complete representation of an equilibrium diagram. One phase-boundary surface can be defined by several isothermals on one diagram in the form of a contour map. Figure 7.21 shows this for the liquidus of the alloy system of fig. 7.19.

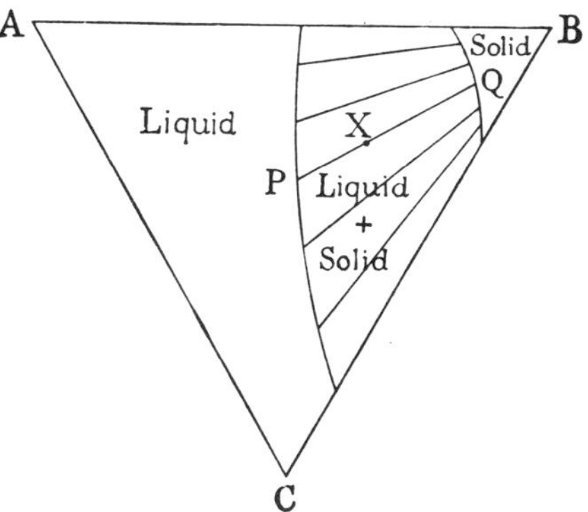

Fig. 7.20.—Isothermal section of fig. 7.19

Vertical sections are useful for showing how a binary system is affected by the presence of a third metal. It is usual to take such sections parallel to one side of the base triangle, that is at a constant proportion of one of the elements, as in fig. 7.22. The liquidus and solidus lines have the same significance as in binary diagrams, but the compositions of the phases

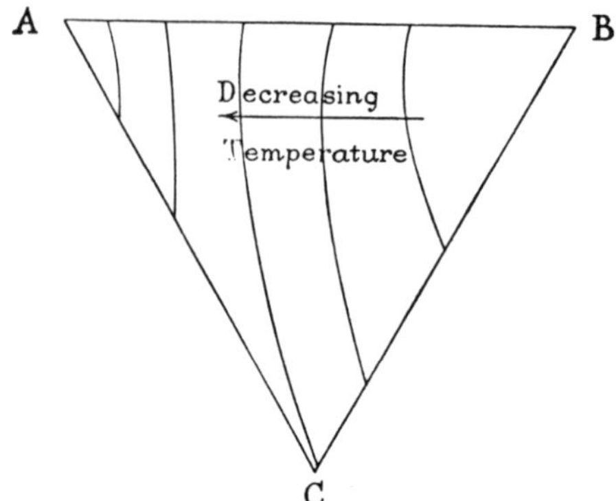

Fig. 7.21.—Isothermals of liquidus surface of fig. 7.19 at equal temperature intervals

existing at a point in the two-phase region cannot be read off the section unless the tie-line happens to lie in the plane of the section.

When a section is taken along a tie-line, then the lever rule of binary equilibrium diagrams can be applied.

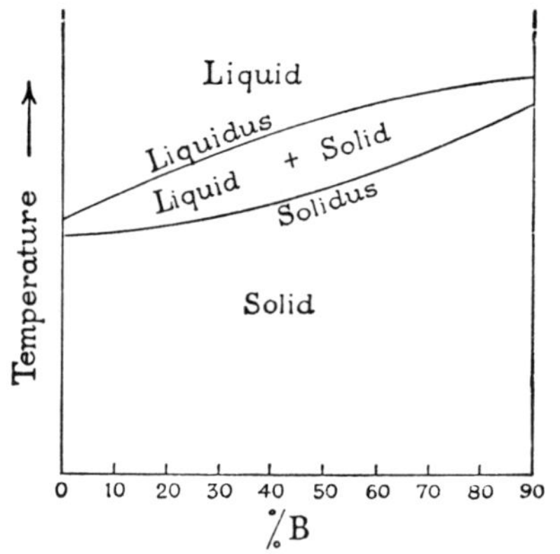

Fig. 7.22.—Vertical section of fig. 7.19 for 10% C

7.17. Ternary eutectic system

In the case of a ternary system where there is complete immiscibility in the solid state, the sides of the model will be binary eutectic diagrams of the type of fig. 7.4. Between these, the liquidi will be smooth curved surfaces which intersect to form three valleys KE, LE, and ME, with all three surfaces meeting at E (fig. 7.23). The view of the model from above is shown in fig. 7.24. A liquid with composition represented by X will, on cooling, first intersect the A-liquidus and pure A will solidify. The

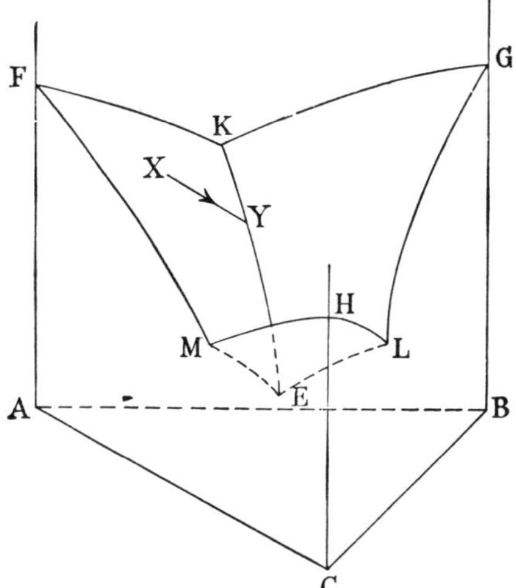

Fig. 7.23.—Ternary equilibrium diagram for complete solid insolubility

composition of the remaining liquid must move in a direction directly away from A along AX produced. Along this line the ratio of B to C will remain constant.

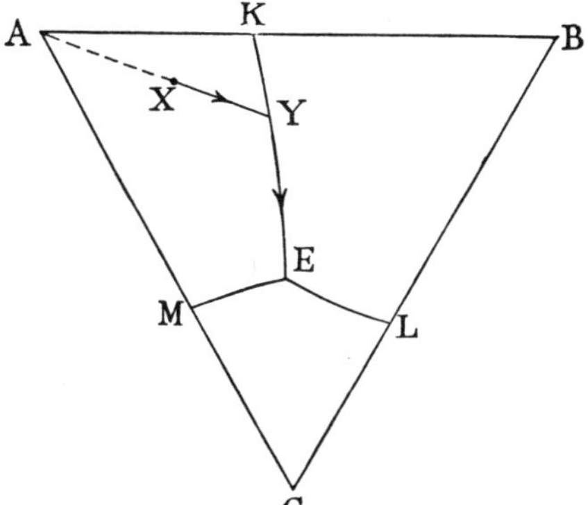

Fig. 7.24.—View of liquidi surfaces of fig. 7.23

Ultimately the liquid composition reaches KE at Y. The proportion of the whole that is solid A is XY/AY. At Y the B-liquidus is reached and the *binary eutectic reaction*

$$\text{liquid} \rightleftharpoons \text{A}+\text{B}$$

occurs. As the binary eutectic continues to form, the composition of the liquid remains on the intersection of the two surfaces, eventually reaching

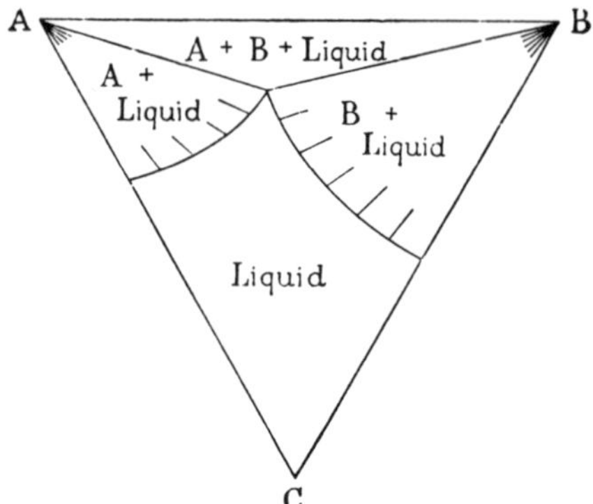

Fig. 7.25.—Isothermal section of fig. 7.23 below temperature of AB binary eutectic and above melting-point of C

E. At E the amounts of A, B, and liquid present will be such that their centre of gravity lies at X. On further extraction of heat, the *ternary eutectic reaction*

$$\text{liquid} \rightleftharpoons A+B+C$$

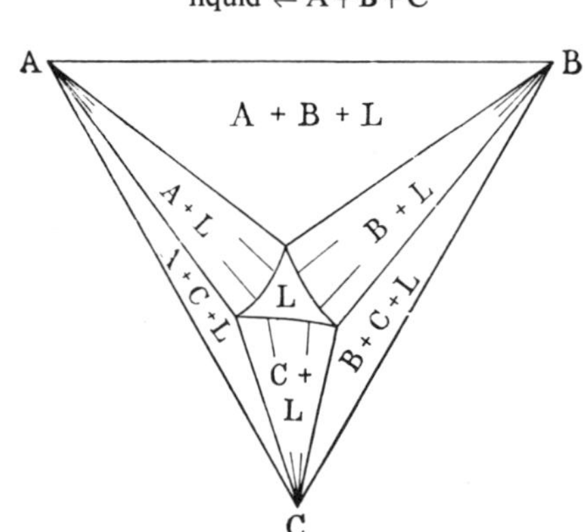

Fig. 7.26.—Isothermal section of fig. 7.23 below temperatures of binary eutectics and above temperature of ternary eutectic

occurs. The composition of the liquid remains at E until solidification is complete.

Two isothermal sections of the intermediate stages of cooling are shown in figs. 7.25 and 7.26. From these the following points may be observed:

(i) Boundaries between single and two-phase regions are curved.

(ii) Boundaries between two and three-phase regions are straight lines, and in fact are the limiting tie-lines of the two-phase regions.

(iii) Three-phase regions are triangles.

7.18. Rules for phase compositions and quantities in a ternary equilibrium diagram

In a two-phase region the rules for binary diagrams (p. 84) apply along tie-lines.

In a three-phase region the compositions of the three phases are given by the corners of the triangle, and the quantities of each are such that if placed at the respective corners they would balance about the point representing the alloy composition, i.e., an extension of the lever rule.

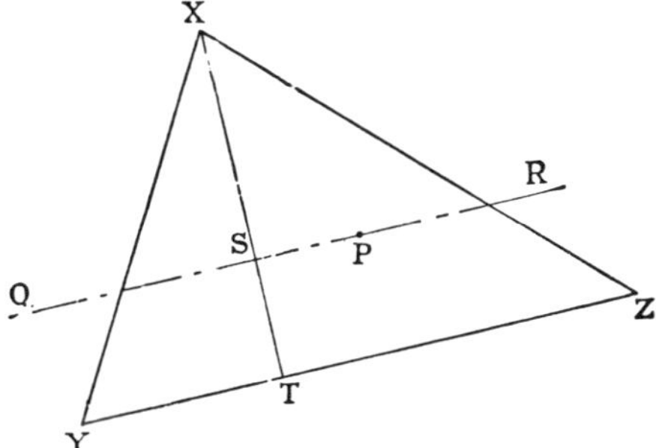

Fig. 7.27.—Lever rule applied to three-phase region

Point P in fig. 7.27 lies in the three-phase region XYZ. If x, y, and z, respectively, represent the weights of the phases present, then they must be in equilibrium about any horizontal axis through P. Taking axis QR parallel to YZ and a perpendicular from X cutting QR and YX in S and T, respectively, then by moments about QR,

$$x.\mathrm{XS} = (y+z).\mathrm{ST}$$

The proportion of phase X is given directly from

$$\frac{x}{x+y+z} = \frac{\mathrm{ST}}{\mathrm{XT}}$$

The proportions of Y and Z are found in a similar manner.

More complicated ternary diagrams are beyond the scope of this work, although no new principles are involved.

QUESTIONS

1. From the copper-nickel equilibrium diagram (fig. A.9) determine the compositions and proportions of the phases present at 1200 °C in an alloy containing 30% nickel.

2. From the lead-magnesium equilibrium diagram (fig. A.13) determine the chemical formula of the intermediate compound. Describe the changes which occur in a mixture containing 30% lead and 70% magnesium as it is slowly cooled from 650 °C to 200 °C. Give the quantities and compositions of the phases present at 500 °C and 300 °C. [MST]

3. From the equilibrium diagram of the sodium-potassium system (fig. A.12) calculate the formula of the intermetallic compound. On a copy of the diagram indicate the constitution of the material in each region.

Describe in detail the changes that occur as a 50–50 alloy is slowly cooled from 50 °C to −50 °C. What is the distribution of phases at 0 °C?

4. Construct on graph paper, using the data given below, the equilibrium diagram of the zirconium-molybdenum system to the following scales: 1 mm = 2% by weight and 1 mm = 40 K.

Zirconium (Zr) exists as α-phase up to 862 °C, when it changes to β-phase which melts at 1860 °C.

Molybdenum (Mo) exists as γ-phase up to its melting-point at 2620 °C.

An intermediate compound $ZrMo_2$ is formed at 1880 °C by the peritectic reaction

liquid (58% Mo) + Mo rich solid solution ± $ZrMo_2$

Limits of solubility are 22% Mo in βZr, negligible solubility of Mo in αZr, and 10% Zr in Mo. The solubility of Zr in Mo falls to zero at 0 °C.

A eutectic between the β solid solution and $ZrMo_2$ exists at 31% Mo and 1520 °C.

A eutectoid decomposition of β occurs at 7·5% Mo and 780 °C.

Typical curves for phase boundaries between the points given may be assumed. [P]

5. In an alloy system between elements A and B, A melts at 700 °C and B melts at 500 °C. The γ-phase melts at 800 °C, at which temperature its composition is 60% B. The following isothermal reactions occur:

Peritectic:	γ (70% B) + liquid (90% B) ⇌ δ (85% B) at 550 °C.
Eutectic:	liquid (25% B) ⇌ α (16% B) + γ (40% B) at 450 °C.
Eutectoid:	γ (50% B) ⇌ α (12% B) + δ (90% B) at 300 °C.
Peritectoid:	α (9% B) + δ (93% B) ⇌ β (35% B) at 200 °C.

The solid solubilities of B in A and A in B at 0 °C are negligible. The β-phase exists at 0 °C from 32% B to 40% B.

Draw, to scale, the equilibrium diagram, assuming the usual shapes for phase boundaries, and indicate the phase or phases present in each region. [MST]

6. Two hypothetical metals, A and B, whose melting-points are 700 °C and 500 °C respectively, are miscible in all proportions in the liquid state and are partially soluble in one another in the solid state, the maximum solubilities being 5% B and 25% A by weight. The solubilities are 2% and 5%, respectively, at 0 °C.

The two metals form a compound A_2B which melts at 750 °C and in which neither metal is soluble. The atomic weights of A and B are 30 and 50, respectively. Eutectics are formed at 22% and 60% by weight of B and at temperatures of 450 °C and 320 °C, respectively.

Construct and label the equilibrium diagram, assuming that all the lines on it are straight.

Find, for an alloy containing 45% by weight of A, (*a*) the temperatures at which melting begins on heating and at which melting is complete, and (*b*) the composition and distribution of the phases at 100 °C. [MST]

7. When are the rules used in the interpretation of binary equilibrium diagrams applicable to vertical sections of ternary equilibrium diagrams?

8. From the isothermal section at 650 °C of the chromium-iron-nickel equilibrium diagram (fig. A.14) estimate the compositions of and percentages of the various phases present at

(*a*) 30% Cr, 30% Fe, 40% Ni;

(*b*) 70% Cr, 10% Fe, 20% Ni;

(*c*) 50% Cr, 30% Fe, 20% Ni.

In the two-phase region, assume that the tie-line is parallel to the nearest side of the equilateral triangle. [MST]

9. Three elements A, B, and C are soluble in all proportions in the liquid phase but only partially soluble in the solid phase. For an isothermal section of their ternary equilibrium diagram, at a certain temperature T below the ternary eutectic point, the boundary of the three-phase area may be represented by the three lines joining the following composition points:

60% A	25% B	15% C
10% A	80% B	10% C
20% A	15% B	65% C

When the three elements are combined in pairs to form simple binary alloys, the limits of solubility at temperature T of the solid solutions formed are as follows:

α solid solution: 60% A, 40% B and 80% A, 20% C.

β solid solution: 85% B, 15% A and 80% B, 20% C.

γ solid solution: 75% C, 25% A and 75% C, 25% B.

Assuming all phase boundaries to be straight lines, draw the isothermal section of the ternary equilibrium diagram for temperature T, marking the phase or phases present in each area, and determine the composition and quantities of the phases present at this temperature for the following alloys:

(*a*)	30% A	10% B	60% C
(*b*)	30% A	30% B	40% C

In two-phase regions the tie lines may be assumed to run parallel to the nearest side of the quadrilateral. [MST]

10. In an alloy system between two elements A and B, there is some solubility of B in A and this solid solution is one component of a eutectic. On the equilibrium diagram for this system, the liquidus and solidus for the solid solution of B in A may be approximated by straight lines. T_m is the melting point of pure A; C_E and C_α are the compositions of the eutectic and of the A-rich solid solution, respectively, at the eutectic temperature T_E. If C_L and C_S are the respective compositions of the liquid and solid phases in a two-phase mixture at a temperature T, where $T_E < T < T_m$, find the segregation coefficient K, i.e., the ratio of C_S to C_L. All compositions are expressed as fractions by weight of component B.

A bar of A containing B as an impurity of concentration C_0, where $0 < C_0 < C_E$, may be purified by the zone-refining method. A molten zone, of length a, is propagated along the bar from one end. At the front of the zone, solid of composition C_0 melts and at the rear, the solid that separates is of composition determined by the segregation coefficient K. Assume that the liquid–solid interfaces are perpendicular to the axis of the bar and move at equal rates so that the zone length remains constant. Also assume that the liquid in the molten zone is homogeneous in composition at all times. Prove that the final impurity distribution is given by

$$C = C_0[1-(1-K)e^{-Kx/a}],$$

where C is the composition at a point distant x from the end at which melting started. [MST]

CHAPTER 8

The Iron-carbon System

8.1. Allotropic forms of pure iron

On cooling molten pure iron, it solidifies at 1539 °C into a body-centred cubic structure. On further cooling, this structure changes at 1400 °C to a face-centred cubic structure and again at 910 °C back to body-centred cubic. These structures have been designated α, γ, and δ from room temperature upwards. The α-iron loses its ferromagnetic properties when heated above 770 °C—a temperature known as the Curie

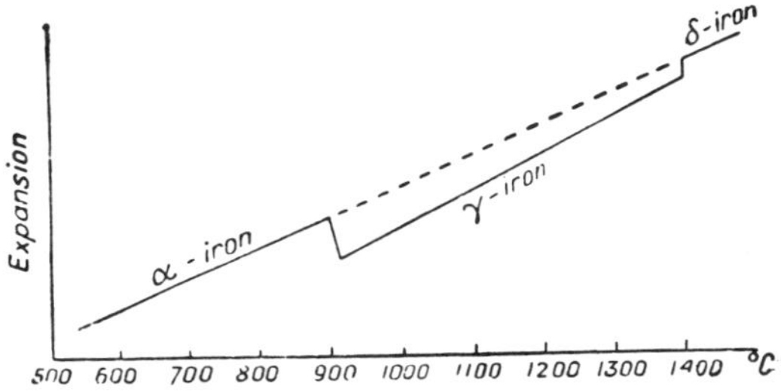

Fig. 8.1.—Change of length with temperature of pure iron [Gregory]

point. The non-magnetic α-iron was formerly known as β-iron until X-ray crystallographic methods showed that there was no change of crystal structure at 770 °C. The ferromagnetism is due to the strong interaction between the magnetic moments of adjacent atoms causing them to become oriented parallel to one another. The magnetic moments are due to the spins of unpaired $3d$ electrons (see Section 2.9). Above the Curie temperature, thermal agitation overcomes this interaction so that the metal has no net ferromagnetism.

The γ-iron is a structure of closest packing, whereas the α- and δ-iron are not. Hence there will be an abrupt change of dimensions at the temperatures of the α-γ and γ-δ transitions. A graph of change of length with temperature is shown in fig. 8.1. This phenomenon is known as *dilatation.* Dilatation can be expressed quantitatively either as the *volume dilatation*, i.e., the change of volume per unit volume, or as the *linear dilatation*, i.e., the change of length per unit length. The allotropic changes also involve a change of heat content—analogous to the latent heat of fusion, so that if a specimen is cooled with a steady rate of heat loss, the temperature–time graph shows arrests at the change points as in fig. 8.2.

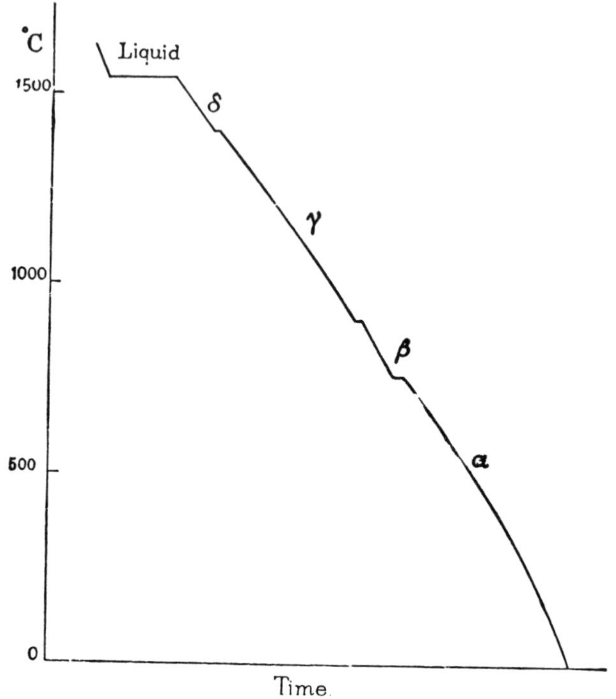

Fig. 8.2.—Cooling curve of pure iron with steady rate of heat loss

8.2. Solubility of carbon in iron

The carbon atom is smaller than the iron atom (the diameters are 0·154 nm and 0·256 nm, respectively) and dissolves interstitially in all three phases. The solubility in γ-iron is at a maximum of 1·7% at 1130 °C, the solid solution being known as *austenite*. The solubilities in the body-centred cubic phases are considerably smaller, the maxima being 0·1% at 1492 °C in δ-iron and 0·03% at 723 °C in α-iron. The solid solution of carbon in α-iron is known as *ferrite*, while that in δ-iron has not been given a special name.

The reason for the difference in solubility may be seen when the crystal structure is examined. Each space in a face-centred cubic structure is bounded by six spherical surfaces symmetrically arranged giving an " octahedral " space (fig. 8.3). Expressed as coordinates in terms of the unit cell, the centres of these spaces are at $(0, 0, \frac{1}{2})$ and $(\frac{1}{2}, \frac{1}{2}, \frac{1}{2})$. The three dimensions that govern the size of a sphere that could be inserted into the space are equal. If D is the diameter of the iron atom, the diameter of the largest sphere that could be put into the space is $0{\cdot}414D$, i.e., 0·106 nm. This may be seen from the section in fig. 8.4. The diameter of the carbon

Fig. 8.3.—Model of "space" between spheres packed in face-centred cubic structure

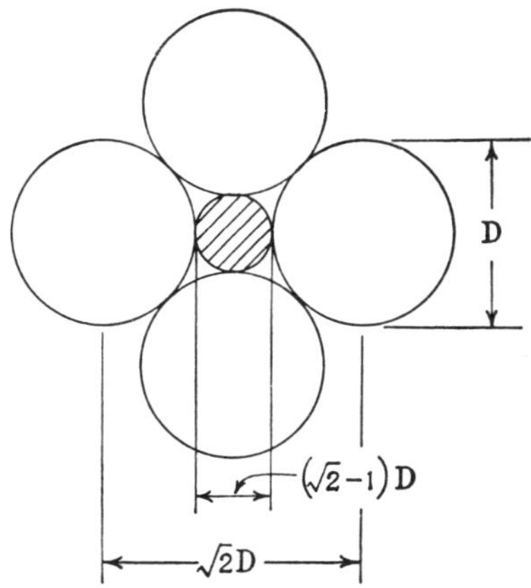

Fig. 8.4.—Section through centre of octahedral "space" in face-centred cubic structure, showing largest sphere that can be inserted

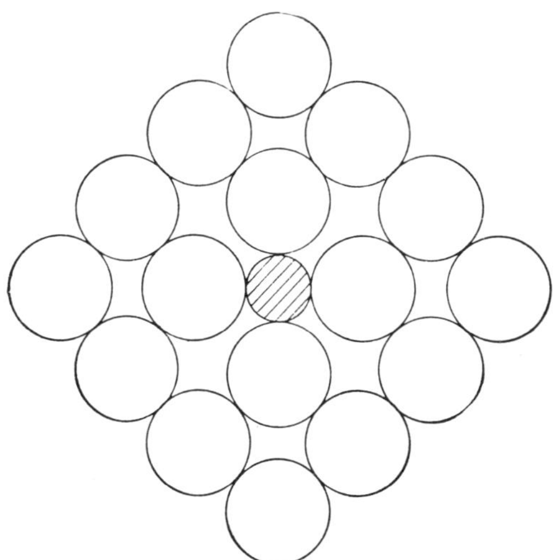

Fig. 8.5.—Hypothetical distortion around a carbon atom in interstitial solid solution in γ-iron

atom is 0·154 nm, so that it can get into the space only by causing local distortion, which may be pictured as in fig. 8·5. There will be an overall distortion of the lattice, causing an increase in the measured lattice parameter, but also quite a lot of distortion in the immediate vicinity of the carbon atom. The limit of solubility, which is about 1·7% by weight, shows that the ratio of carbon atoms to iron atoms is

$$\frac{1 \cdot 7/12}{98 \cdot 3/56} \approx \frac{1}{12}$$

that is, the maximum number of carbon atoms that can go into interstitial solution in the γ-iron is about one to every twelve available spaces, since there will be one space per iron atom. Now each space has twelve equidistant nearest neighbouring spaces. The distortion due to one carbon atom can therefore be assumed to be such that, on average, eleven of the twelve nearest spaces cannot also take a carbon atom.

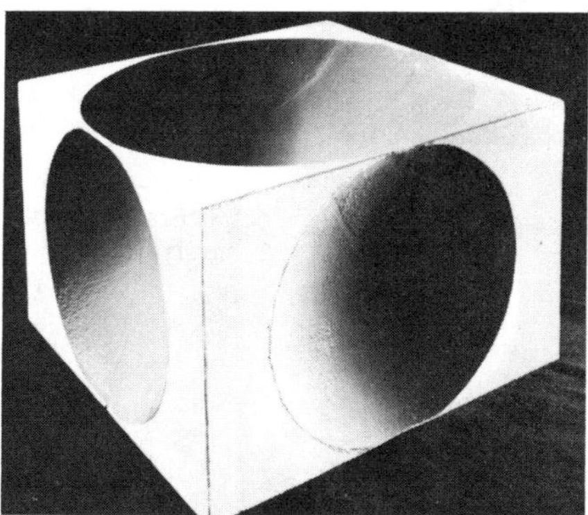

Fig. 8.6.—Model of " space " between spheres packed in body-centred cubic structure

The body-centred cubic lattice is not so densely packed as the face-centred cubic lattice, and hence the spaces are larger. They are not, however, of the cubical shape of the face-centred cubic lattice, but have one smaller and two larger dimensions. A picture of the " space " is shown in fig. 8.6. The position into which the largest sphere could be fitted in this space has the coordinates ($\frac{1}{2}$, $\frac{1}{4}$, 0). In this position it would lie with its centre in one face of the unit cell and touching four atoms (fig. 8.7). The diameter of the sphere that would just fit is $[\sqrt{(5/3)}-1]D$ or 0·074 nm for the α-iron lattice. Obviously, this space would not take a carbon atom as easily as the spaces in γ-iron, and if an atom did go into such a space, the distortion would be so great that another carbon atom

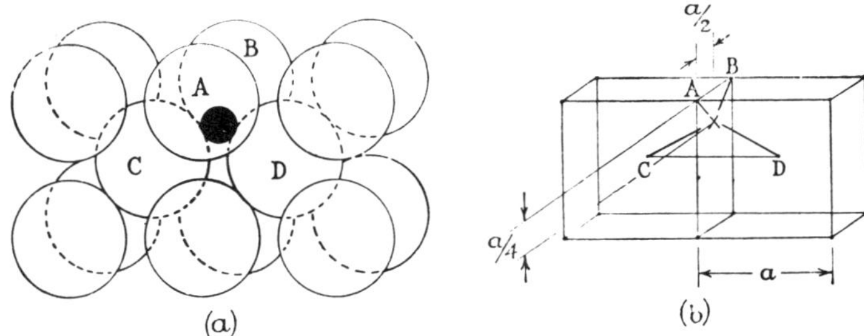

Fig. 8.7.

(*a*) Interstitial atom (black) that just fits into body-centred cubic structure, touching atoms A, B, C, and D

(*b*) Centres of atoms.
$AC = AD$, etc. $= \sqrt{3}a/2 =$ diameter D of larger atom.
$AX = BX = CX = DX = \sqrt{[(\frac{1}{2}a)^2+(\frac{1}{4}a)^2]} = \sqrt{5}a/4 = \sqrt{5}D/2\sqrt{3}$
Radius of interstitial atom $= AX-\frac{1}{2}D = \frac{1}{2}D[\sqrt{(5/3)}-1]$

could not get in for some considerable distance. The site of the carbon atoms in ferrite is not this position, but that shown in fig. 8.8, and is also one that will not take many atoms, actually about one carbon atom per 700 spaces at maximum solubility.

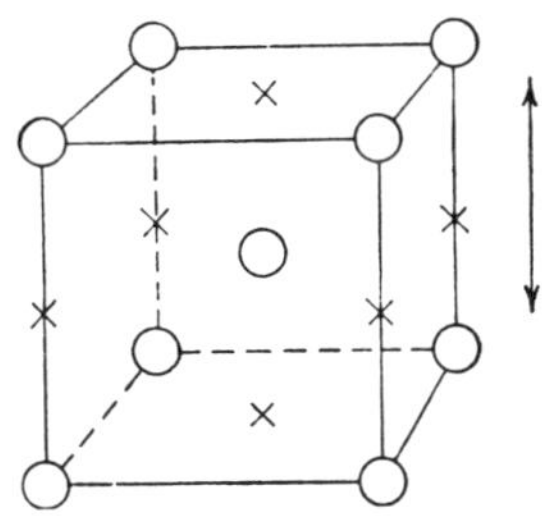

Fig. 8.8.—Actual positions occupied by carbon atoms in ferrite lattice shown by crosses. Arrow shows direction of resulting distortion of lattice parallel to one axis only

8.3. Cementite and ledeburite

Iron and carbon form an intermediate compound which contains 6·67% carbon. This compound, known as *cementite*, and having the formula Fe_3C has been considered in Section 6.8.

The equilibrium diagram (fig. 8.9) shows that between iron and cementite there is a eutectic-forming series of alloys with the eutectic reaction occurring at 4·3% carbon and 1130 °C. The eutectic is between cementite and austenite and is given the name *ledeburite*. When iron ores are smelted with coke (i.e. carbon) in a blast furnace, the iron oxide is reduced to iron, the oxygen forming carbon monoxide. When the temperature is above 1130 °C, an iron alloy containing 4·3% dissolved carbon would melt, so that any alloy of this composition would first run to the base of the furnace. The crude iron-carbon alloy formed in a blast furnace will thus be of about eutectic composition. It is known as *pig iron*.

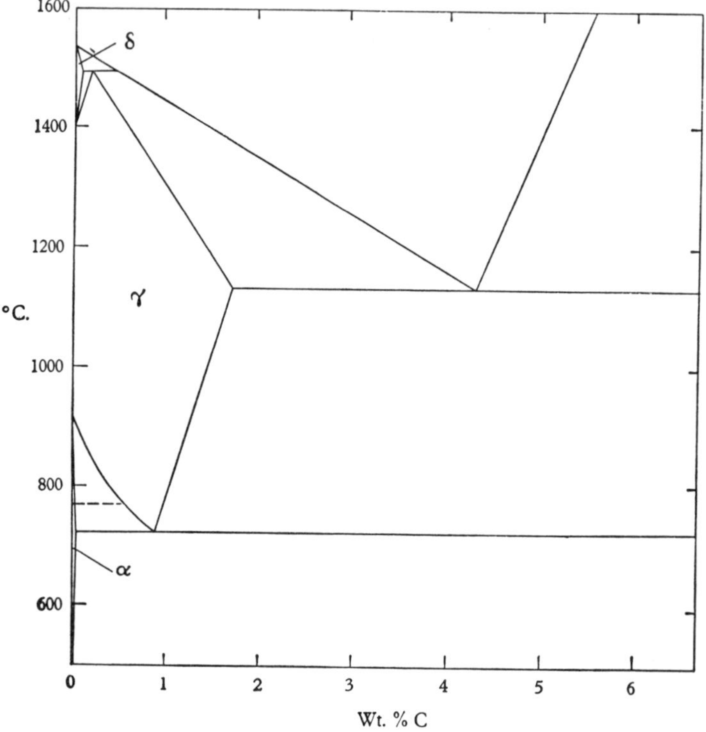

Fig. 8.9.—Equilibrium diagram of iron-carbon system

The greatest solubility of carbon in austenite is 1·7% at the eutectic temperature. Above and below this temperature, the solubility decreases. Thus for compositions above about 0·2% carbon the limit of solubility is reached, on heating, when the solidus line is reached and the two-phase (austenite + liquid) region entered. Also all compositions above about 0·9% on cooling will precipitate cementite when the limit of solubility is reached.

8.4. Modifications of allotropic change temperatures

Austenite has a much higher solubility for carbon than either ferrite or δ-iron, the result of which is that the presence of carbon raises the temperature of the $\gamma \rightarrow \delta$ change and lowers that of the $\gamma \rightarrow \alpha$ change. The temperature range of austenite thus increases initially with an increase of carbon content.

On cooling, say, a 0·3% carbon alloy from the austenite phase, the change to ferrite does not begin until the temperature falls to 850 °C. At this temperature, ferrite crystals nucleate at positions on the austenite grain boundary.

The lower boundary of the austenite region is formed by two lines, one from 910 °C at pure iron—the ferrite separation line—and the other from 1130 °C at 1·7% carbon—the cementite separation line. These intersect, showing that a eutectoid reaction will occur. The equilibrium temperature for this reaction is given as 723 °C, and the composition of the eutectoid is given in various works as anything from 0·80% to 0·89%. In the absence of conclusive evidence as to the correct value, the value of 0·89% has been used throughout this book.

8.5. Eutectoid and peritectic decomposition of austenite

The eutectoid forms by simultaneous precipitation of ferrite and cementite and is known as *pearlite*. The structure is often laminated, consisting of alternate thin plates of cementite and ferrite. These evenly-

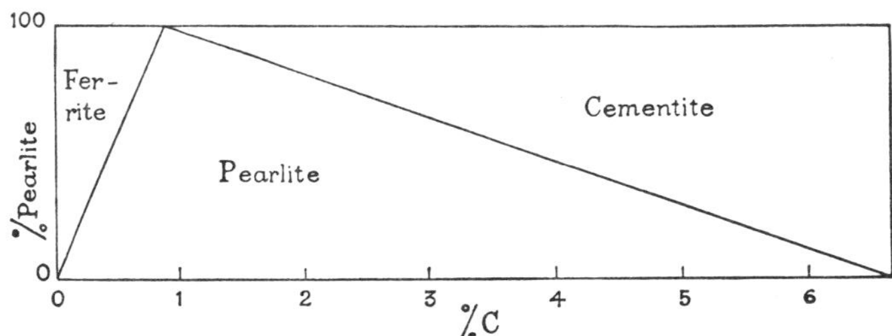

Fig. 8.10.—Proportions of eutectoid and pro-eutectoid structures in microstructures of iron-carbon alloys at room temperature

spaced lines sometimes produce an irridescent appearance after polishing and etching—an appearance resembling mother-of-pearl—hence the name. The composition of pearlite is $(0{\cdot}89/6{\cdot}67)\times 100 = 13{\cdot}3\%$ cementite and 86·7% ferrite.

The structure of any alloy at room temperature will be part ferrite and part pearlite if the carbon content is below 0·89%, and part pearlite and part cementite if above. The proportions of these constituents for any composition can be determined from fig. 8.10.

The upper part of the austenite region terminates in a peritectic reaction at 0·18% carbon and 1492 °C.

8.6. Nomenclature of iron-carbon alloys

Any alloys with less than 0·03% carbon (i.e., entirely ferrite at 723 °C) are known as *pure irons*.

Those alloys with carbon contents between 0·03% and 1·7% (i.e., entirely austenite at 1130 °C) are known as *steels* and are further divided into *hypo-eutectoid* and *hyper-eutectoid* steels, being those with carbon contents respectively less and more than the eutectoid composition. Alloys containing more than 1·7% and less than 6·67% carbon are known as *cast irons*

8.7. Microstructures of steels

Pure irons containing ferrite only, show sharp grain boundaries as in fig. 5.3. With increase of carbon content, pearlite appears in increasing quantity. Under the microscope the pearlite appears dark or striped and may be distinguished clearly from the ferrite as in fig. 8.11*a*. In hyper-eutectoid steels, the primary or pro-eutectoid* cementite is concentrated on the former austenite grain boundaries and appears white between the dark pearlite areas as in fig. 8.11*d*.

The cementite layers of the pearlite, which appear dark at intermediate magnifications, are actually etched at their boundaries only, and under high magnification the boundaries may be resolved, showing white cementite between.

8.8. Phase-transformation diagrams

The phases and their proportions at any temperature for any particular alloy can be read directly from the equilibrium diagram. Also the manner in which the phases are distributed can be deduced. The results are most easily presented on a *phase-transformation diagram*.

Consider the slow cooling from a temperature of 1500 °C of an alloy containing 0·5% carbon. At a temperature of about 1500 °C austenite commences to separate from the liquid and continues to do so as the temperature falls until at 1430 °C all the liquid has transformed. The austenite remains unchanged until at about 775 °C, when ferrite nucleates and continues to separate until 723 °C when the composition is 0·47/0·86 austenite containing 0·89% carbon and 0·39/0·86 ferrite containing

* *Pro-eutectic and pro-eutectoid* refer to phases formed before the eutectic and eutectoid structure, respectively.

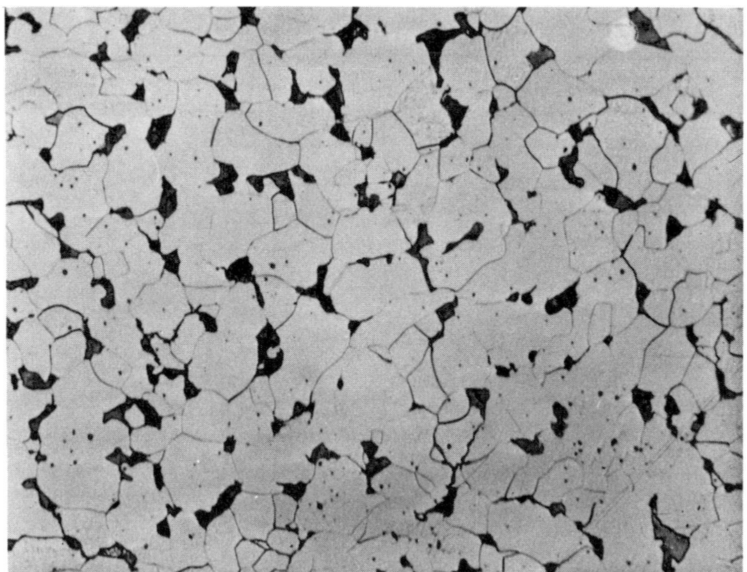

Fig. 8.11(*a*).—Microstructures of steels: 0·15% C (× 400)

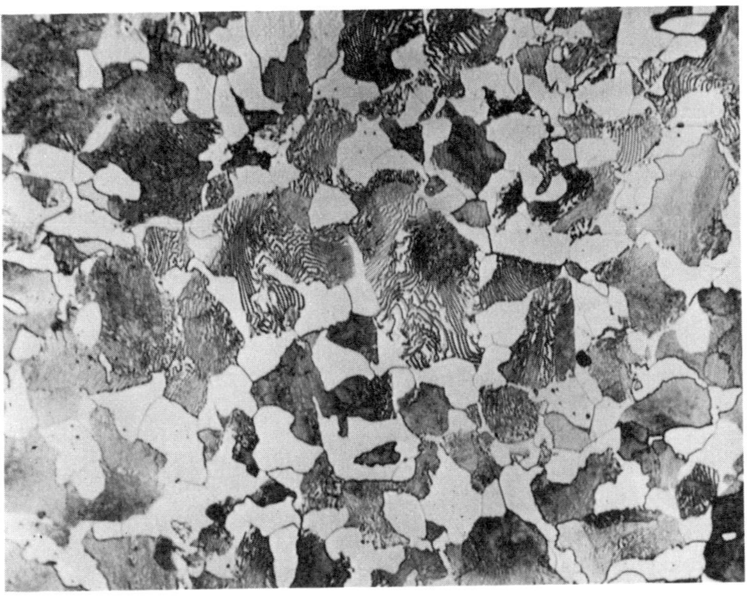

Fig. 8.11(*b*).—Microsctructures of steels: 0·53% C (× 400)

Fig. 8.11(*c*).—Microstructures of steels: eutectoid lamellar pearlite (× 500)

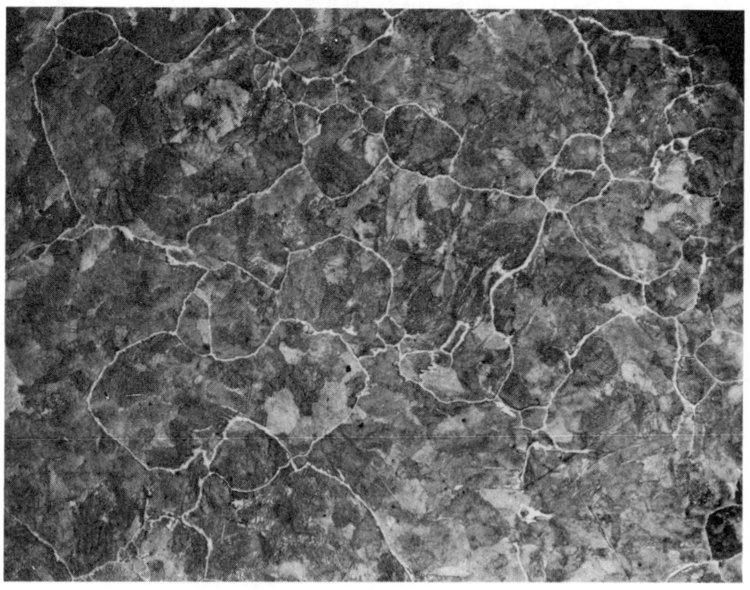

Fig. 8.11(*d*).—Microstructures of steels: 1·2% C (× 200)

0·03% carbon. At 723 °C this austenite changes to the eutectoid pearlite. The final composition is

$$\frac{0{\cdot}39}{0{\cdot}86} = 45{\cdot}3\% \text{ pro-eutectoid ferrite}$$

$$0{\cdot}867 \times \frac{0{\cdot}47}{0{\cdot}86} = 47{\cdot}4\% \text{ eutectoid ferrite}$$

$$0{\cdot}133 \times \frac{0{\cdot}47}{0{\cdot}86} = 7{\cdot}3\% \text{ eutectoid cementite}$$

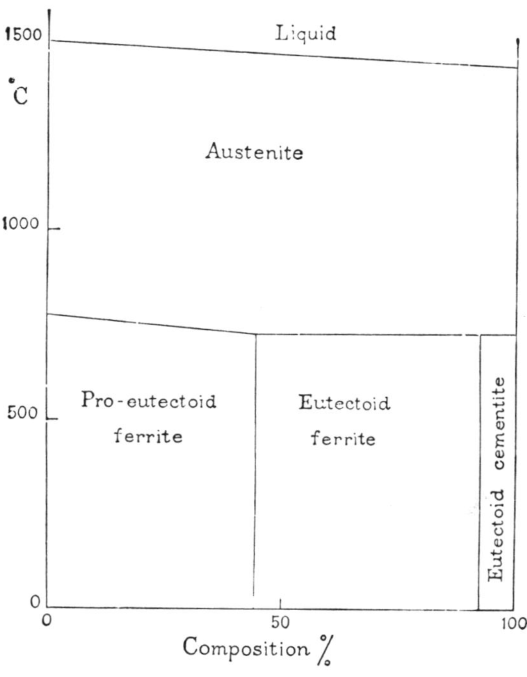

Fig. 8.12.—Phase-transformation diagram for a 0·5% C steel

The phase-transformation diagram showing the changes is given in fig. 8.12. It will be seen that this form of diagram shows for each temperature the proportion of each structural constituent and the phase from which it was formed.

Changes involving the δ-phase or ledeburite would be more complicated, as may be seen from the examples in figs. 8.13 and 8.14.

The changes associated with the δ-phase are of little practical importance, because no commercial processing of alloys is carried out in that range. The austenite-ferrite change is of great importance, because it is on this change that all the heat treatment of steels depends.

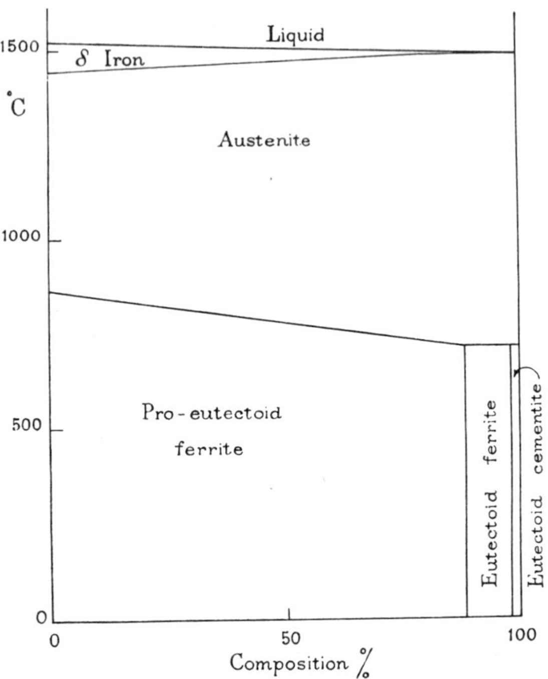

Fig. 8.13.—Phase-transformation diagram for a 0·1% C steel

8.9. Nomenclature of change points

To facilitate reference to the various changes and the temperatures at which they occur, they are referred to by the letter A (*arrêt*—referring to the arrest in a time-temperature curve) with various subscript numerals to identify the changes as follows:

	Change on heating	*Change on cooling*
A_1	Pearlite to austenite	Austenite to pearlite
A_2	Loss of ferromagnetism	Gain of ferromagnetism
A_3	Last ferrite absorbed in austenite	First ferrite nucleates from austenite
A_{cm}	Last cementite absorbed in austenite	First cementite nucleates from austenite
A_4	First δ-iron forms from austenite	Last δ-iron dissolves in austenite

The lines of the equilibrium diagram with their appropriate labels are

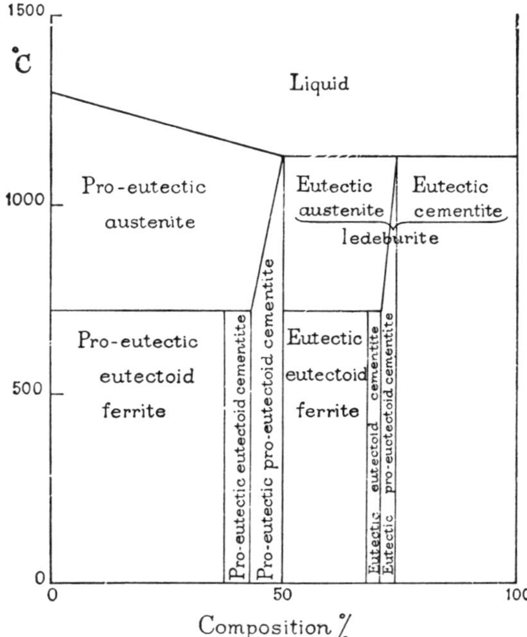

Fig. 8.14.—Phase-transformation diagram for a 3·0% C iron-carbon alloy

shown in fig. 8.15. For carbon contents above *a*, ferromagnetism is lost when the last ferrite disappears, so that A_2 and A_3 coincide along *ab*. Also for hypereutectoid steels all the ferrite present at low temperatures is in the pearlite, so that the A_1, A_2, and A_3, temperatures are the same.

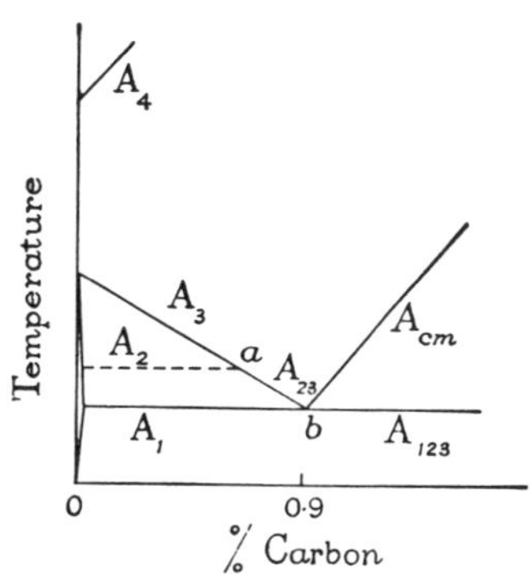

Fig. 8.15.—Nomenclature of arrest points on iron-carbon equilibrium diagram

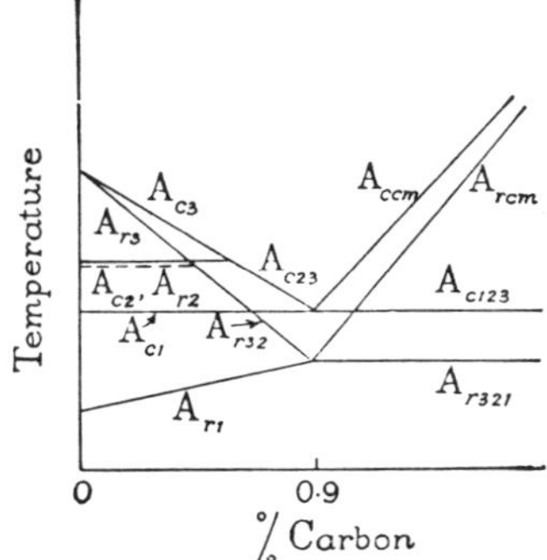

Fig. 8.16.—Nomenclature of change points in steel for heating and cooling

This equilibrium diagram shows the temperatures and compositions for changes carried out under equilibrium conditions, i.e., at infinitely slow rates of heating and cooling. In practical cases, these rates are finite and, because the changes depend upon the diffusion of carbon, the rate of which is limited, there is a thermal lag. Each change temperature is higher during heating than during cooling with the exception of the A_2-temperature. The loss of magnetism at this temperature is not due to change of structure and does not involve migration of atoms.

A further subscript is used to distinguish between heating and cooling change temperatures; *c* (from the French *chauffage*) indicates the transformation temperature on heating and *r* (from *refroidissement*) on cooling. The difference between A_{c1} and A_{r1} in plain carbon steels is of the order of 25–50 K and is much greater in alloy steels. The constitution diagram for actual changes is then as shown in fig. 8.16.

8.10. Determination of change points

The temperatures of the changes for a particular alloy may be found from heating and cooling curves. At the A_1-temperature there will be considerable absorption or evolution of heat giving a plateau on the curve. Between A_1 and A_3 heat of change has to be supplied as well as specific heat. Above A_3 only specific heat has to be supplied, so that there will be a change of slope at A_3. Change of slope also occurs at the A_2-temperature. Heating and cooling curves for some steels are given in fig. 8.17.

Alternatively the change temperatures can be found by observing the change of length with temperature. An instrument for doing this is a dilatometer, one form of which is shown in fig. 8.18. The specimen is heated by an electric furnace, its temperature being measured by a thermocouple. The extension of the specimen is transmitted to a dial gauge via silica rods and Invar* bars. Both of these have low coefficients of thermal expansion, and the silica, being an insulator, does not conduct much heat from the ends of the specimen, so that a fairly uniform temperature is maintained throughout its length. The main frame of the instrument is also made of Invar bars, so that if these do change their temperature the effect on the readings will be small. By choice of suitable currents in the electric furnace, controlled rates of heating and cooling can be achieved. During heating and cooling, simultaneous readings of temperature and length change are made. Some typical results are shown in fig. 8.19.

* Invar is a trade name for an iron-nickel alloy containing about 36% nickel. Its coefficient of thermal expansion is approximately 6% of that of a carbon steel at room temperature.

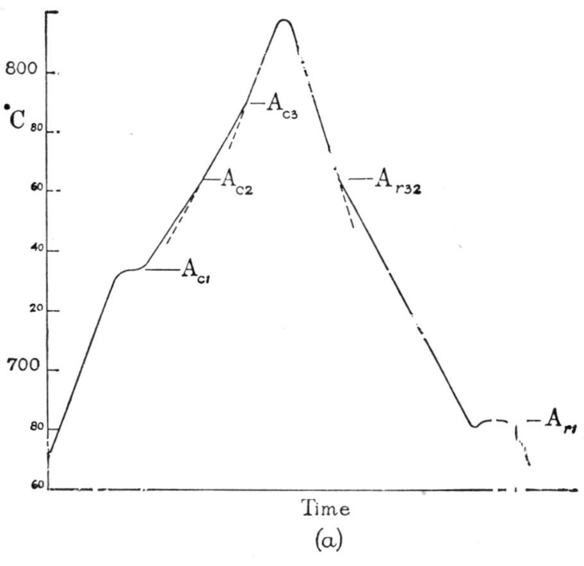

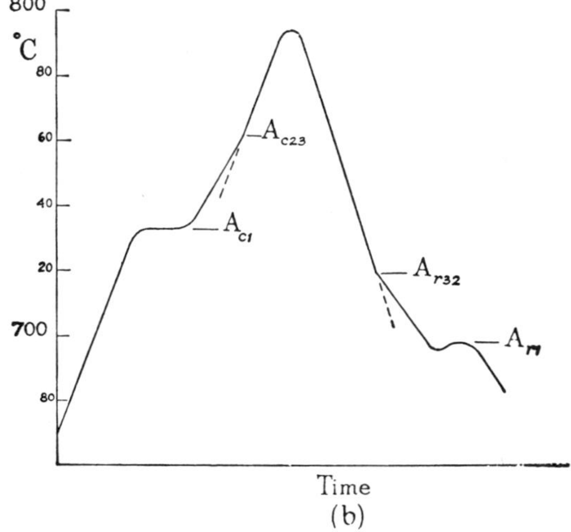

Fig. 8.17.—Time-temperature curves for heating and cooling of steels containing (*a*) 0·34% C, (*b*) 0·53% C

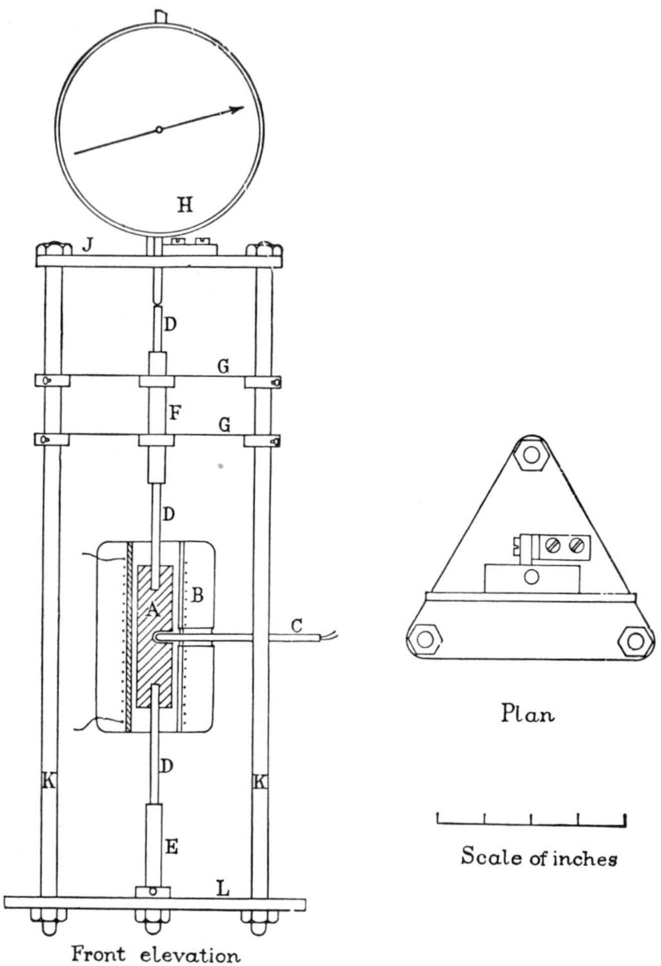

Fig. 8.18.—Typical form of dilatometer

A—specimen (sectioned) inside furnace B (sectioned). C—thermocouple. D—silica rods to reduce conduction heat losses, fitted into lower Invar bar E in base L and upper Invar bar F restrained by spring steel guides G. Dial gauge H secured by bracket on top plate J. K—Invar bar main frame

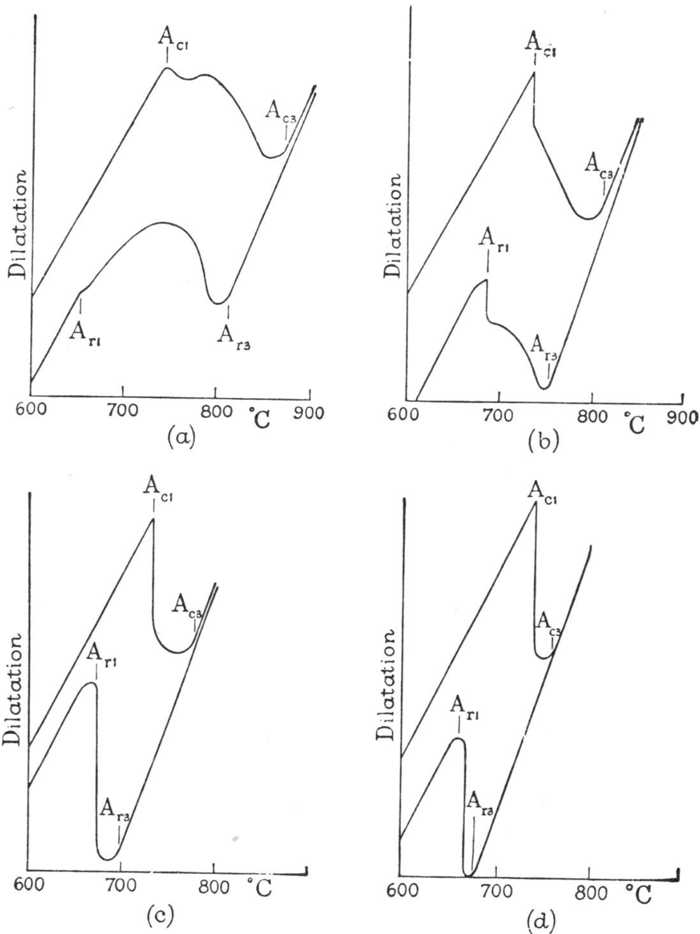

Fig. 8.19.—Temperature-length curves for heating and cooling of steels containing (*a*) 0·12% C, (*b*) 0·36% C, (*c*) 0·52% C, (*d*) 0·69% C

8.11. Cast irons

As stated on p. 114 the term *cast iron* is applied to all iron-carbon alloys containing more than 1·7% carbon. Thus it will be seen from the equilibrium diagram that they are never single-phase on first solidification, but always contain some cementite. The changes on cooling are complex as shown in the phase-transformation diagram in fig. 8.14. The final structure is a mixture of ferrite and cementite.

The formation of cementite is, however, a metastable condition and under certain conditions the cementite breaks down to ferrite and free carbon in the form of graphite. In hypereutectic alloys, some graphite forms from the liquid. A cast iron containing all the carbon as cementite is known as a *white cast iron*, while a graphitized one is known as *grey cast iron*. Graphitization is favoured by a slow rate of cooling and by the presence of silicon.

White cast iron (fig. 8.20), being largely cementite, is extremely hard, but brittle. It gives excellent wear-resisting properties to the surfaces of castings. In grey cast iron (figs. 8.21 and 8.22) the graphite is distributed as flakes, which break up the continuity of the metal matrix—a carbon steel of itself possessing considerable strength and ductility. In ferritic

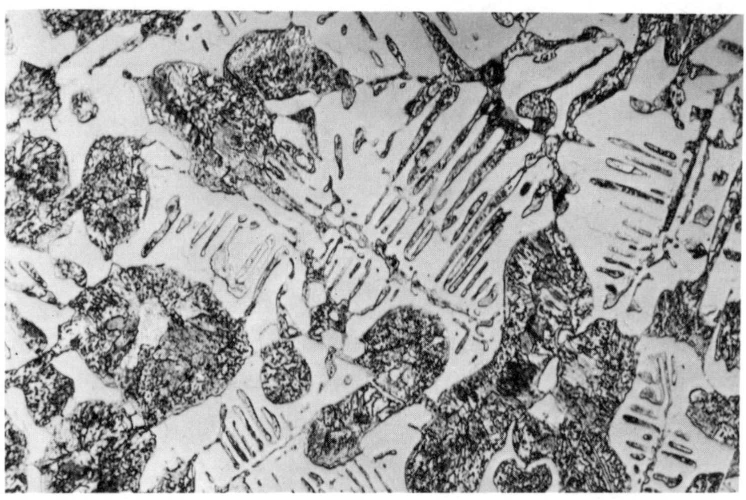

Fig. 8.20.—White cast iron (×200) showing pearlite and cementite

grey cast iron all the carbon is graphite. In pearlitic grey cast iron, some is in pearlite. The graphite flakes have no strength and act as internal cracks making the material weak and brittle and extremely liable to fracture under shock loads.

Malleable cast iron is made by annealing white iron castings for periods of several days under controlled conditions, when the graphite will separate as nodules of approximately spherical shape (fig. 8.23). The weakening effect of the graphite is reduced so that malleable cast iron has a higher strength, and greater ductility and shock resistance. The objection is the cost of the prolonged annealing treatment.

It was found about 1948 that by additions of cerium or magnesium under controlled conditions to the molten metal before pouring into the moulds, the graphite forms spherical particles on casting (fig. 8.24). This

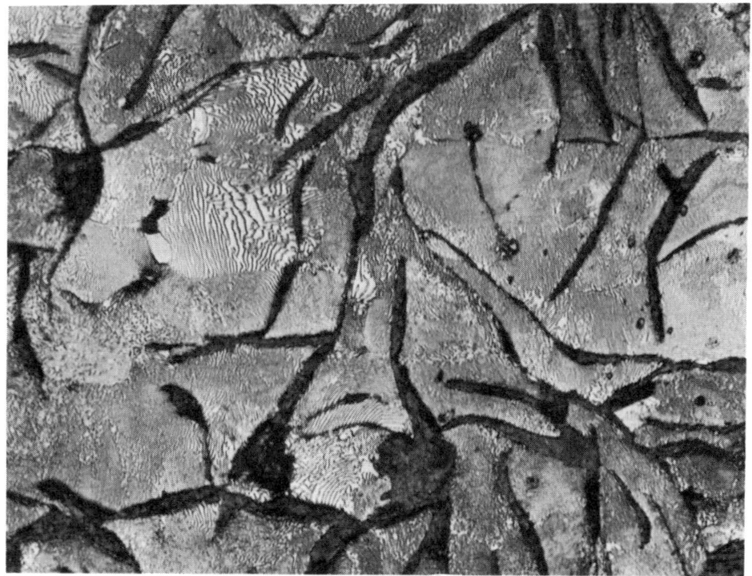

Fig. 8.21.—Pearlitic grey cast iron (×200)
Graphite flakes in pearlite matrix

Fig. 8.22.—Ferritic grey cast iron (×200)
Graphite flakes in ferrite matrix

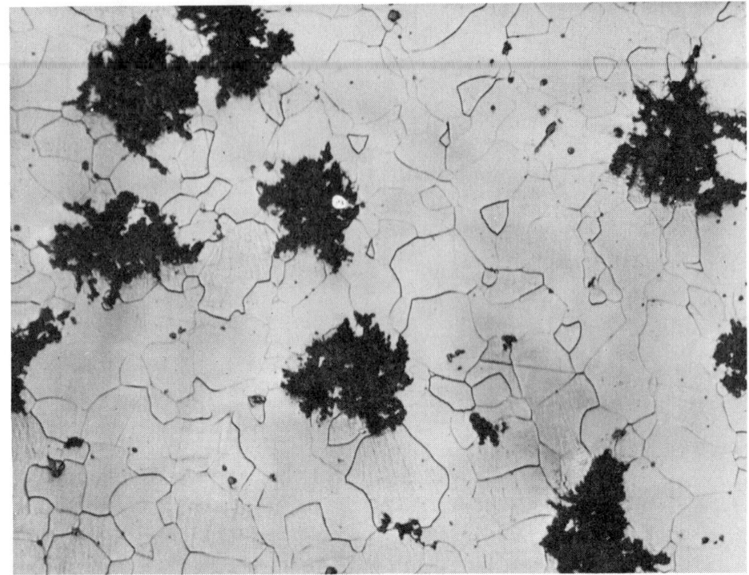

Fig. 8.23.—Blackheart malleable cast iron (×200)
Graphite nodules in ferrite matrix

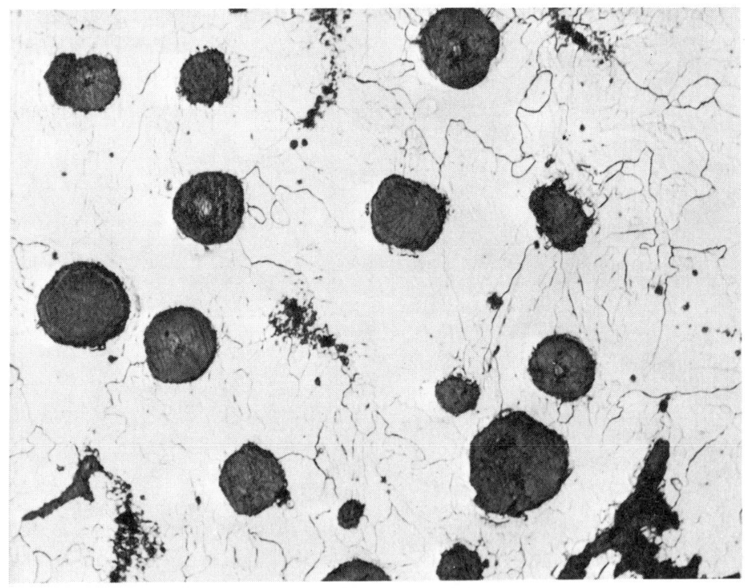

Fig. 8.24.—Spheroidal graphite cast iron (×200)
Graphite spheroids in ferrite matrix

spheroidal graphite cast iron is found to have a strength superior to malleable cast iron and, with suitable annealing, which precipitates further graphite from the pearlite matrix, the ductility is as good.

QUESTIONS

1. Show that the denser way of stacking equal-sized spheres in space is achieved in the face-centred cubic rather than the body-centred cubic arrangement. How then do you account for the fact that the solubility of carbon is greater in γ-iron than in α-iron? Comment on the technological importance of this fact. [MST]

2. Calculate the linear dilatation as a polycrystalline specimen of α-iron changes to γ-iron, assuming that the atoms are rigid spheres.

3. Calculate the linear dilatation as a polycrystalline specimen of titanium changes from close-packed hexagonal to body-centred cubic at 880 °C. Assume that the closest distance between atom centres is the same in each structure.

4. Explain, referring to suitable examples, the meanings of the terms *phase*, *change of state*, *allotropic change*, *crystal lattice*, *unit cell*.

From the lattice parameter of α-iron, find the radius of the iron atom, assuming that the atoms behave as hard spheres in contact. Compute the lattice parameter of γ-iron, assuming that the atoms behave as hard spheres of this same radius. [MST]

5. Draw a phase-transformation diagram to show how the proportions and distribution of the phases present vary with temperature, as a 0·35% carbon steel is cooled under equilibrium conditions from the liquid state to 0 °C.

Sketch the resultant microstructure.

Sketch a graph of the variation of length with temperature for a specimen of this steel during heating from 0 °C to 1000 °C and subsequent cooling to 0 °C. [MST]

6. Draw, to a scale of 1 mm = 10 K and 1 mm = 1%, the phase transformation diagram of an iron-carbon alloy containing 1·25% carbon for the temperature range 500 °C to 1500 °C. Hence calculate the ratio of the proeutectoid cementite to the eutectoid cementite in this alloy at 500 °C. The change in solubility of the α-phase may be ignored. [MST]

7. An Fe-Fe_3C alloy containing 3% carbon has been slowly cooled from the liquid state. At 500 °C what is the ratio of the cementite in the pearlite derived from the eutectic austenite to the proeutectoid cementite derived from the proeutectic austenite? [MST]

8. Draw, to a scale of 1 mm = 5 K and 1 mm = 1%, a phase transformation diagram for a lead-magnesium alloy containing 50% magnesium, during slow cooling from 600 °C to 0 °C. [P]

9. Describe briefly two experimental methods for the determination of change points in a mild steel as it is heated from room temperature to 900 °C and then cooled to room temperature again.

Sketch and explain the form of experimental results which should be obtained. [P]

10. A sample of a medium carbon steel has a linear dilatation of + 1·33% on cooling from the austenite region. Determine the carbon content of the sample given that the unit cell of Fe_3C is orthorhombic, having dimensions

$$a = 0{\cdot}452 \text{ nm}, \qquad b = 0{\cdot}508 \text{ nm}, \qquad c = 0{\cdot}637 \text{ nm},$$

and contains 16 atoms. Assume that the unit cell of austenite has sides of length 0·356 nm and that of ferrite sides of 0·286 nm. [E]

CHAPTER 9

Thermal Energy

9.1. Kinetic energy of a gas molecule

The kinetic energy of translation of the molecules of an ideal gas was discussed in Chapter 3 and shown to be equal to $\frac{1}{2}mn\overline{C^2}$. Also it was shown that

$$pV = RT = \tfrac{2}{3} \times \tfrac{1}{2}mn\overline{C^2}$$

for one mole, so that the kinetic energy was $\frac{3}{2}RT$ per mole or $\frac{3}{2}kT$ per molecule on the average.

Now a gas molecule has three degrees of freedom for translational motion (e.g. the components of its velocity in three mutually perpendicular directions define its velocity). Maxwell has shown that if a system which has several degrees of freedom obeys the ordinary laws of mechanics, then the total energy of a system is equally divided among the different degrees of freedom. This is termed the principle of the *equipartition of energy*. Hence the kinetic energy of translation of a molecule in one degree of freedom is on the average $\frac{1}{2}kT$.

The next problem is the total number of degrees of freedom of a molecule. (So far we have considered only translational degrees of freedom and ignored any possible rotational or vibrational freedoms.)

1. *A monatomic gas*

In a monatomic gas each molecule can have three translational degrees of freedom, but no rotational freedom. It can be said simply that an atom is spherically symmetrical and therefore any rotation could not be detected.*

2. *A diatomic gas*

Suppose each molecule consists of two atoms at a fixed distance apart as shown diagrammatically in fig. 9.1. The molecule can rotate about two perpendicular axes which are themselves at right angles to the line

* The correct explanation is that the angular momentum obeys the Bohr quantization rule (p. 11). The moment of inertia of a single atom or of a diatomic molecule about the axis passing through both atoms is so small that the angular velocity for the lowest value of angular momentum is very high, and the kinetic energy would be much greater than $\frac{1}{2}kT$ at ordinary temperatures. It is therefore unlikely that this rotation could be excited.

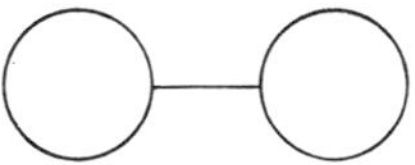

Fig. 9.1

joining the atoms, but there will be no rotation about the axis that passes through both atoms. This last assumption is similar to the one made in the case of the monatomic molecule. Each molecule therefore has three translational and two rotational degrees of freedom, making five in all. By the principle of equipartition of energy, each degree of freedom will have an average energy of $\frac{1}{2}kT$, so that the total energy is

$$\tfrac{5}{2}kT$$

More complicated molecules may also have a third rotational degree of freedom and possibly other degrees if relative movement or vibration between atoms can occur.

9.2. Specific heat of a gas

Now the specific heat of a gas at constant volume is the quantity of heat required to raise the temperature of unit quantity of the gas by one degree, while the volume remains unchanged. This heat goes only to increase the total kinetic energy of the molecules. Therefore for one mole of a monatomic gas it should be $\frac{3}{2}R$ and for one mole of a diatomic gas it should be $\frac{5}{2}R$. A few examples to illustrate this are given in Table 9.1.

TABLE 9.1—EXPERIMENTAL AND THEORETICAL VALUES OF SPECIFIC HEAT AT ROOM TEMPERATURE FOR SOME GASES (in J kmol^{-1} K^{-1})

Gas	Experimental value of C_v	Theoretical value of C_v
Argon	12475	$\frac{3}{2}R = 12470$
Helium	12475	
Air	20720	$\frac{5}{2}R = 20780$
Carbon monoxide	20680	
Nitrogen	20635	

9.3. Specific heat of a solid

Whereas the molecules of a gas were free to move, the atoms in a solid are much less mobile. Each can vibrate about its mean position, and sometimes diffusion will occur whereby atoms may migrate through the bulk of the metal. Further reference to diffusion will appear later.

Now the vibration of each atom can be considered as being compounded from three vibrations, each parallel to one of three perpendicular axes. The energy of an atom in one of its modes of vibration can be expressed at any instant as the sum of its kinetic and its potential energy. The kinetic energy is due to its having a velocity, and it has potential energy because it is moving in a potential field of force. These two energy terms are independent. Hence six independent terms are necessary to specify completely the motion of the atom, i.e., it has six degrees of freedom.

By the principle of equipartition of energy, the energy associated with each degree of freedom will be $\frac{1}{2}kT$ per atom or $\frac{1}{2}RT$ per mole. The total energy is therefore $6 \times \frac{1}{2}RT = 3RT$, and

$$C_v = 3R$$

This was expressed in Dulong and Petit's law in 1819 which was based on experimental results for solid elements at ordinary temperatures. This law states that the product of the specific heat at constant volume and the atomic weight, that is, the *atomic heat*, is the same for all substances, and is equal to 24946 J kmol^{-1} K^{-1}.

9.4. Experimental results for simple substances

Dulong and Petit's law does not represent the behaviour of metals at all temperatures. The actual behaviour can be summarized as follows:

(i). Near absolute zero, the atomic heat is proportional to the cube of the absolute temperature. (This is the Debye T^3 law.)

(ii). At ordinary and higher temperatures, the atomic heat at constant volume converges to $3R$.

(iii). The curves for atomic heat against absolute temperature for all substances are exactly similar in shape and can be made to coincide by altering the scale of the temperature axis, i.e.,

$$C_v = F\left(\frac{T}{\Theta}\right)$$

where the function F is the same for all substances and Θ is a constant which has a particular value for each substance. The shape of the curve is shown in fig. 9.2.

A theoretical analysis by Debye based on quantum mechanical principles applied to the possible vibrational modes in a crystal led to the formula*

$$C_v = 3R\left[12\frac{T^3}{\Theta^3}\int_0^{\Theta/T}\frac{x^3\,dx}{e^x-1} - \frac{3\Theta/T}{e^{\Theta/T}-1}\right]$$

* For a derivation of this formula and others relating to specific heats, see J. de Lauray, "The Theory of Specific Heats and Lattice Vibrations," *Solid State Physics*, Vol. 2, 1956, p. 219. Academic Press Inc., London and New York.

which gives curves in very close agreement with the observed shape. At very low temperatures this reduces to

$$C_v = \frac{12\pi^4 R}{5}\left(\frac{T}{\Theta}\right)^3$$

the Debye T^3 law.

Θ is known as the Debye temperature. Its exact meaning is beyond the scope of this work. It happens to be the temperature at which C_v is within 3·4% of the value of $3R$.

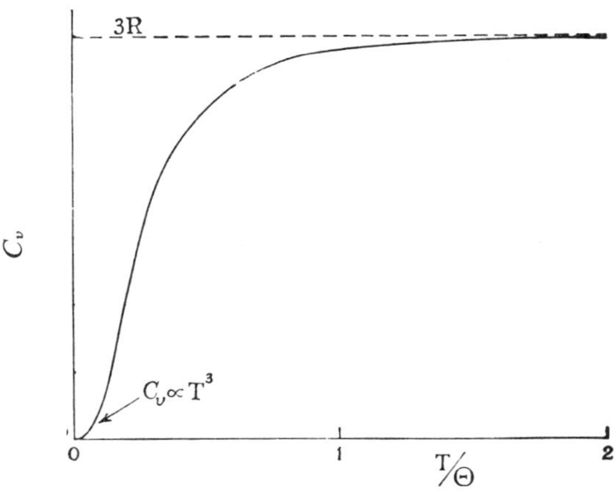

Fig. 9.2.—Variation of specific heat of a solid with temperature. By expressing temperature in the form T/Θ, where Θ is a characteristic temperature for each solid material, one curve applies to all materials

Normally, for solids, the specific heat is measured at constant pressure. This differs from that at constant volume by the energy necessary to compress the solid sufficiently to counter the effects of thermal expansion. It can be shown that

$$C_p - C_v = \alpha^2 VTK$$

where α is the coefficient of volumetric thermal expansion, V is the molar volume and K is the bulk modulus.

9.5. Thermal expansion

The curve of the variation of potential energy of an atom in a solid with internuclear spacing, as discussed in Section 4.12, is of the form shown in fig. 9.3. If an atom possesses an energy greater than the minimum of this curve, it can move between the limiting positions at which the

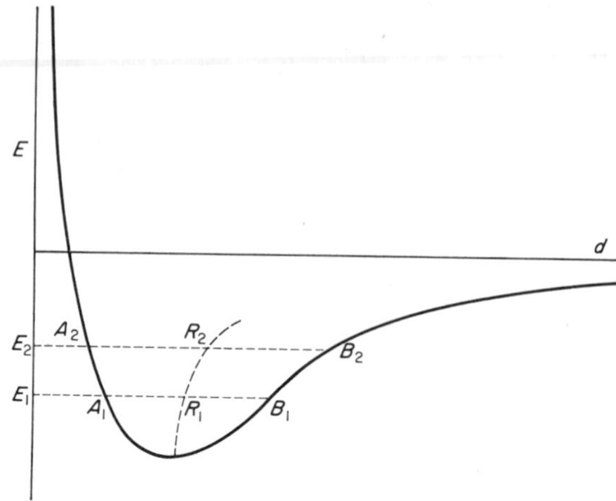

Fig. 9.3.—Variation of energy E with intermolecular spacing d

potential energy equals the total energy of the atom. When between these positions, the potential energy is less and the excess energy of the atoms becomes kinetic energy. The atom will then vibrate between these limits. At a temperature T_1, for which the energy is E_1, the atoms are vibrating so that the spacing varies from A_1 to B_1 with a mean position R_1. At a higher temperature T_2, the vibration is between A_2 and B_2 with a mean position R_2. Since $R_2 > R_1$ due to the asymmetry of the curve, the solid expands. The amplitude of vibration is of the order of 1/10 of the atomic spacing at ordinary temperatures. Melting occurs when the amplitude is about 12%. The smaller the coefficient of thermal expansion at room temperature, then the higher the melting point.

9.6. Stable and metastable states

A stable state of equilibrium of a system may be defined as one in which the free energy is a minimum; that is, to disturb the system slightly from this state some work must be done. Now it is possible for many physical or chemical systems to exist more or less indefinitely in states which are not those of minimum free energy. Hence they cannot be regarded as being stable states

As an example, consider a system consisting of oxygen and hydrogen at room temperature. The stable state is that in which they are combined in the molecular form. But if the gases are mixed at this temperature, they do not combine chemically. Once a spark has passed, however, they will combine, and in doing so give out much heat energy, showing that the system has now attained a state of lower free energy and hence a more stable one.

Such a system is said to be in a *metastable* state. True equilibrium can be reached by raising the temperature. Also the speed of chemical reactions can be raised by increasing the temperature.

Thus a higher temperature favours a more rapid rate of approach to equilibrium. A metastable state is possible when the temperature is too low to enable changes to occur at an appreciable rate.

9.7. A simple example of a metastable state

A box resting on a flat surface in the position shown in fig. 9.4*a* is stable because the centre of gravity G is in the lowest position possible. In the position shown in fig. 9.4*b*, it is not in stable equilibrium because G is not in the lowest position. But on tilting it slightly, the potential energy rises, showing that the immediately neighbouring positions are

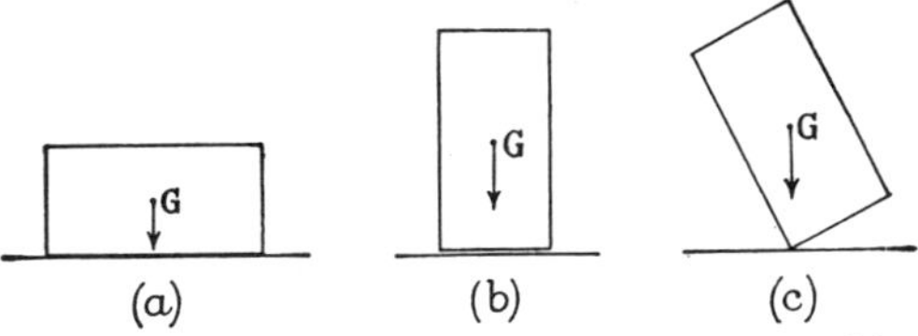

Fig. 9.4.—(*a*) Stable, (*b*) metastable, and (*c*) unstable positions of box

less stable. Hence, unless disturbed, the box will remain indefinitely in this position which is one of metastability. For the box to pass from the metastable to the stable position, it must pass through the position shown in fig. 9.4*c*, which is that of the maximum potential energy it can have while still in contact with the surface. It could balance here but

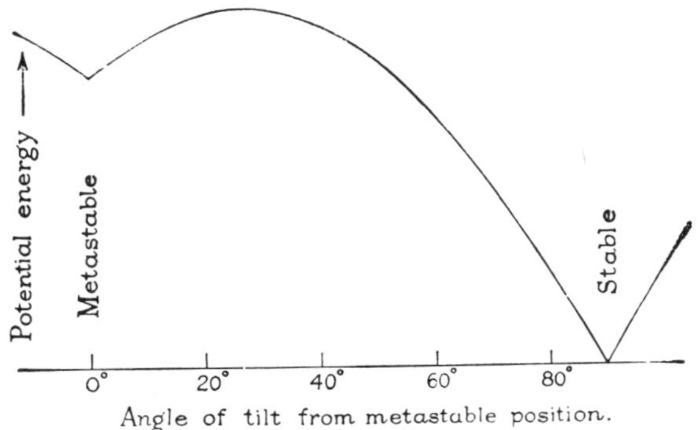

Fig. 9.5.—Variation of potential energy of box in fig. 9.4 with angle of tilt

would topple one way or the other if given the slightest disturbance. This position is therefore one of *unstable* equilibrium.

A graph of potential energy against angle of tilt from the metastable position for a box in which one dimension is twice the other is shown in fig. 9.5.

9.8. Activation energy

The same principle operates in many physical and chemical changes, where an atom or other particle can move from a metastable position to another of greater stability by passing through intermediate positions of higher energy. This is represented diagrammatically in fig. 9.6.

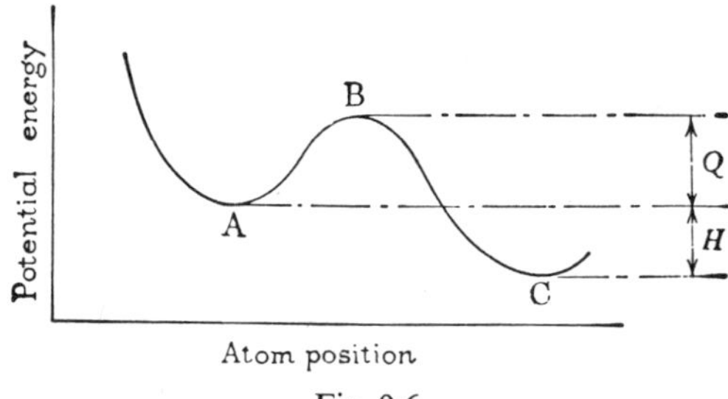

Fig. 9.6

An atom at position A can reach a more stable position C only by passing through the unstable position B. It will be able to do this only if it can receive the necessary additional energy to take it to the level B. The additional energy Q, i.e., the difference in the energy levels of positions A and B is the *activation energy*.

The final position C has a lower energy level so that in passing from B to C a total energy of $Q+H$ is released. Thus there is a net energy release of H. This is termed the *heat of reaction* and may appear in any of the forms that energy can take—heat, light, sound, etc. Since atoms may have energy due to thermal excitation and there is a scatter of energy values, some atoms will at times have enough energy to overcome the barrier.

The speed at which such a change can occur in a large population of such atoms will depend on the magnitude of Q and also on the number of atoms that possess energy equal to or greater than Q at any instant.

When Q is small, other things being equal, the reaction can proceed faster than when Q is large.

The number of atoms possessing energy Q at any instant depends upon the distribution of energy in a system. Owing to collisions and the forces between the atoms, there is a constant interchange of energy and fluctuations of energy of any particular atom. In a system with a large population of atoms, which has reached a state of thermodynamic equilibrium, there is a steady-state distribution of energy with a certain fraction of the population in each energy range.

9.9. Maxwell–Boltzmann law

For such a distribution, the proportion of atoms in each energy range is given by the Maxwell–Boltzmann law. The number of atoms which have energies in the range E to $E+dE$ at a particular temperature T is given by

$$dN = A\, e^{-E/kT}\, dE$$

where k is Boltzmann's constant and A is a constant. The form of the distribution curve is shown in fig. 9.7.

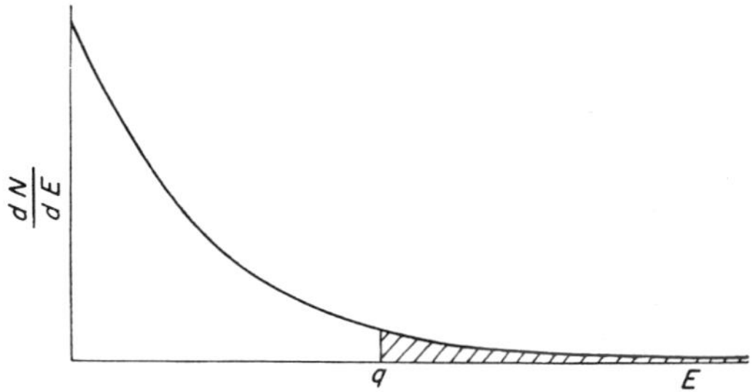

Fig. 9.7.—Maxwell–Boltzmann distribution law. Shaded area is proportional to number of atoms with energy greater than q

If N_T is the total number of atoms in the system, then

$$N_T = \int_0^\infty A\, e^{-E/kT}\, dE$$

$$= AkT$$

so that

$$A = N_T/kT$$

The number of atoms with energies greater than a particular value q is

$$N = \int_q^\infty \frac{N_T}{kT} e^{-E/kT}\, dE$$

$$= N_T e^{-q/kT}$$

If the energy is expressed as Q per kilomole of atoms or molecules (i.e., $Q = N_0 q$), then the expression becomes

$$N = N_T\, e^{-Q/RT}$$

where R is the universal gas constant.

9.10. Arrhenius' rate law

For a reaction for which the activation energy is Q per kilomole of atoms or molecules, the number of atoms with sufficient energy to overcome the potential barrier will be proportional to $e^{-Q/RT}$. The rate of reaction is proportional to this number so that

$$\text{rate of reaction} = v = Ae^{-Q/RT} = Ae^{-q/kT}$$

where A is a constant.

Writing this in the form

$$\log v = \log A - \frac{Q}{RT}$$

it will be seen that $\log v$ varies linearly with $1/T$. If, for a particular reaction, experimental values of $\log v$ when plotted against $1/T$ exhibit a straight line, then the reaction is a thermally-activated one and the slope of the line will be $-Q/R$, from which the activation energy Q may be calculated.

Some practical cases of thermally-activated processes will be discussed in the following sections.

9.11. Diffusion

The atoms in a crystal, in addition to vibrating about their mean positions, can also change positions by moving into adjoining vacant sites, a process known as diffusion. Such vacant sites are formed during crystal growth or later by thermal activation, the fraction of sites that are vacant being a function of temperature. In a homogeneous pure material diffusion cannot be detected. If, however, some atoms of a radioactive isotope of the same element are introduced, these are in effect labelled atoms and diffusion can be observed. Atoms of a second kind can also diffuse into a pure metal, a process which can be observed and measured.

The rate at which diffusion can occur has been formulated into Fick's laws, the first of which is:

If C is the concentration of the diffusing atoms at any point, and $\partial C/\partial x$ is the concentration gradient in the x-direction, then the rate at which atoms cross unit area perpendicular to the x-direction is

$$\frac{dm}{dt} = D\frac{\partial C}{\partial x}$$

where dm is the mass crossing unit area in time dt. Here D is the diffusion coefficient and has dimensions of (length^2/time).

In general C is not constant at any one point, and the variation of C with time is required. This can be expressed by Fick's second law:

$$\frac{\partial C}{\partial t} = \frac{\partial}{\partial x}\left(D\frac{\partial C}{\partial x}\right)$$

of if D is assumed to be constant

$$\frac{\partial C}{\partial t} = D\frac{\partial^2 C}{\partial x^2}$$

An atom can change its position only by crossing a potential barrier. Hence diffusion is a thermally-activated process, and the diffusion coefficient is found to obey Arrhenius' rate law.

Example.—The diffusion coefficient of carbon from a gas into the surface of a certain steel is given by

$$D = 0{\cdot}49\times 10^{-4}\, e^{-1{\cdot}53\times 10^8/RT}\text{m}^2\ \text{s}^{-1}$$

If carburizing is carried out at 927 °C (1200 K) then

$$\begin{aligned} D &= 0{\cdot}49\times 10^{-4}\, e^{-1{\cdot}53\times 10^8/8314\times 1200} \\ &= 1{\cdot}17\times 10^{-10}\ \text{m}^2\ \text{s}^{-1} \end{aligned}$$

At a temperature 50 K higher,

$$\begin{aligned} D &= 0{\cdot}49\times 10^{-4}\, e^{-1{\cdot}53\times 10^8/8314\times 1250} \\ &= 1{\cdot}99\times 10^{-10}\ \text{m}^2\ \text{s}^{-1} \end{aligned}$$

The rate has been approximately doubled by raising the temperature by 50 K.

9.12. Chemical reactions

Many chemical reactions have activation energies of approximately 4×10^7 J kmol^{-1}. The ratio of the rates of reaction at 27 °C and 37 °C is then

$$\frac{e^{-4\times 10^7/8314\times 310}}{e^{-4\times 10^7/8314\times 300}} = \frac{e^{-15{\cdot}53}}{e^{-16{\cdot}04}} = e^{0{\cdot}51} \approx 1{\cdot}7$$

i.e., the rate of reaction is approximately doubled when the temperature is raised by 10 K. This explains why a little gentle heat from a Bunsen burner will promote a reaction in a test tube.

9.13. Thermionic emission

It has already been stated that the valency electrons in a metal are free to move—not being attached to any particular atom. They are, however, confined within the boundary of a metal except when energy is supplied in some form. It therefore appears that some form of potential barrier exists at the surface, and that the electrons must be given enough energy to pass this barrier before they can escape.

The release of electrons when a substance is heated may be studied in a diode—a form of electronic valve containing an emitter, which can be heated, and an electrode which is given a positive potential relative to the emitter to collect the electrons. With a suitable geometry of construction of the diode and a sufficient voltage, all electrons leaving the emitter will reach the positive electrode—or anode—and the thermionic current i may be measured.

Some experimental results for the emission from clean tungsten are given in Table 9.2. It is found in this as in other cases that i increases as

TABLE 9.2—VARIATION WITH TEMPERATURE OF EMISSION FROM CLEAN TUNGSTEN

(values from S. Dushman, H. N. Rowe, J. W. Ewald and C. A. Kidner, *Phys. Rev.*, Vol. 25, 1925, p. 338).

T (K)	i(A mm^{-2})
1470	$7{\cdot}63 \times 10^{-10}$
1543	$4{\cdot}84 \times 10^{-9}$
1640	$4{\cdot}29 \times 10^{-8}$
1761	$4{\cdot}62 \times 10^{-7}$
1897	$4{\cdot}31 \times 10^{-6}$
2065	$4{\cdot}72 \times 10^{-5}$
2239	$4{\cdot}66 \times 10^{-4}$

the absolute temperature T of the emitter is raised, and that a plot of log i against $1/T$ gives a straight line within the limits of experimental accuracy as in fig. 9.8. This suggests a relationship of the form

$$i = f e^{-\omega/kT}$$

where ω is the activation energy necessary for an electron to leave the metal. Both f and ω will have values characteristic of the emitter material.

It is found by theoretical treatment* of the problem that

$$i = AT^2 e^{-\chi/kT}$$

* Reimann, *Thermionic Emission* (Chapman & Hall, 1934).

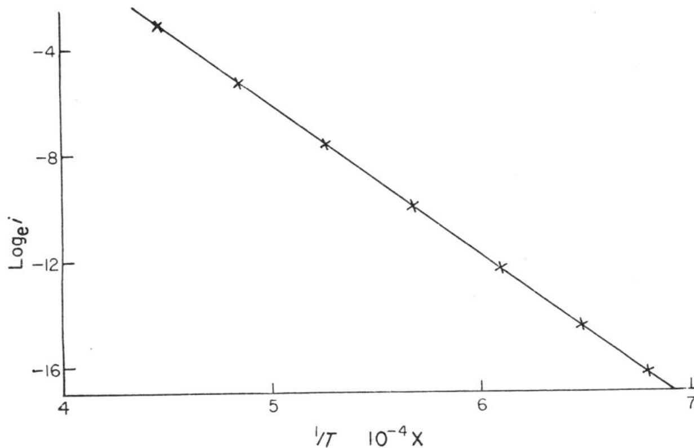

Fig. 9.8.—Plot of $1/T$ against $\log_e i$ for values given in Table 9.2

This differs from the previous value in containing the term T^2. The emission varies so greatly with temperature that it is not possible to distinguish by experiment between these two equations. The quantity χ is called the *work function* and is slightly less than ω.

The work function also turns up in the case of photoelectric emission. When a metal is irradiated with electromagnetic radiation of a sufficiently short wavelength, electrons are spontaneously emitted from the metal and may be detected with suitable electronic equipment. The energy of the photons at the critical wavelength for emission is found to equal the work function found from thermionic-emission experiments. This phenomenon is utilized in photoelectric cells.

QUESTIONS

1. Given that the kinetic energy of translation of a molecule of a gas at an absolute temperature T is $\frac{3}{2}kT$, where k is Boltzmann's constant, derive expressions for the specific heats of monatomic and diatomic gases.

Discuss the variation in specific heat of simple solids as they are cooled from room temperature towards absolute zero. [P]

2. Given that for a mass M of an ideal monatomic gas $pV = MRT$ where R is a constant, show that the specific heat at constant volume is $\frac{3}{2}R$.

Explain why at very low temperatures the specific heat at constant volume of a diatomic gas, such as hydrogen, falls below the room temperature value of $\frac{5}{2}R$.

Show that for particles of dust, each having a mass of $7{\cdot}5 \times 10^{-18}$ kg, in thermal equilibrium with the atmosphere at 15 °C, the root-mean-square velocity is about $0{\cdot}04$ m s^{-1}. [MST]

3. Calculate the specific heat of (*a*) copper, and (*b*) iron at temperatures above their Debye temperature.

4. Derive a value for the specific heat of a solid based on the principle of equipartition of energy between the various degrees of freedom.

In what ways do the observed specific heats of metals vary with temperature? [P]

5. State Arrhenius' rate law and explain what is meant by activation energy.

The rate v of linear growth of new crystals in a sample of cold-worked aluminium at different temperatures T is given in the following table:

T (°C)	200	250	300	350	400
v (m s^{-1})	$5{\cdot}62\times10^{-13}$	$1{\cdot}38\times10^{-10}$	$1{\cdot}35\times10^{-8}$	$6{\cdot}76\times10^{-7}$	$1{\cdot}82\times10^{-5}$

Show that this is a thermally-activated process, and calculate the activation energy per kmol of atoms.

6. Many physical and chemical thermal rate processes have activation energies of about $\frac{1}{2}$eV per atom. Show that the rate is approximately doubled by raising the temperature by 10 K near room temperature.

7. State Arrhenius' rate law for a thermally-activated process. Give examples of processes of technological interest which obey this law.

Some experiments on the diffusion of hydrogen in nickel have given the following results:

Temperature (°C)	162·5	237	355	496
Diffusion coefficient (m^2 s^{-1})	9×10^{-12}	$4{\cdot}6\times10^{-11}$	$3{\cdot}1\times10^{-10}$	$1{\cdot}34\times10^{-9}$

Show that these data are consistent with Arrhenius' law and determine the activation energy. [MST]

8. Some values quoted for the diffusion coefficient of carbon into iron determined from tests on an almost pure iron with a gaseous carburizing agent are

T (°C)	800	900	950	1000	1050	1100
D (m^{2}s^{-1}) $10^{-12}\times$	1·5	7·5	11·8	20	28	45

Examine whether these values follow an Arrhenius rate law relationship. Suggest a reason for any that do not.

Estimate a value for the activation energy for the diffusion of carbon in γ-iron. [E]

9. From considerations of Arrhenius' rate law, show that, if a system can exist in a metastable state with energy E_1 and a stable state with energy E_2 ($E_1 > E_2$), then, under conditions of thermal equilibrium, the numbers of units (e.g. atoms, molecules) in the two states are given by

$$\frac{N_1}{N_2} = \exp\left[-(E_1 - E_2)/kT\right]$$

For a hydrogen atom, the total energy of an electron is given by

$$E = 2{\cdot}18 \times 10^{-18}/n^2 \text{ J}$$

where n is the principal quantum number. In a flame at 3300 K containing 10^{20} hydrogen atoms, approximately how many atoms are in the first and in the second excited states? [E]

10. State what is meant by the terms *activation energy* and *work function* of a process.

In an experiment to determine the thermionic emission behaviour of tungsten the following results were recorded:

Current (A)	$2{\cdot}4 \times 10^{-9}$	$4{\cdot}0 \times 10^{-8}$	$7{\cdot}0 \times 10^{-7}$	$1{\cdot}2 \times 10^{-5}$	$9{\cdot}0 \times 10^{-5}$
Temperature (°C)	1155	1265	1394	1545	1727

Assuming that the current density and temperature are related by an Arrhenius type law, determine, from the above data,

(*a*) the activation energy required for electrons to leave an emitter, and

(*b*) which of the five tests should be repeated because of a probable error in a measurement. [E]

CHAPTER 10

The Deformation of Metal Single Crystals

10.1. Introduction

An examination of the behaviour of single crystals during deformation is necessary to an understanding of the deformation of polycrystalline aggregates. Single crystals of metals in suitable form for carrying out mechanical tests may be prepared by several methods, some of which are outlined in Section 5.9.

If such a crystal is pulled, there is first a very small elastic deformation proportional to the load, and then at a certain load plastic deformation commences and continues with a certain amount of increase of load until the specimen fractures.

10.2. Geometry of deformation by slip

When the crystal deforms plastically, it remains a crystal with the same structure. In the more common manner of deformation, namely *slip*, whole layers of atoms have moved over one another, so that atoms in a layer that has moved take up positions where other atoms were before. Each kind of crystal is found to have definite planes on which slip occurs and definite directions of sliding in each plane. They are known as *slip* or *glide planes* and *slip* or *glide directions*, respectively, and together they constitute the *glide elements.* A list of the glide elements for some of the commoner metals is given in Table 10.1.

The slip direction is always the direction of closest packing* and the slip planes are usually the ones of most dense packing. These most densely packed planes are also the most widely separated and so would require a lower shear stress to produce slip than would be necessary on planes of less dense packing.

One slip plane with one slip direction constitute a *slip system.* Slip will occur on the particular system for which the resolved shear stress is greatest.

In the hexagonal close-packed metals, zinc and cadmium, the ratio of the lattice parameters c/a is larger than the ideal and the most densely

* The direction of closest packing is the direction for which the strain energy of a dislocation is smaller than for any other direction because the Burgers vector is the smallest possible (see Section 10.11).

packed plane is the (001) plane—the basal plane. The slip direction is the [100] direction, which is along any of the diagonals of the hexagon, i.e., the close-packed directions. (The [100], [010], and [110] directions are all equivalent in this lattice.) The relationships of the glide elements for this type of lattice are shown in fig. 10.1.

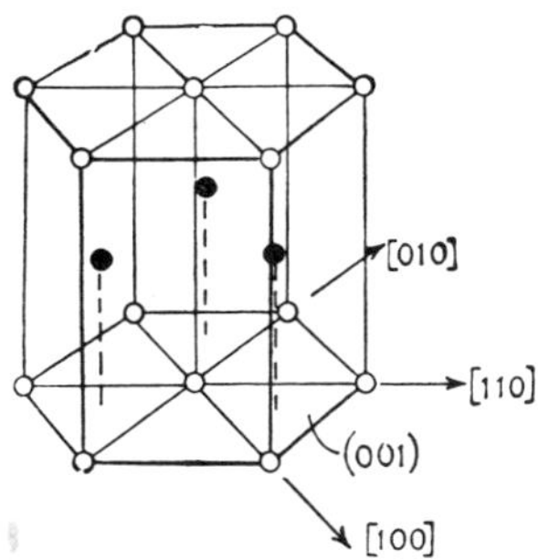

Fig. 10.1.—Glide elements for hexagonal close-packed metals that slip only on basal plane

In all other hexagonal close-packed metals c/a is smaller than the ideal and the (001) plane is not always the most densely packed plane. Some of these metals have other slip planes, as for example, titanium, which can slip on (100) planes as well as (001). The slip direction is always a [100] direction. The possible slip systems for titanium are shown in fig. 10.2.

Slip does not occur on every plane by an equal amount, but only on certain planes which may be hundreds of atom layers apart, and by large amounts on those planes. If the specimen were originally circular, with the glide planes lying at an angle to the axis, then the boundaries of the glide planes would be ellipses and after sliding over one another would give the effect shown in fig. 10.3. The glide steps are so large and so far separated in terms of atomic distances that they are visible to the naked eye. An actual photograph of a deformed crystal showing the slip band markings on the surface is shown in fig. 10.4. Work with the electron microscope has shown that there is a fine structure within the slip bands as shown in fig. 10.5.

For the face-centred cubic structure, the slip plane is the close-packed plane (111) and the slip direction is [10$\bar{1}$]. A model exposed on the slip plane is shown in fig. 10.6. There are four such planes of the form {111} and three possible slip directions for each, giving twelve possible slip systems. As in the previous case, the slip system that operates is the

TABLE 10.1—GLIDE AND TWIN ELEMENTS OF COMMON METAL CRYSTALS AT ROOM TEMPERATURE

Structure	Metal	Glide plane	Glide direction	Critical stress MPa	Twinning plane	Twinning direction
	Al			0·8		
	Cu			0·5		
Face-centred cubic	Ag	(111)	[10$\bar{1}$]	0·4	(111)	[11$\bar{2}$]
	Au			0·9		
	Ni			3·3		
Body-centred cubic	α-Fe	(101) (112) (123)	[11$\bar{1}$]	28		
	W	(112)		—	(112)	[11$\bar{1}$]
	Mo	(112)		73		
	K	(123)	[11$\bar{1}$]	—		
	Na	(112)		—		
	Mg			0·4		
Hexagonal close-packed	Zn	(001)	[100]	0·15	(102)	[10$\bar{1}$]
	Cd			0·09		
	Be			1·4		
	Ti	(100)	[100]	14		

Various values of critical stresses have been determined by different workers on materials of different purity. The lowest value has been quoted in each case.

[Data taken from C. S. Barrett, *Structure of Metals*, 2nd Ed. (McGraw-Hill, 1952) and from W. J. McGregor Tegart, *Elements of Mechanical Metallurgy*, (Macmillan, 1966).]

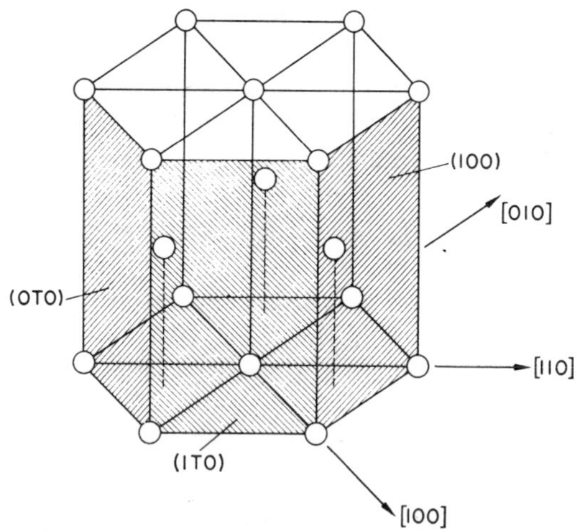

Fig. 10.2.—Glide elements of titanium. Slip systems are (100) [010], (010) [100], and (1$\bar{1}$0) [110]

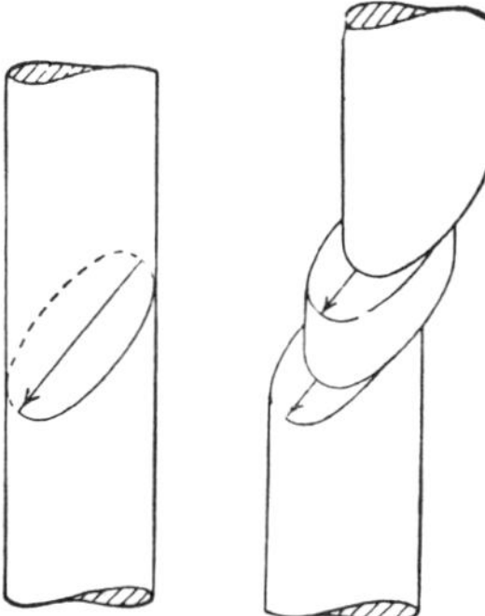

Fig. 10.3.—Diagram of single crystal of circular cross-section before and after slip on two parallel planes

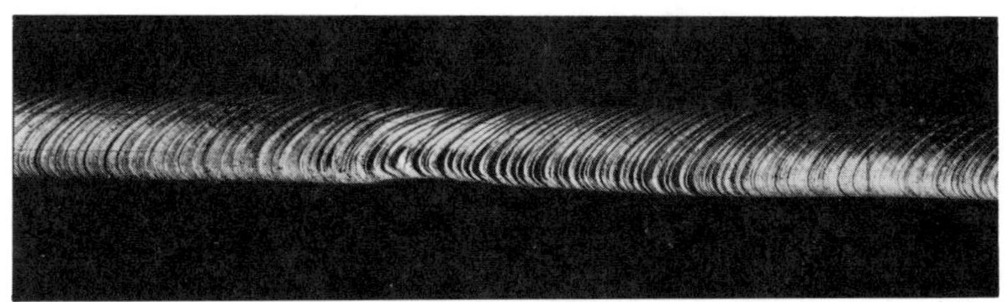

Fig. 10.4.—Photograph of slip bands on extended copper-aluminium single crystal [Elam]

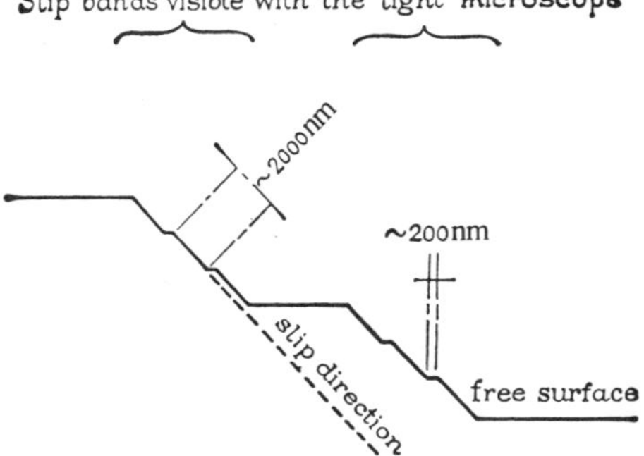

Fig. 10.5.—Schematic representation of fine structure of slip bands [Brown]

one on which the resolved shear stress is greatest, and there is a high probability of two slip systems operating at the same time.

For the body-centred cubic metals listed in Table 10.1 there are three families of slip planes listed for α-iron and one or another of these for

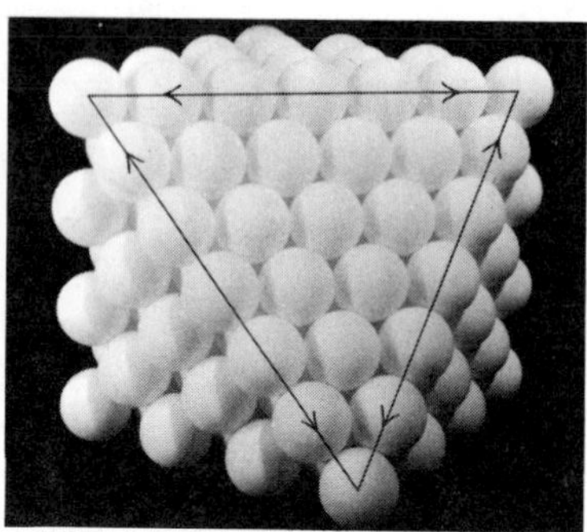

Fig. 10.6.—Model of face-centred structure showing (111) plane and three slip directions

each of the other metals. Models sectioned on the possible slip planes are shown in fig. 10.7. In the body-centred cubic crystals, there are no close-packed planes. There is, however, a close-packed direction which is [111], the cube diagonal, and this is always the slip direction.

10.3. Deformation by twinning

Another manner in which crystals can transform is by *twinning*. The displaced portion is separated from the undisplaced portion by a crystallographic plane characteristic of the particular crystal lattice and is symmetrically disposed to the undisplaced portion about this plane of separation. This may be most easily visualized as one portion being a mirror image of the other. The effect is shown in fig. 10.8. The twinning planes and directions of some commoner metals are listed in Table 10.1.

Twinning commonly occurs during the deformation of body-centred cubic and hexagonal close-packed metals but rarely, if ever, occurs in face-centred cubic metals due to deformation. In this last class, it occurs frequently due to crystal growth, especially during the recrystallization of cold-worked metals. A twin in this class of crystal is, in effect, a stacking fault, the close-packed layers (see p. 54) being arranged ABCABACBA, etc.

Twinning due to deformation of a single crystal is usually accompanied by a sharp noise, indicating an abrupt process. The stress necessary for

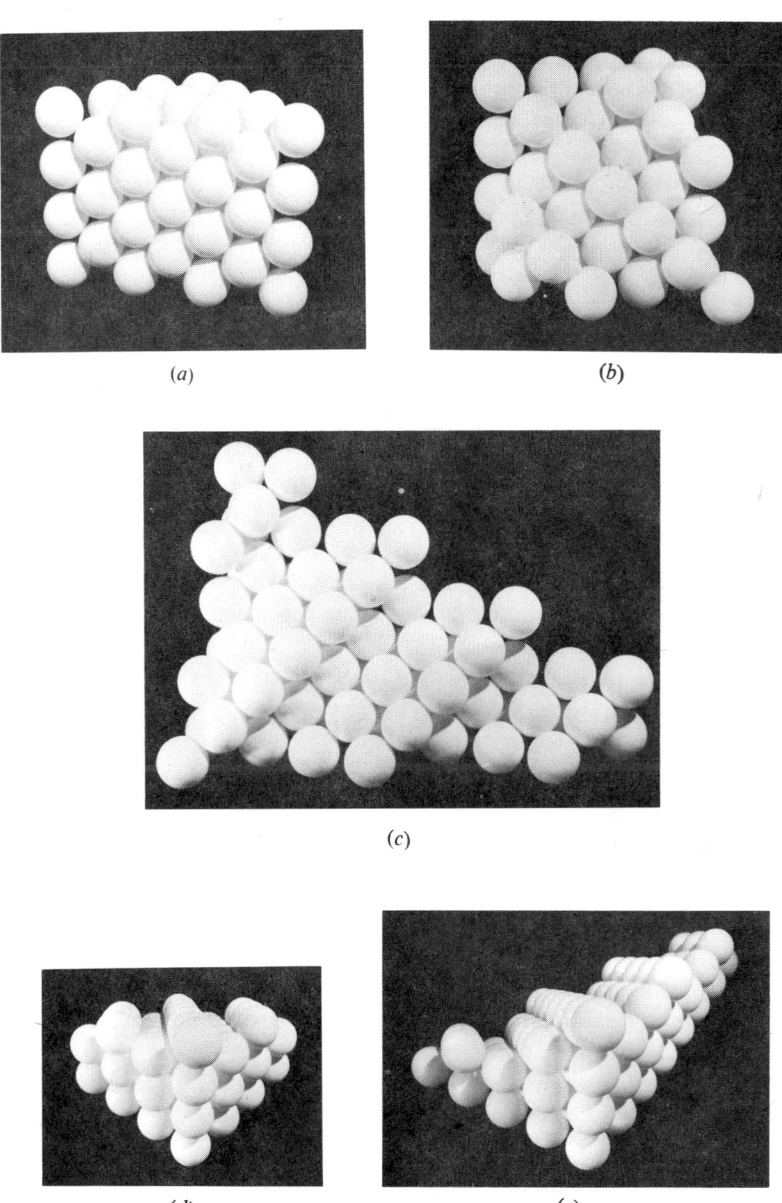

(*a*) (*b*)

(*c*)

(*d*) (*e*)

Fig. 10.7.—Models of body-centred structure showing slip planes

(*a*) view on (110) plane, (*b*) view of (112) plane, (*c*) view of (123) plane, (*d*) view of (112) plane along slip direction, (*e*) view of (123) plane along slip direction

twinning in cadmium has been observed to be of the order of 1·5–4·0 MPa, which is a much higher order than the stress necessary to cause slip in an undeformed cadmium crystal (see Table 10.1).

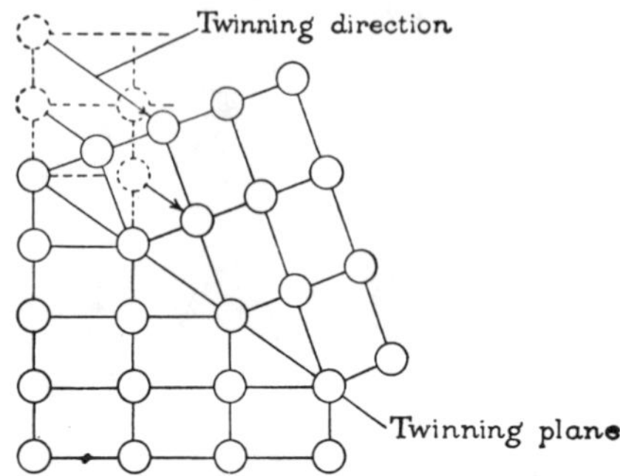

Fig. 10.8.—Atomic movements on twinning. Section on (110) plane of body-centred cubic lattice perpendicular to (112) twin plane

10.4. Resolved shear stress

The shear stress on the slip plane in the slip direction depends upon the load applied to the crystal and upon the orientation of the glide elements to the direction of the load.

Consider a load P applied axially to a crystal as shown in fig. 10.9. Let

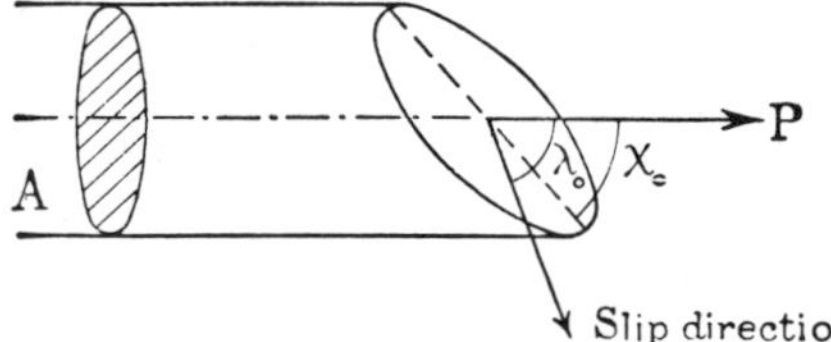

Fig. 10.9.—Angle relationships of slip plane and slip direction to crystal axis

A be the area of cross-section of the specimen,
χ_0 be the angle between the axis of the specimen and the slip plane,
λ_0 be the angle between the axis of the specimen and the slip direction.
The area of the slip plane is

$$\frac{A}{\sin \chi_0}$$

The resolved component of force in the slip direction is

$$P \cos \lambda_0$$

Hence the resolved shear stress in the slip direction is

$$\tau_0 = \frac{P}{A} \cos \lambda_0 \sin \chi_0$$

10.5. Load–extension curves

A series of stress–elongation curves for cadmium are shown in fig. 10.10. Each curve relates to a crystal of different initial orientation of the slip plane to the crystal axis, the values of χ_0 being given in each case. The origins of successive curves have been displaced in the direction of the elongation axis to reduce confusion. It will be observed that the initial nominal stress (i.e., P/A) is greatly different for the different orientations. If, however, the resolved shear stress in the slip direction is calculated, it is found to be constant and independent of the initial orientation. This is the *Law of Critical Shear Stress* which is obeyed by all metal crystals.

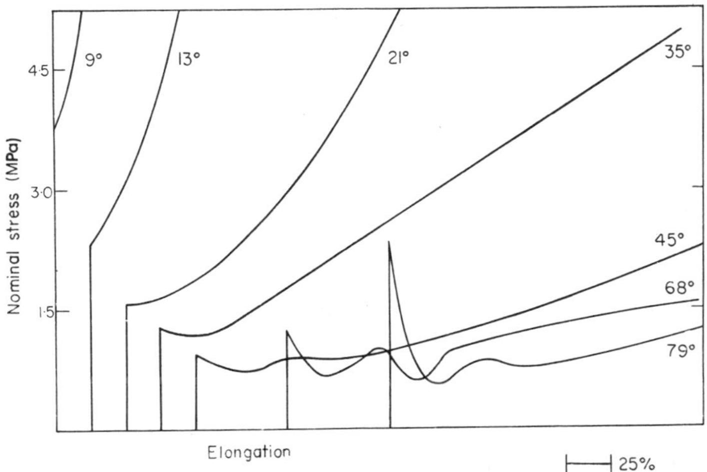

Fig. 10.10.—Nominal stress–elongation curves for cadmium single crystals. Figures on curves indicate angle between slip plane and direction of tensile stress. [After Boas]

Some values of the critical shear stress are given in Table 10.1 (p. 143). It will be observed that they are mostly of the order of 0·1– 1MPa. Tests with material of higher purity give lower values of critical shear stress.

When χ_0 is small, the area of the slip plane is large, and when χ_0 approaches 90° the resolved component of the force is a smaller fraction of the applied force, so that the nominal stress in either case is greater than when χ_0 has values in the region of 45°.

As slip proceeds, the area of the slip plane ($A_0/\sin \chi_0 = A_1/\sin \chi_1$) remains almost constant, but $\cos \lambda$ increases because λ decreases. For crystals so oriented that χ_0 and hence λ_0 is large, $\cos \lambda$ increases rapidly as slip proceeds, more rapidly than τ increases due to work hardening (see Section 10.7), so that the load P necessary to cause further slip will decrease. This is shown in fig. 10.10 for the crystals with large values of χ_0.

10.6. Relationship of elongation to strain

The amount of shear (or shear strain) is defined as the relative displacement of two planes which are parallel to the slip plane and which are at unit distance apart measured in a direction perpendicular to the plane.

Let the slip plane be the xy-plane of a coordinate system, and let the y-axis be the slip direction as in fig. 10.11. Let OP_0 and OP_1 be the positions of the axis before and after deformation, the coordinates of P_0 and P_1 being (x_0, y_0, z_0) and (x_1, y_1, z_1), respectively. Also let l_0 be

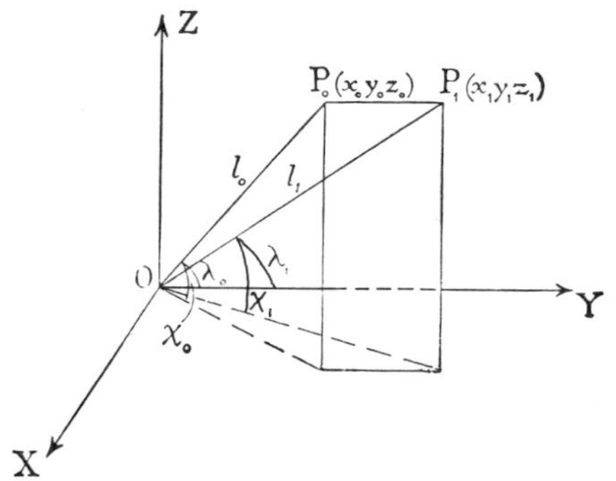

Fig. 10.11.—Coordinate system for analysis of deformation of crystal from OP_0 to OP_1

the initial length OP_0 of the crystal and the angles χ_0 and λ_0 have the meanings assigned in Section 10.4. Also let the same symbols with the subscript 1 denote the values after deformation.

It is assumed that the scale is such that the dimensions are much larger than the size of the slip bands, so that the deformation can be regarded as homogeneous.

Since the deformation is a pure shear, P_0 moves to P_1 where P_0P_1 is parallel to OY and the amount of this movement P_0P_1 is sz_0, where s is the magnitude of the shear strain.

Then

$$x_1 = x_0$$

$$y_1 = y_0 + sz_0$$

$$z_1 = z_0$$

Also, since

$$l_0^2 = x_0^2+y_0^2+z_0^2$$

$$l_1^2 = x_1^2+y_1^2+z_1^2$$

then

$$\frac{l_1^2}{l_0^2} = \frac{x_1^2+y_1^2+z_1^2}{x_0^2+y_0^2+z_0^2}$$

$$= \frac{x_0^2+y_0^2+z_0^2+2sz_0y_0+s^2z_0^2}{x_0^2+y_0^2+z_0^2}$$

$$= 1+\frac{2sz_0y_0+s^2z_0^2}{l_0^2}$$

But

$$y_0 = l_0 \cos \lambda_0 \quad \text{and} \quad z_0 = l_0 \sin \chi_0$$

so that

$$\left(\frac{l_1}{l_0}\right)^2 = 1+2s \sin \chi_0 \cos \lambda_0 + s^2 \sin^2 \chi_0$$

Hence if the initial orientation of the crystal is known, the amount of shear s for any measured extension of the crystal axis can be calculated.

Also because $z_1 = l_1 \sin \chi_1$ and $z_0 = l_0 \sin \chi_0$, and because $z_1 = z_0$, then if e is the fractional elongation of the axis,

$$(1+e) = \frac{l_1}{l_0} = \frac{\sin \chi_0}{\sin \chi_1}$$

Also $l_0 \sin \lambda_0 = l_1 \sin \lambda_1$, so that

$$(1+e) = \frac{l_1}{l_0} = \frac{\sin \lambda_0}{\sin \lambda_1}$$

Thus as the crystal elongates, χ and λ become smaller. The orientation of the crystal changes during extension so that the slip plane and the slip direction approach the specimen axis. Also the total volume will remain constant so that $A_0l_0 = A_1l_1$, where A_0 and A_1 are the respective cross-sectional areas.

10.7. Shear hardening curves

From the relations given in the equations derived in Sections 10.4 and 10.6 it is possible to calculate for each point on a load-extension curve for a single crystal the resolved shear stress and the shear strain, provided that the initial orientation of the crystal axis is known. The *shear hardening* curve so obtained is characteristic of the metal. It is found to depend upon the purity of the metal and the temperature, but is independent of the crystal orientation. Figure 10.12 shows the shear hardening curves for several metals at room temperature.

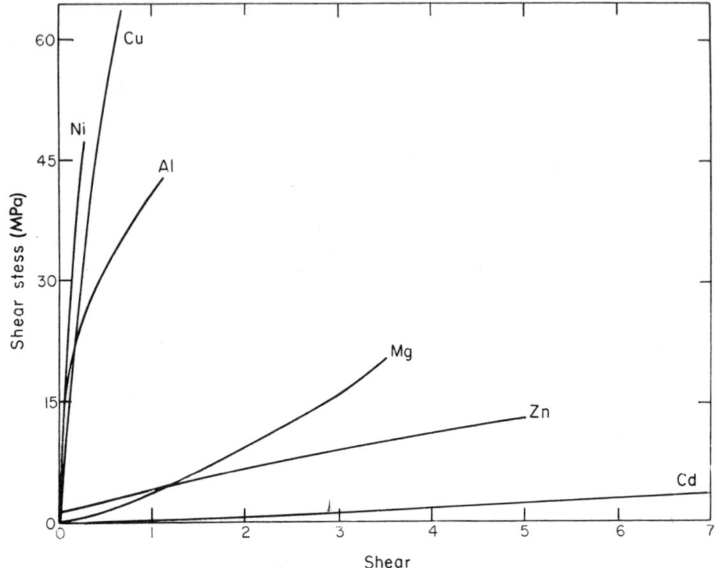

Fig. 10.12.—Shear hardening curves for single crystals of some face-centred cubic and hexagonal close-packed metals. [After Boas]

In every case there is *shear hardening*, that is to say, the shear stress necessary to cause further shear increases as the amount of deformation increases. The hexagonal close-packed metals, zinc, cadmium, and magnesium have low strain hardening rates. The general form of the shear hardening behaviour of a single crystal of a face-centred cubic metal in the early stages of deformation depends upon the orientation of the applied load to the crystal axis. For certain orientations, the resolved shear stress on two slip systems will be almost equal and slip will occur on both. This causes the rapid strain hardening shown in fig. 10.12. For other orientations, the resolved shear stress will reach the critical value on one slip system only. The behaviour is then very similar to that of a hexagonal close-packed metal and is described as *easy glide*. Because of the rotation of the lattice during straining, the resolved shear stress on the other slip systems will change and a second system later becomes operative—causing rapid strain hardening.

10.8. The nature of slip

The simplest assumption that can be made about the manner in which slipping occurs is that the whole of one slip plane slides simultaneously over the next, behaving as a rigid entity. This would mean that all the atoms move simultaneously, which would be possible only if the shearing force causing the movement had a completely uniform distribution over the slip plane, i.e., the shear stress is constant at all points. Thermal vibrations would, however, make this an impossible condition to achieve, even if the force could be applied to give an otherwise uniform shear stress. Therefore some areas must try to slip before others. There is no reason why this should not happen, since there is some flexibility in the coupling between the atoms in any plane. Slip can therefore start at one place and spread outwards.

At any stage of the slipping, a boundary could be drawn between the slipped region and the unslipped regions, and there would be a region of misfit along this boundary. This misfit is called a *dislocation* and the boundary is a *dislocation line.* It must always be a closed loop inside a crystal or have its ends at free surfaces of the crystal.

10.9. Theoretical shear strength

It should be possible to calculate the shear stress necessary to cause all the atoms in one plane to slip simultaneously.

Consider the shearing of two layers of atoms past each other in a homogeneously strained crystal. Let the spacing between atom centres in the direction of slip be a and the spacing of the rows be b as in fig. 10.13*a*. Let the shear displacement of the upper row over the lower be x when the shear stress is τ.

The shear stress is obviously zero when the upper row is in any equilibrium position, i.e., $x = 0$, a, $2a$, etc. Also, it will be zero when the upper layer is displaced by $\frac{1}{2}a$, $\frac{3}{2}a$, etc., because these are unstable positions at which each atom of the upper layer is sitting on top of an atom in the lower layer, and a slight movement either way in the absence of a shear force would cause the upper layer to move to one of the stable positions. For sufficiently small displacements from any of the stable equilibrium positions, Hooke's law would be obeyed, the shear stress being the product of the modulus of rigidity G of the material and the shear strain:

$$\tau = G\frac{x}{b}$$

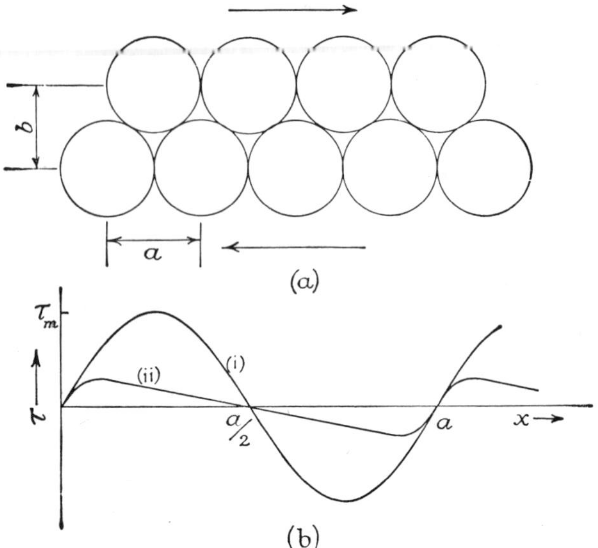

Fig. 10.13.
(*a*) Two rows of atoms being sheared by shear stress τ.
(*b*) Variation of τ with amount of shear

The curve of τ against x will change sign for every increase of x by $\frac{1}{2}a$, and its precise shape depends upon the nature of the interatomic forces. Frenkel made the assumption that the curve might be represented by a sinusoidal function of period a, that is

$$\tau = \tau_m \sin \frac{2\pi x}{a}$$

as curve (*i*) in fig. 10.13*b*.

The value of the constant τ_m is determined by the condition that the slope for small displacements must agree with the relation above. For the sinusoidal relationship near the origin,

$$\tau = \tau_m \frac{2\pi x}{a}$$

and from the Hooke's law relation,

$$\tau = G \frac{x}{b}$$

Hence

$$\tau_m = G \frac{x}{b} \cdot \frac{a}{2\pi x}$$

$$= \frac{G}{2\pi} \cdot \frac{a}{b}$$

This maximum value τ_m of τ is the shear stress that would have to be applied before slip would occur, i.e., the critical shear stress.
Now since a $\approx b$, then $\tau_m \approx G/2\pi$, which is of the order of 10 GPa for most materials.

As stated in Section 10.5, the observed values of the critical shear stresses for most metal crystals is of the order of 0·1–1 MPa. Hence there is a discrepancy of 10^4–10^5 between the theoretical result derived above and the observed values.

The assumptions made in the calculation must therefore be examined. Calculations involving the nature of the inter-atomic forces show that the shape of the curve should be more nearly as curve (*ii*) in fig. 10.13*b*. The maximum shear stress for this curve is found to be of the order of $G/30$ and to occur for a displacement of about 0·1 *a*. The major part of the discrepancy still exists.

Hence the conclusion must be reached, as discussed in Section 10.8, that the whole of a crystal does not slip simultaneously but that slip spreads progressively across a slip plane, actual slip motion taking place only at the dislocation line.

10.10. The geometry of dislocations

When a dislocation passes across a crystal, the atoms behind it have sheared relative to the adjacent layer by the unit of slip, which is of a definite amount and in a definite direction. The vector which defines this displacement is called the *Burgers vector*. In a hexagonal close-packed crystal, for example, the Burgers vector has a magnitude equal to the unit cell side *a* and a direction parallel to one of the hexagon diagonals.

Dislocations may be classed according to the direction of the line of misfit relative to the Burgers vector. If the line is perpendicular to the Burgers vector, as in fig. 10.14, the dislocation is an *edge dislocation*. As the slipped region extends, the dislocation line moves in the direction of the Burgers vector.

If the dislocation line is parallel to the Burgers vector, the dislocation is a *screw dislocation*, as shown in fig. 10.15. A set of parallel but separate planes, when intersected normally by a screw dislocation line, undergo a shear which transforms them into a single helicoidal or screw surface spiralling around the line. Thus beginning on any plane, a circuit taken once around the dislocation line will result in a translation by an amount

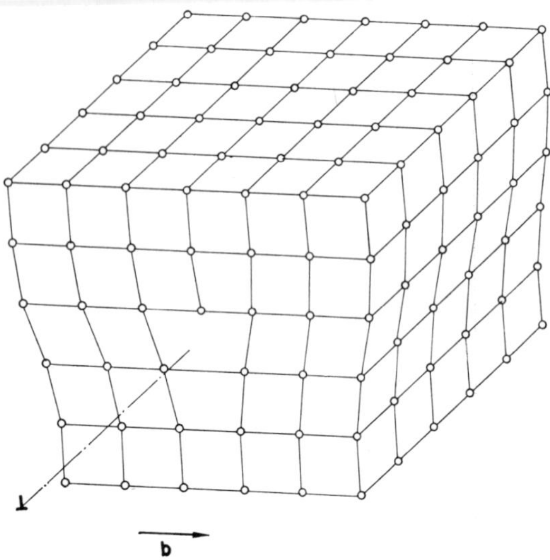

Fig. 10.14.—Edge dislocation in a simple cubic crystal. **b** is Burgers vector. ⊥ is symbol used to denote an edge dislocation

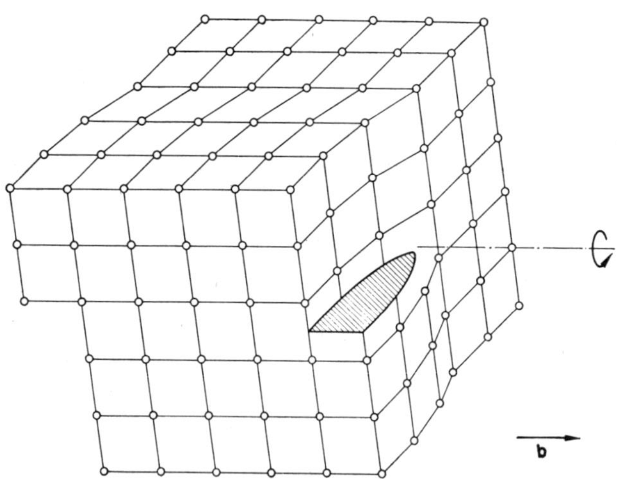

Fig. 10.15.—Screw dislocation in a simple cubic crystal

equal to the Burgers vector along the direction of the screw dislocation line. As slip continues, the dislocation line moves perpendicular to the Burgers vector.

In general a dislocation line can be curved, being partly edge and partly screw dislocations. Also if two slip planes exist with a common slip direction, the dislocation line can lie partly in each plane, so that it is not necessarily confined to two dimensions. An edge dislocation can normally move only in the plane which is defined by its Burgers vector and the dislocation line, while a screw dislocation is free to move on any of the planes in which the Burgers vector lies.

An edge dislocation can move to a parallel plane, a process known as *dislocation climb*, either by the diffusion of vacancies already present in the crystal to the end of the half plane of atoms, thereby removing the extreme row (fig. 10.16), or by atoms joining the half plane to make an extra row and leaving vacancies behind.

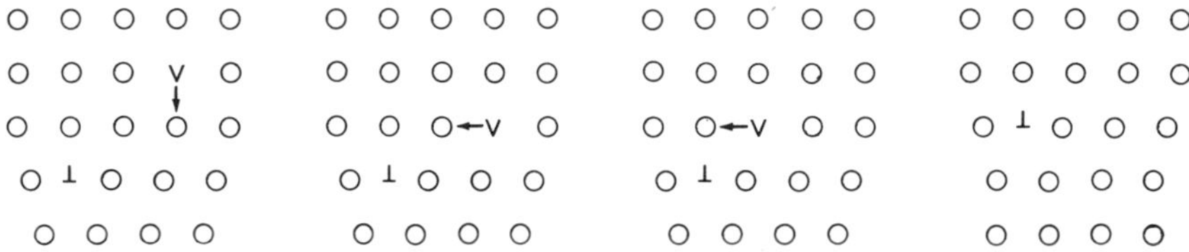

Fig. 10.16.—Dislocation climb by absorbing vacancies

As stated above, a screw dislocation is free to move on any of the planes in which the Burgers vector lies. Hence a screw dislocation can move out of the slip plane by this process of *cross-slip*. Apart from the need to create extra dislocation lines at the ends of the moving part, this requires no extra force or energy, in contrast with dislocation climb where thermal activation is necessary for vacancy diffusion. Further reference to these processes of cross-slip and dislocation climb will be made in Section 11.6.

When a slip dislocation has passed completely through an initially perfect region, it leaves a perfect crystal structure behind it.

There is strain energy stored up in the dislocation due to the elastic distortion, but by comparison it requires very little energy to cause it to move. Hence if a dislocation is present in a slip plane it will move under small forces.

10.11. Strain field around a screw dislocation

The crystal lattice is distorted in the vicinity of a dislocation. Hence there is strain energy associated with its presence. The magnitude of this energy may be estimated for a screw dislocation as follows.

Consider a thin hollow cylinder, length l, radius r, and thickness δr, which is cut along a plane that passes through the axis and is sheared by an amount b (fig. 10.17, a and b). The deformation is typical of that in a

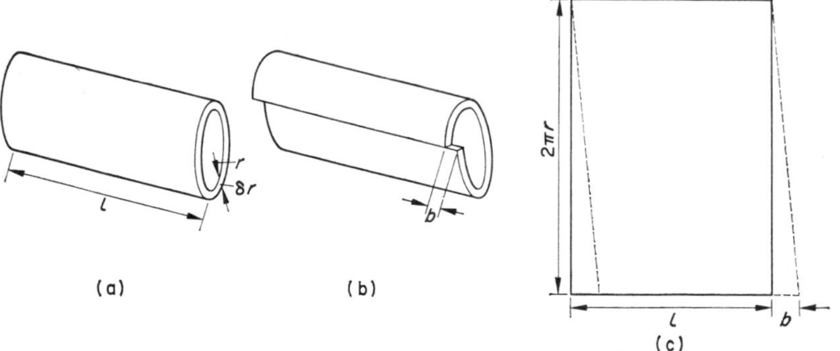

Fig. 10.17.—Elementary cylinder around a screw dislocation (a) before, (b) after slip: (c) cylinder flattened to show shear deformation

region around a screw dislocation, the dislocation line coinciding with the axis of the cylinder. The nature of the deformation can be seen to be the same as shear in a flat sheet (fig. 10.17c), in which

$$\text{shear strain} = \gamma = \frac{b}{2\pi r}$$

$$\text{shear stress} = \tau = \frac{Gb}{2\pi r}$$

where G is the elastic shear modulus.

The elastic strain energy stored in this cylinder is

$$\delta E = \tfrac{1}{2} \times \text{stress} \times \text{strain} \times \text{volume}$$

$$= \frac{Gb^2 l\, \delta r}{4\pi r}$$

The total strain energy is given by the integral

$$E = \int_{r_2}^{r_1} \frac{Gb^2 l}{4\pi r}\, dr$$

$$= \frac{Gb^2 l}{4\pi} \log_e r_1/r_2$$

where r_2 and r_1 are the inner and outer radii of the region over which the integration is taken. These limits of integration cannot be stated precisely. The outer limit may be set by the size of the crystal or the distance to the next neighbouring dislocation which also has its own strain field, and is likely to be of the order of 10^{-3} m or less. In the core of the dislocation at radii of the order of a few atom spacings, the medium cannot be regarded as a continuous Hookean solid, and the actual strain energy due to the forces between individual atoms will be small compared with the total. The inner limit can therefore be taken as three or four atom spacings, i.e., about 10^{-9} m.

Then

$$\log_e r_1/r_2 = \log_e 10^6 \approx 4\pi$$

so that the strain energy per unit length of a screw dislocation is approximately,

$$Gb^2$$

This is about 5 eV per atom length of dislocation line.

10.12. Strain field around an edge dislocation

Again, this is most easily considered in terms of thin shells which have the dislocation line as axis and which are cylindrical before the slip takes place. These become distorted after slip as shown in fig. 10.18. There is no circular symmetry and the strain is a function of the angle θ. On the side of the extra half plane there is a region of compression and on the other side a region of tension. A detailed analysis which is too lengthy to consider here, gives a value for the strain energy that is approximately

$$\frac{Gb^2}{1-\nu}$$

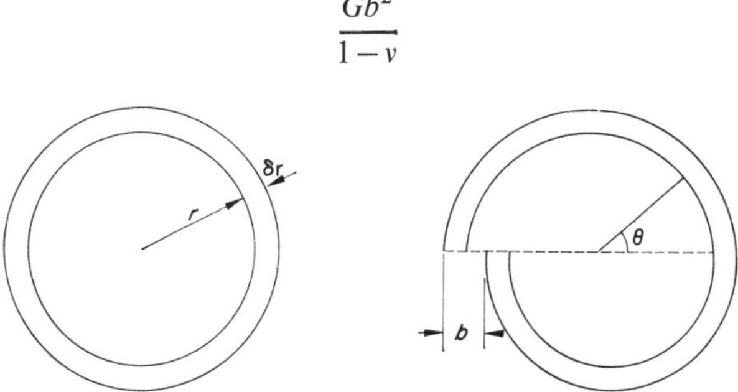

Fig. 10.18.—Distortion of elementary cylinder around an edge dislocation

per unit length, where ν is Poisson's ratio. For $\nu = \frac{1}{3}$, the energy of an edge dislocation is about $\frac{3}{2}$ of that of a screw dislocation per unit length.

Because the energy is proportional, in each case, to b^2, the most stable dislocations, i.e., the ones of minimum strain energy, will be those of minimum Burgers vector, i.e., those with this vector in a close-packed direction.

10.13. Force on a dislocation line

Consider a block of a single crystal, as shown in fig. 10.19, containing a slip plane ABCD of dimensions l and w. Suppose the block is subjected to an external stress such that the component of shear stress on a plane parallel to the slip plane and in a direction parallel to the Burgers vector is τ. The passage of an edge dislocation along the slip plane from AB to DC will produce a shear displacement of magnitude b.

The force on the upper surface, which is τlw has done work τlwb. Let F_e be the force per unit length on the edge dislocation due to the shear stress τ. The total force on the dislocation is $F_e w$ and is moved a distance l so that the work done is $F_e wl$. This is equal to the external work so that

$$F_e = \tau b$$

A similar argument can be applied to a screw dislocation which moves from AD to BC and produces the same overall displacement as in the previous case. The force F_s per unit length is also of magnitude τb. In each case, the force acts perpendicular to the dislocation line and parallel to the direction of motion.

Also, for a mixed dislocation, it will again be τb per unit length and act perpendicular to the dislocation line. Hence a dislocation loop would behave as if subject to an internal pressure (fig. 10.20).

10.14. Forces between dislocations

When dislocations approach one another, their strain fields will intersect. If this leads to an increase of the total energy, then there will be a repulsion between them. Conversely, if the energy decreases, then there will be an attraction. The magnitude of the force is equal to the rate of change of energy with distance.

For two parallel screw dislocations of equal strength (i.e., having Burgers vector of equal magnitude) and opposite signs separated by a distance r, the strains due to them at large distances will cancel so that strain effects will matter only within a distance of the order of r. The strain energy per unit length of the pair will be

$$E \approx \frac{Gb^2}{2\pi} \log_e r/r_2$$

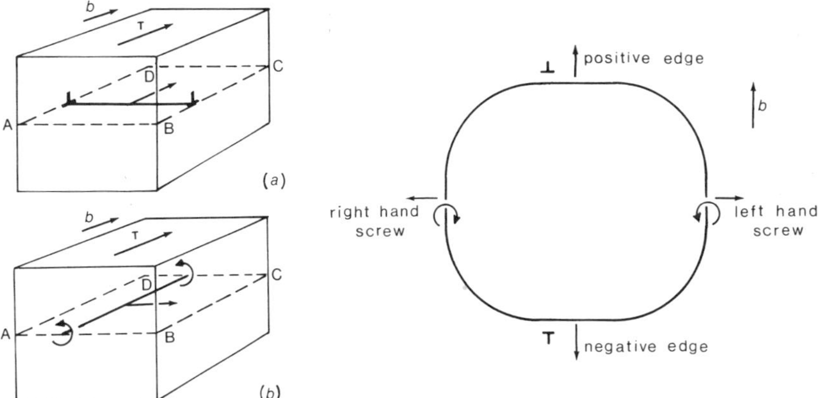

Fig. 10.19.—Slipping of single crystal due to passage of (*a*) an edge dislocation, (*b*) a screw dislocation

Fig. 10.20.—Directions of forces on dislocation loop in crystal subject to applied shear stress

which decreases as r decreases. Hence there will be an attractive force per unit length equal to

$$\frac{dE}{dr} = \frac{Gb^2}{2\pi r}$$

If one dislocation can move under the action of this force until it meets the other they will annihilate one another.

Conversely, for two screw dislocations of the same sign the strains would be additive so that they would repel each other. If the two like dislocations are far apart and the energy of each is E, the total energy is $2E$. Bringing them together is equivalent to having a dislocation with a Burgers vector $2b$, so that the energy increases to $4E$.

For a dislocation at a distance $r/2$ from a free surface, the case is similar to that of putting a negative dislocation in a continuous medium as a mirror image (image dislocation) as in fig. 10.21. The forces due to the dislocation and its image will cancel at the surface. Hence the dislocation is attracted to the free surface by a force $Gb^2/2\pi r$ per unit length.

In edge dislocations, the stress field does not have radial symmetry and the force between dislocations does not act along the line between them except in special cases.

Two parallel edge dislocations on the same plane or nearby planes will repel one another. If they are on more distant planes, then the direction of the force depends upon the angle between the line joining them and the slip plane. If this is less than 45°, then they repel each other. If it is greater than 45° then they attract one another. Because movement is

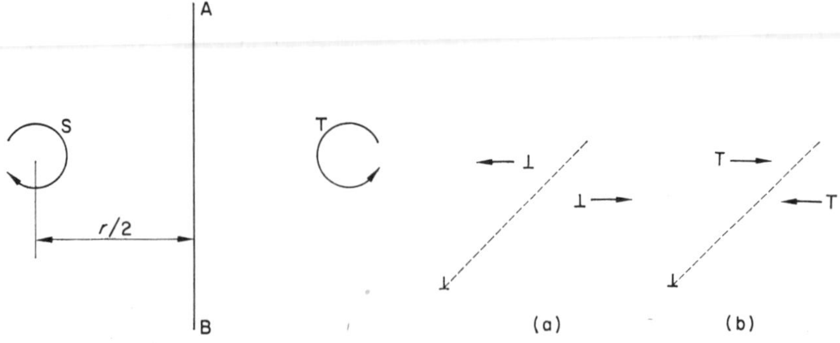

Fig. 10.21.—A screw dislocation *S* distant *r*/2 from a free surface AB produces the same stress-free effect at AB as would the dislocation *S* together with an image dislocation *T* of opposite sign in a continuous medium

Fig. 10.22.—Direction of forces on (*a*) like (*b*) unlike edge dislocations

restricted to the slip planes, they will move until the line between them is perpendicular to the slip planes (fig. 10.22). Unlike dislocations on the same slip plane would attract and annihilate one another, while those on different planes would be attracted to and move to be on a line making 45° to the slip plane.

In a bent crystal there would be a collection of like dislocations scattered at random. Given time they would move into positions where they are lined up thereby decreasing the total energy. The crystal would then have a polygonal structure, regions of perfect lattice being separated by *low-angle grain boundaries* known as *tilt boundaries* (fig. 10.23).

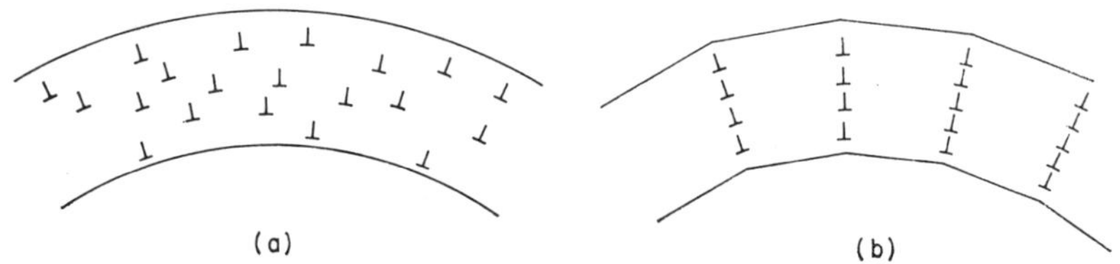

Fig. 10.23.—(*a*) Dislocations in a bent crystal, (*b*) dislocations lined up in tilt boundaries producing polygonization

Another type of low-angle boundary known as a *twist boundary* is that produced by an intersecting grid of screw dislocations which give a relative angular rotation of the portion about an axis perpendicular to the boundary (fig. 10.24).

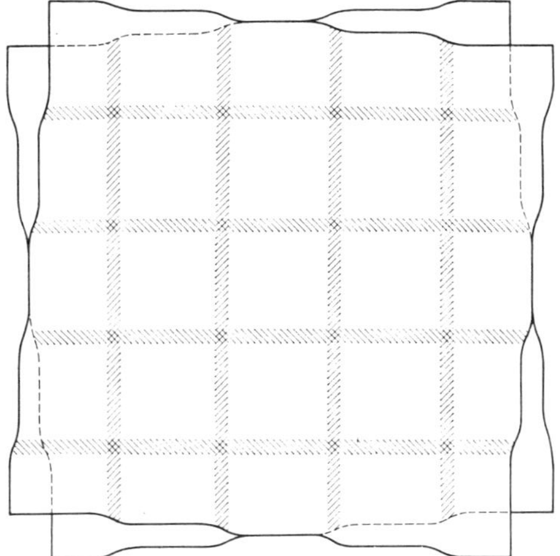

Fig. 10.24.—Twist boundary formed by two series of screw dislocations

10.15. Tension of a dislocation line

The energy of a dislocation line is proportional to its length, so that work has to be done to increase the length. This can be considered as work against a line tension T, the magnitude of which is $T = dE/dl \approx Gb^2$

A dislocation line can be bent to a curved shape only if acted upon by a shear stress which balances a component of the line tension. Consider an arc δs of a dislocation line of strength b curved to a radius r by a shear stress τ (fig. 10.25). The angle subtended by the arc is

$$\delta\theta = \delta s/r$$

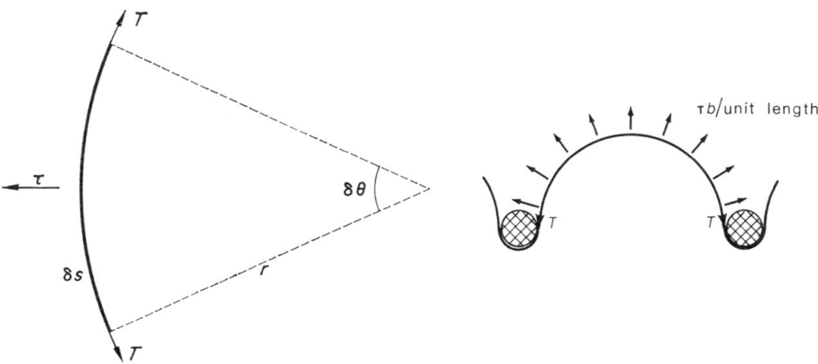

Fig. 10.25.—Curving of dislocation line due to shear force

Fig. 10.26.—Dislocation loop bowing between two obstacles

The force acting on the arc due to the applied stress is $\tau b\ \delta s$ and is balanced by the components of T, i.e., by

$$2T \sin \frac{\delta\theta}{2} \approx T\ \delta\theta$$

so that

$$\tau = T/br$$

Hence the sharper the curvature of the dislocation, the greater is the shear force necessary to maintain that curvature.

Suppose a dislocation moving under the effect of an external shear stress, τ, meets two obstacles a distance L apart. It will be bent to a smaller and smaller radius of curvature as it tries to pass between them until it is a semi-circle (fig. 10.26), after which any further movement will result in an increase of the radius of curvature. Hence the critical state is when the dislocation loop is semicircular with radius $L/2$. The components of the force on the dislocation line in the direction perpendicular to the line joining the obstacles is τbL and must equal $2T$. Hence

$$\tau = \frac{2T}{bL} = \frac{2Gb}{L}$$

That is, in a material containing obstacles which are, on average, L apart, the yield shear stress will be $2Gb/L$.

10.16. The origin of dislocations

It has been calculated that the average stress necessary to create an edge dislocation in a perfect crystal is of the same order of magnitude as Frenkel's theoretical stress for the simultaneous slip of a layer of atoms. Hence we are led to the belief that dislocations must exist in crystals when they are formed.

Frank* has drawn attention to the fact that it is practically impossible for a crystal to grow from vapour or a dilute solution unless dislocations are present. An atom arriving at the crystal surface from the liquid or vapour side will not attach itself permanently to the crystal unless a step or ledge is present. Hence a single atom on a plane face as in fig. 10.27*a* will not remain attached, but if an atom reaches a point such as A in fig. 10.27*b*, the attractive forces are sufficient to hold it there.

Continual growth of a crystal will be possible if the structure is such that the crystal always has steps on its faces however much it grows. This kind of structure exists around a screw dislocation, the crystal

* F. C. Frank, *Advances in Physics*, Taylor and Francis, London, Vol. 1, 1952, p. 91.

building up as a spiral plane. Spiral growth has been observed on many crystals, some examples being shown in fig. 10.28.

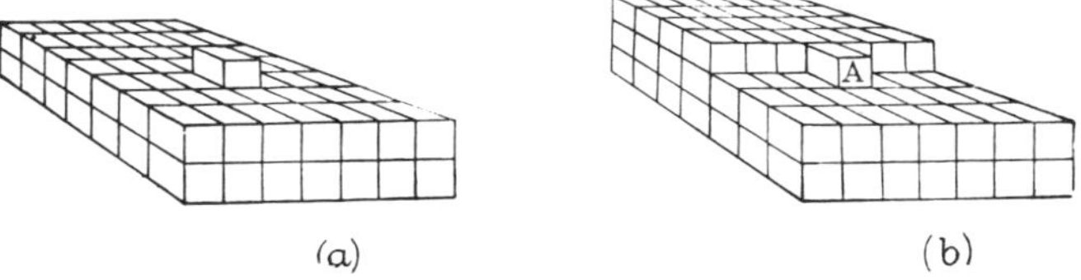

Fig. 10.27.—Atoms, represented by cubes, during crystal growth
(*a*) single atom will not attach itself to surface,
(*b*) step provides sufficient forces to hold atom A

Hence unless a dislocation forms during nucleation, a crystal is unlikely to grow.

Also during crystal growth, atom sites may be left vacant. It is possible for vacancies to move together by diffusion and, having done so, the larger vacancy then formed may collapse creating a dislocation ring, of which a section should show two dislocations as in fig. 10.29.

Since in general there are several possible slip systems in a crystal, dislocations can occur in planes that are not parallel to one another. In general, growth dislocations are likely to be in different directions. Since a dislocation line can end only on an outer surface, these dislocations must run together except at free surfaces. It is possible for three disloca-

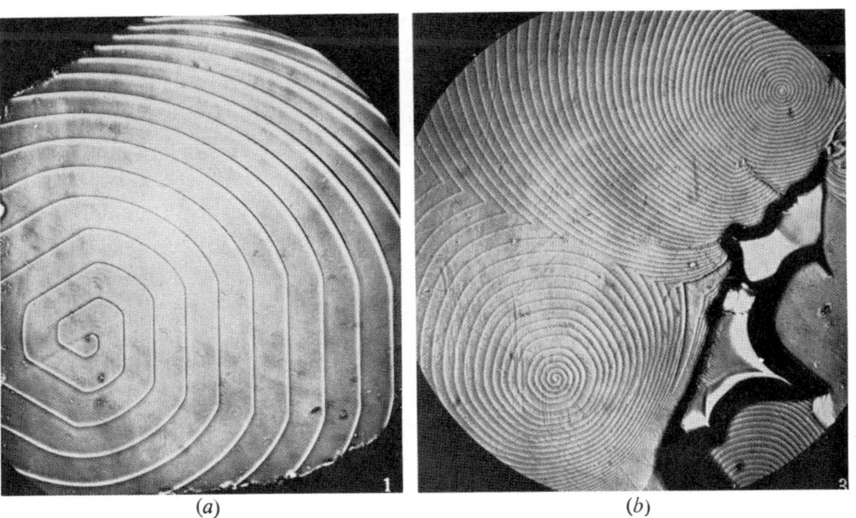

(*a*) (*b*)

Fig. 10.28.—Spiral growth on silicon carbide crystals revealed by phase-contrast microphotography (× 60)
(*a*) single spiral, (*b*) spirals from two dislocations of opposite hand. [Verma]

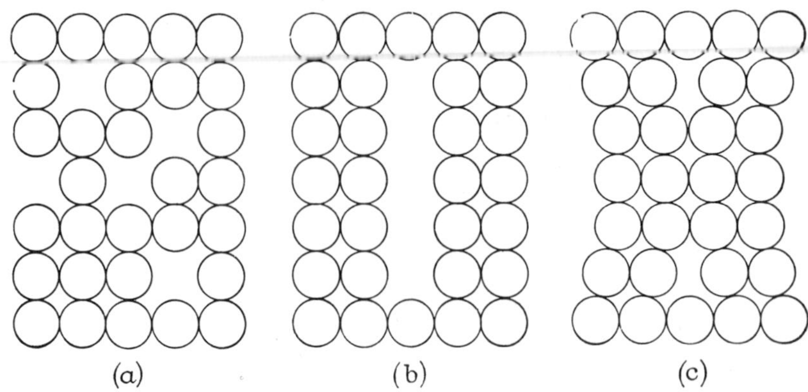

Fig. 10.29
(*a*) Vacancies randomly scattered in crystal (the number of vacancies is exaggerated)
(*b*) Vacancies have diffused into a single layer
(*c*) Collapse of vacancy giving two dislocations

tions to meet at a point if the three Burgers vectors add up to zero as in fig. 10.30. Hence the dislocations can build up a three-dimensional network. Experimental evidence for the existence of such networks has been found,* an example being shown in fig. 10.31.

Of the dislocations already in a crystal, some may be unable to move under the applied stress, either because the resolved stress is not sufficient or because they are anchored by some mechanism. These *sessile dislocations* may anchor the ends of another dislocation line for which slip would be favoured, giving a *Frank–Read source of dislocations*.† Figure 10.32*a* shows a dislocation which is anchored at its ends A and B and which has its slip plane in the plane of the diagram. When a shear stress is applied in the direction shown, the dislocation line will form a loop moving in the direction shown by the arrows in fig. 10.32*b* and *c*. The loop grows until it sweeps around A and B, and eventually the two sides meet forming a complete loop and another line dislocation between A and B as in fig. 10.32*d*.

Fig. 10.30.—Three Burgers vectors which have zero sum
$$\mathbf{b}_1 + \mathbf{b}_2 + \mathbf{b}_3 = 0$$

* P. B. Hirsch, " Direct Experimental Evidence of Dislocations," *Met. Rev.*, Vol. 4, 1959, p. 101.

† F. C. Frank and W. T. Read, *Phys. Rev.*, Vol. 79, 1950, p. 772.

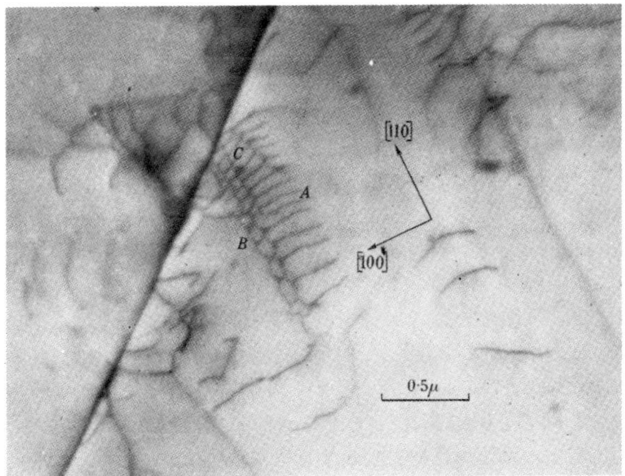

Fig. 10.31.—Hexagonal network of dislocations in stainless steel deformed by rolling, revealed by electron microphotography at × 22000 magnification. [Whelan, Hirsch, Horne, and Bollmann]

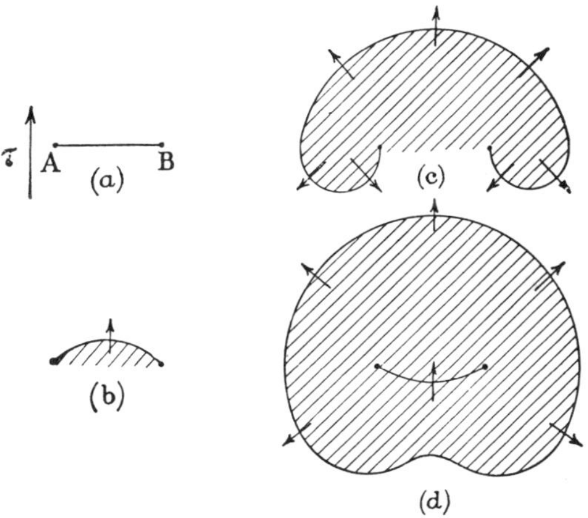

Fig. 10. 32.—A Frank–Read source forming a dislocation loop under the action of the applied shear stress τ

The critical stage, which requires the maximum shear force, is that when the loop is approximately semicircular, the radius of curvature then being a minimum. Sources of greater length, i.e., greater distance between A and B, will operate under lower shear stresses. If the distance between the fixed points A and B is l then the stress required to push the dislocation through this critical stage is $2\,Gb/l$. Sources of greater length will operate under lower shear stresses.

For a single crystal of pure metal, typical values of 1 MPa for the critical shear stress, 3×10^{-10} m for b and 10^5 MPa for G give a value for l of 6×10^{-10} m. If this is taken to be the typical distance between dislocations, then their density is of the order of 10^8 lines m^{-2}, a value in agreement with that commonly found in annealed single crystals.

Under the applied shear force, the loop will move out to the boundaries of the crystal, and the source can form a succession of loops. It would appear that such a source could give an indefinite number of dislocations.

This mechanism can explain the observed behaviour that slip is a cascade process, slip of about 1000 atom spacings occurring on one plane, and also that slip occurs on certain planes only, these being the planes on which there are possible Frank–Read sources. The shear stress necessary to activate a Frank–Read source depends upon the distance AB, being greater for smaller values of AB, and as this distance is not necessarily constant for different sources, the stresses to cause slip on different planes will be of different magnitude.

QUESTIONS

1. Describe the modes of deformation due to stress of single crystals of (*a*) cadmium, (*b*) copper. What reason may be given for the difference in the form of the shear hardening curves of single crystals of these two metals? [P]

2. What is meant by a " slip system " of a crystal?
Cadmium slips on {001} planes in ⟨110⟩ directions. Titanium slips on {100} planes in ⟨110⟩ directions. Suggest reasons for this difference.
Sketch a structure cell of each metal showing a slip system. How many possible slip systems does a crystal of each metal have? [MST]

3. A single crystal specimen of pure zinc of 65 mm² cross-sectional area yielded in tension at a load of 160 N. The tensile load was then increased to a maximum value of 220 N, and the resulting plastic extension of the specimen on a 50 mm gauge length was found to be 17 mm. If the inclinations of the active slip plane and the slip direction to the specimen axis in the unloaded specimen were 22° 21′ and 34° 42′, respectively, calculate: (*a*) the critical shear stress for zinc; (*b*) the inclinations of the slip plane and the slip direction to the specimen axis at the conclusion of the test; (*c*) the value of the shear stress on the slip plane in the slip direction when the specimen was under maximum load. [P]

4. A specimen, 32 mm² in cross-sectional area, was made from a single crystal of pure magnesium. The inclinations of the active slip plane and the slip direction to the specimen axis were 27° 00′ and 38° 24′, respectively. If, when tested in tension, the specimen yielded at a load of 75 N, calculate the critical shear stress for magnesium. [P]

5. In a series of single-crystal specimens of zinc the inclinations of the active slip plane and the slip direction to the specimen axis are equal and in the range 10° to 80°. If the critical shear stress for zinc is 760 kPa, draw a graph showing the relationship between this angle of inclination and the yield stress (i.e., load/cross-sectional area) of the specimen in simple tension. Sketch and explain the shapes of the complete stress–strain curves for the specimens having angles of inclination of 10°, 45°, and 80°. [MST]

6. In the face-centred cubic system find:
   (*a*) the angle between [110] and [211];
   (*b*) the angle between [110] and (211);
   (*c*) the stress resolved on the slip system $(1\bar{1}1)$ [110], when a stress of 350 kPa is applied in the [211] direction. [MST]

7. Derive Frenkel's approximate expression for the theoretical yield strength in shear of a single crystal in terms of the shear modulus G.

The critical shear strength of copper determined by experiment is 1·0 MPa. What is the ratio of theoretical to experimental shear stress? What mechanism has been postulated to explain this discrepancy?

(For copper, take $G = 4{\cdot}2 \times 10^4$ MPa.) [P]

8. Calculate the average spacing of dislocations in a 1·5° tilt boundary in a silver crystal. What are the Miller indices of the plane of this boundary? [E]

9. Explain what is meant by a screw dislocation.

Show that in a typical metal, such as copper for which $G = 4{\cdot}2 \times 10^4$ MPa, the stored strain energy around a screw dislocation of length n times the interatomic distance is of the order of $5n$ e V and discuss any assumptions made.

If the energy stored in a piece of heavily cold-worked metal is 5×10^7 J m^{-3} and this energy is entirely due to dislocations, determine approximately the density of dislocations. [MST]

10. Calorimetric measurements showed that by cold working the energy stored in an iron specimen increased by $5{\cdot}9 \times 10^3$ J kg^{-1}. Calculate the total length of dislocation line produced by this treatment per mm^3 of the crystal. Assume that screw and edge dislocations are generated in equal numbers and that the ratio of their strain energies is $\frac{2}{3}$.

[Shear modulus for iron $= 8{\cdot}2 \times 10^4$ MPa.] [P]

11. Describe a mechanism whereby dislocations can multiply in a crystal.

Assuming that the line tension of an edge dislocation is constant and approximately equal to Gb^2, where G is the shear modulus of the crystal and b the Burgers vector, show that a critical value of shear stress must be attained in order to activate a Frank–Read source. Calculate the value of this shear stress for a source of length $10^5 b$ if $G = 7 \times 10^4$ MPa.

12. It can be shown that a dislocation starts to move in the slip plane when the applied shear stress $\tau \approx Gb/L$, where G = the shear modulus, b = the unit translation, L = the length of dislocation. What is the minimum value of L for slip to occur in the [110] direction in a crystal of copper under a tensile stress of 50 MPa acting normal to (100)?

(For copper $G = 4{\cdot}2 \times 10^4$ MPa.) [MST]

CHAPTER 11

The Strengthening of Metals

11.1. Introduction

It was shown in the preceding chapter by theoretical considerations that the strength of perfect crystals should be very high, of the order of 1000 MPa. It is possible to grow dislocation-free crystals—known as *whiskers*—which have strengths which approach the theoretical value. However, these are very small, very expensive to produce, and no plant has yet been made for large scale production.

Single crystals of pure metals normally have strengths of the order of 0·1–1 MPa due to the presence of dislocations and few obstacles to their motion. Because of their low strengths, they are obviously of no practical use.

Commercial metals, in common use by engineers, have strengths which lie between 200 and 2000 MPa and must of necessity be cheap. They are mainly produced in the form of polycrystalline alloys and are usually subjected to heat treatment and mechanical working to obtain optimum values of strength and ductility (see Sections 12.5 and 12.7). It is by controlling composition and grain size and by further processing that barriers to dislocation movement are introduced and the enormous increase of strength over that of pure single crystals is obtained relatively cheaply.

It is important for the engineer to appreciate the nature of these factors, so that when other requirements demand a certain class of material, the best method for strengthening that type can be used, consistent with the economic factors involved.

11.2. Dislocations and obstacles

Strengthening is brought about by obstructing the movement of dislocations.

Consider dislocations of strength b created at a Frank–Read source S in a slip band AB (fig. 11.1) due to an applied shear stress τ and which have moved until held up by obstacles P and Q distant L from S, there being n dislocations in each equilibrium pile-up. Further movement of the leading dislocation will be opposed by a local shear stress τ_0 due to the obstacle. The magnitude of this can be estimated by the Principle of Virtual Work.

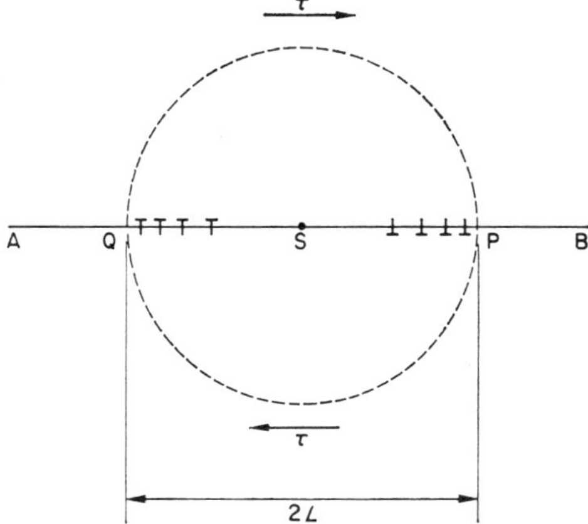

Fig. 11.1.—Piling-up of dislocations at obstacles

Let the leading dislocation move a distance δx. Then all the others will also move δx and the work done on unit length of the dislocations by the external shear force is $n\,\tau\,\delta x$. The work done by unit length of the leading dislocation against the local shear stress is $\tau_0\;\delta x$. Under equilibrium conditions, these are equal so that

$$\tau_0 = n\tau$$

Hence the obstacle causes a stress concentration factor of magnitude n.

If slip did not occur, the whole region between P and Q would have an elastic shear strain equal to τ/G, where G is the shear modulus. After slipping, the elastic strain over the central part of the region drops to a negligible value. (This is a reasonable assumption if the stress necessary to cause the Frank–Read source to operate is small compared with τ.) The area in which the elastic shear strain drops to zero is roughly that enclosed by the broken circle in fig. 11.1, which has a diameter $2L$. The amount of slip necessary at the centre to reduce the shear strain to zero is $2L\tau/G$. The displacement produced by the n dislocation loops is $n\mathbf{b}$ so that

$$n = \frac{2L\tau}{Gb}$$

A more exact treatment gives

$$n = \pi k\,\frac{L\tau}{Gb}$$

where k is a factor having the value 1 for screw dislocations and $(1 - \nu)$ for edge dislocations in an isotropic medium, where ν is Poisson's ratio.

Hence for more closely spaced obstacles, a greater shear stress is necessary to generate the same number of dislocations from a Frank–Read source, i.e., the material shows greater strength against plastic deformation, unless or until the stress concentration is sufficient to break down the barrier.

The various types of obstacles are discussed in the following sections.

11.3. Polycrystalline materials

In a polycrystalline material, neighbouring grains have different orientations, so that slip planes in one crystal are not usually parallel to those in its neighbours and a dislocation cannot pass directly across a boundary. Hence the intercrystalline boundaries will act as obstacles and cause piling up of dislocations. Figure 11.2 shows an actual photograph of piled-up dislocations. Eventually a stage may be reached when the resolved shear stress on the most favourable slip plane in the neighbouring

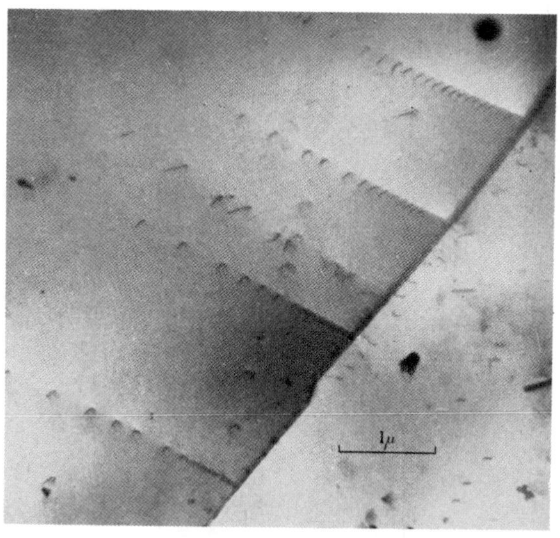

Fig. 11.2.—Pile-up of dislocations against a grain boundary in slightly deformed stainless steel (× 13000). [Whelan, Hirsch, Horne, and Bollmann]

crystal reaches the critical shear stress, so that slip can also occur; then there can be some movement to cause relief of the pile-up of dislocations in the first crystal.

It may be seen from fig. 11.3 that the strengthening effect is much greater in the hexagonal close-packed metal zinc than in the face-centred cubic metal aluminium, which is in contrast to the behaviour of single crystals shown in fig. 10.12. This is because the hexagonal close-packed metal has only one set of slip planes per crystal, and so the slip planes in

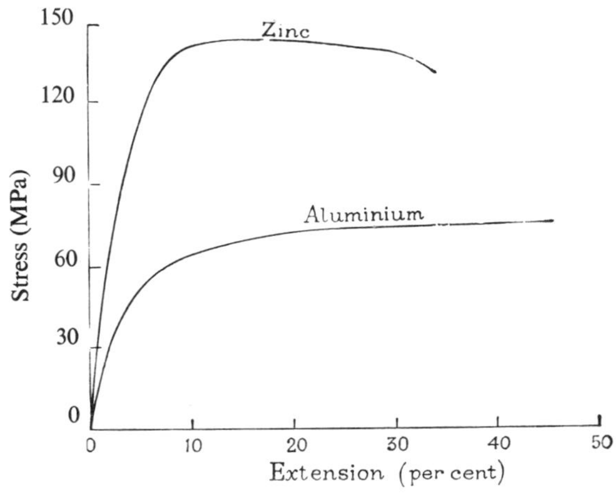

Fig. 11.3.—Comparison of typical stress-strain curves of polycrystalline hexagonal close-packed and face-centred cubic metals

adjacent crystals may have very different orientations. In the example shown in fig. 11.4 no slip in crystal B could accommodate slip in crystal A. The face-centred cubic metal has four possible slip planes in each crystal which are at such angles to each other that there will always be one whose orientation is very close to the active slip plane in the neighbouring crystal.

The smaller the grains, the closer are the obstacles and hence the strength is greater. The effect of grain size on the yield stress σ_y in steel is given by the Petch equation

$$\sigma_y = \sigma_i + kd^{-1/2}$$

where d is the mean grain diameter, k is a constant, and σ_i is the *friction stress* characteristic of the metal. σ_i is a measure of the resistance of the material to dislocation motion due to effects other than grain boundaries.

11.4. Solid solution alloying

When a foreign atom is introduced into a lattice, there is some distortion, which is more marked the greater the difference in the size between the foreign atom and the atoms of the parent lattice. Hence the foreign atom will act as a barrier to the passage of dislocations.

Another form of action of a solute atom in a substitutional solid solution can be understood from fig. 11.5 which shows the form of the atom arrangement around a dislocation. Those atoms above the dislocation are squeezed into holes too small for them and those below are stretched to fill the holes. The energy of the dislocation would be reduced if there were migration of the smaller atoms to the upper side and the larger atoms to the lower side of the dislocation, a migration that would tend to occur by diffusion. If such a dislocation is then to move, the energy of the dislocation would have to be increased because it would move to a position where the atoms were randomly distributed, that is, the force needed to move the dislocation is increased.

The effect of substitutional solid-solution alloying is illustrated in fig. 11.6 which shows the variation of strength with composition for single crystals of silver-gold alloys. Although the difference in size of the atoms of these metals is less than 0·2%, the maximum strength, which occurs at about the 50% alloy, is about seven times that of the stronger pure metal.

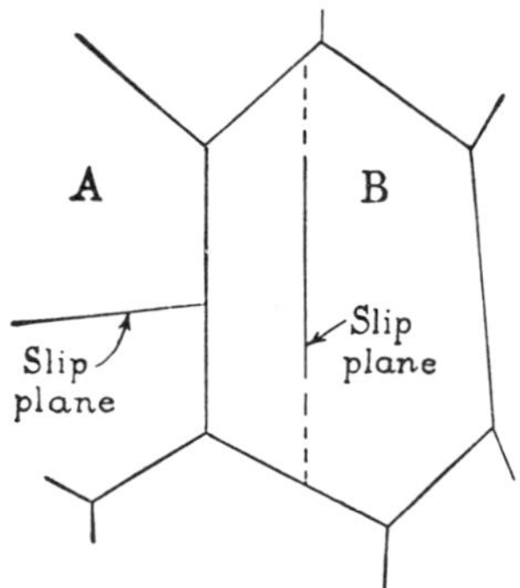

Fig. 11.4.—Slip planes in adjacent crystals such that slip in B cannot accommodate slip in A

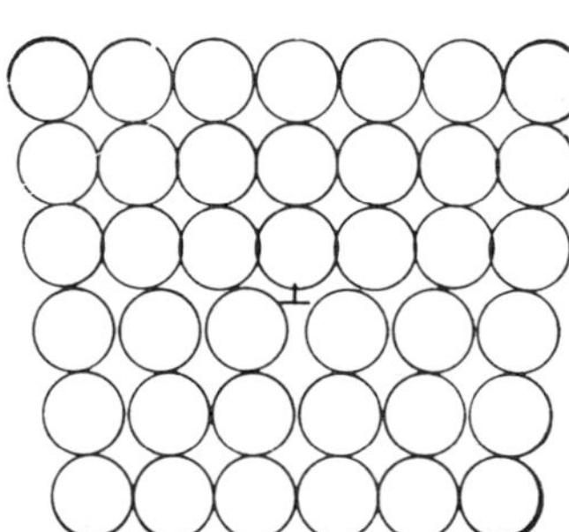

Fig. 11.5—Distortion of atoms around a dislocation

In the case of interstitial solid solutions, the solute atoms would diffuse into the cavity immediately under the dislocation. The dislocation

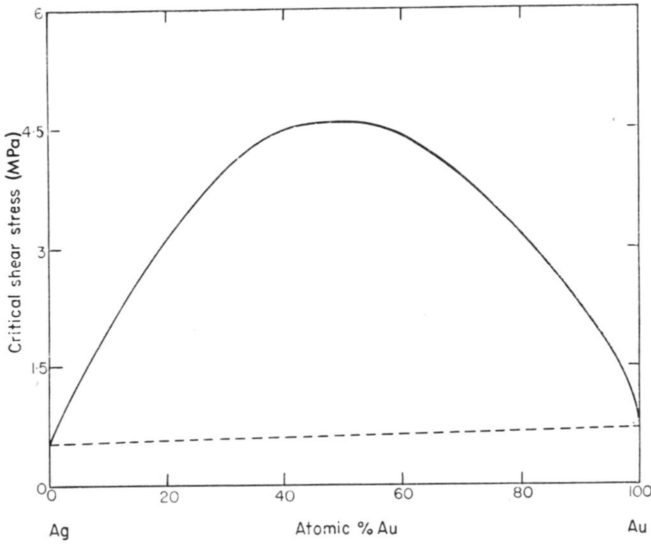

Fig. 11.6.—Variation of critical shear stress with composition in silver-gold single crystals. [After Boas]

would then not move until a sufficient force were applied to break it away from the locking effect of the solute atom. Once the dislocation had moved, it would then continue to move under a lower force. Cottrell has suggested that this is the explanation of the sharp yield point that is observed in mild steel (fig. 12.5) and some other alloys. The carbon and nitrogen in a solid solution would diffuse to the dislocations forming a *Cottrell atmosphere*. This concept, later amplified by Cottrell and Bilby in 1949, was of great help to the understanding of yield point phenomena in general, but there were difficulties in detail. More recent theories suggest that yielding is due to the rapid multiplication of the few dislocations which are not pinned by the Cottrell atmosphere.* Once yielding has occurred, no yield point is seen during a second loading immediately after the first. As the dislocations have been moved away from the solute atoms, a large break-away force is no longer required. Also on resting a specimen that has been yielded, the solute atoms have time to diffuse to the new positions of the dislocations, and this causes a return of a yield-point phenomenon. This action is called *strain ageing*. In some mild steels it will occur in a matter of days at room temperature or in minutes at 100 °C, because solid state diffusion is a thermally-activated process (see Section 9.11).

* For a review, see E. O. Hall, *Yield Point Phenomena in Metals and Alloys* (Macmillan, London, 1970).

11.5. Two phases in equilibrium

In an alloy series of two metals which are insoluble or nearly so in each other in the solid state, the strength varies approximately linearly between the strengths of the pure constituents. In the case of the lead-tin alloys, of which the equilibrium diagram is shown in fig. 11.7, there are regions of solid solubility at each end of the composition range and a eutectic is formed between. Values for the strengths of alloys of this series are also shown. It will be observed that there is a rapid increase of strength with addition of the second element in the regions of solid solubility. This is to be expected as the difference in atomic size is considerable. In the two-phase region, the strength varies in a much more gradual manner. The eutectic grains may be strong due to the large numbers of phase boundaries, and so have a strengthening effect upon the aggregate of pro-eutectic and eutectic grains.

This effect is most marked if the eutectic (or eutectoid) is formed with an intermediate compound, as these tend to have complex structures

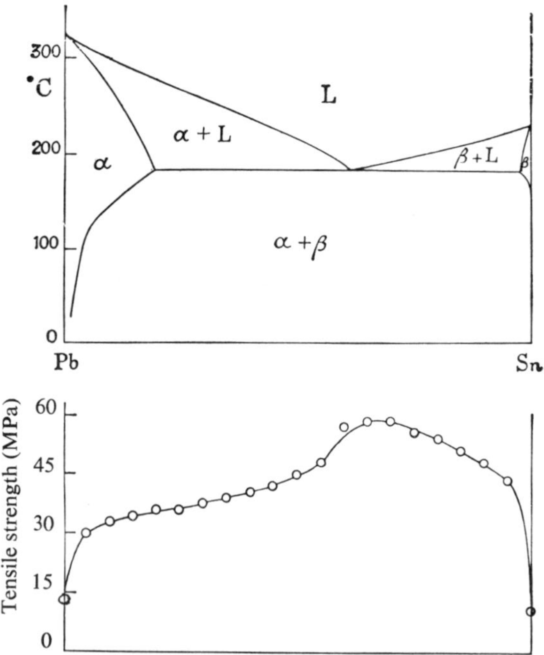

Fig. 11.7.—Lead-tin equilibrium diagram and strengths of lead-tin alloys. [Tensile strength values from *The Properties of Tin Alloys* (Tin Research Institute, 1947)]

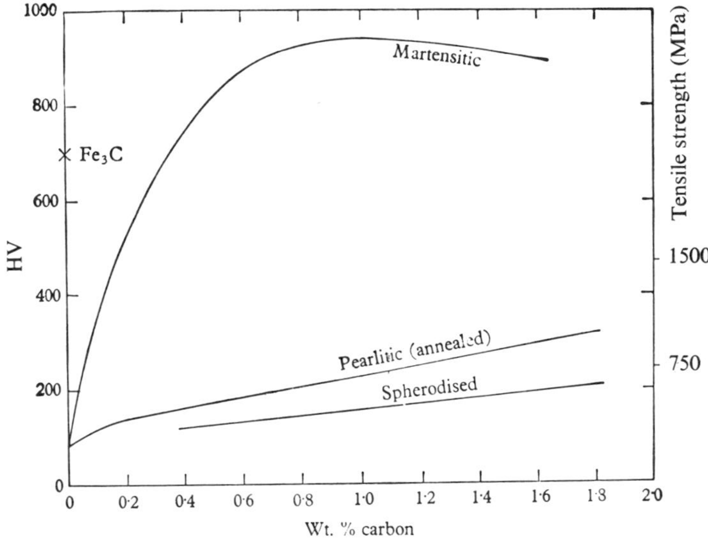

Fig. 11.8.—Variation in hardness of steel with carbon content in annealed, quenched, and quenched + fully tempered condition

with no planes of easy glide. In the case of the iron-carbon alloys, cementite has a hardness of 700 HV.* Pure iron has a hardness of about 90 HV, and an increase of carbon content, thereby increasing the proportion of cementite, increases the hardness in an almost linear manner as shown by the curve marked "pearlitic" in fig. 11.8.

The effects of solid solution and two-phase alloying in varying mechanical properties can also be seen in the copper-zinc alloys or *brasses*, for which the equilibrium diagram is shown on fig. A.11 and the mechanical properties in fig. 11.9. Up to about 37% zinc there is a single-phase solid solution known as α-brass. This has the same crystal structure as copper, i.e., face-centred cubic. As the amount of zinc is increased, the strength is increased and there is some improvement of ductility. Between about 45% and 50% a second phase exists, the body-centred cubic β-brass. This is harder than the α-brass, but it is not so ductile and so tends to break in a brittle manner at low loads instead of showing high strength. In the intermediate region (37% to 45%) the alloys are two-phase α-β brasses, which for the lower zinc contents are stronger, but as the zinc content increases the decrease of ductility makes the strength fall off.

* See p. 202 for a discussion of the meaning of hardness.

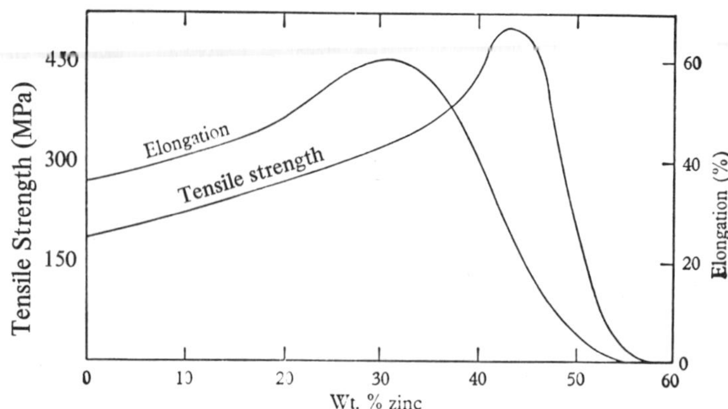

Fig. 11.9.—Mechanical properties of brasses

The types of brass in common engineering use are known as 70/30 or *cartridge brass* and 60/40 brass or *Muntz metal.* The figures denote the approximate copper/zinc ratio in each case. The 70/30 alloy, being α-brass, is easily deformed and can be cold-worked, this being the usual method employed in the final stages of manufacture; cold rolling, cold drawing of tubes, etc. The 60/40 alloy, being an α-β brass, cannot be cold-worked much without fracturing, and is used either cast to shape or hot-worked.

11.6. Dispersion hardening

When a second phase is formed as a precipitate within the matrix of a primary phase, there is generally a difference of size so tnat there is distortion causing a stress field local to each precipitate particle. In general, these precipitates will not have glide planes in common with the matrix and so will act as dislocation barriers. Their strengthening effects depend upon the size and spacing of the particles.

If the precipitate is very coarse with large gaps then the dislocation line can bend sufficiently under low stresses to pass between them leaving a closed loop around each particle as in fig. 11.10. Increasing the degree of dispersion and hence reducing the width of the gaps increases the strength of the material until the particles become sufficiently small for cross-slip or dislocation climb (see Section 10.10) to allow the dislocations to bypass the particle or for the stress concentration due to dislocation pile-up to be sufficient to fracture the particles. Hence the strength depends upon the scale of dispersion and is found to be a maximum for a particle size of about 100b. Three methods of achieving dispersion hardening are described later in this chapter.

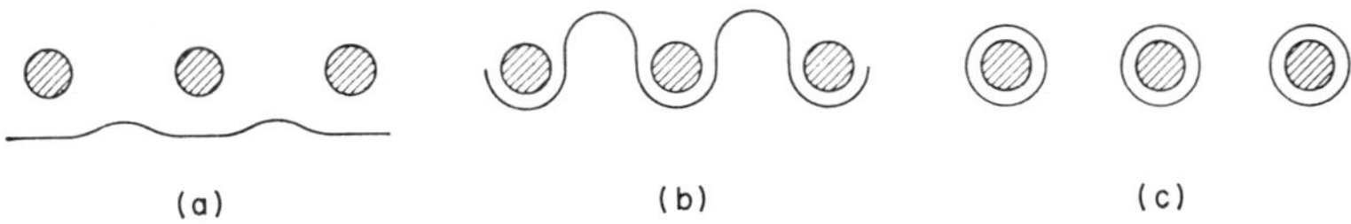

Fig. 11.10.—Dislocation line (*a*) approaching, (*b*) bulging between, (*c*) leaving loops around precipitate particles

11.7. Work hardening

The shear hardening of single crystals as they are deformed was discussed in Section 10.7 and the work hardening of polycrystalline materials will be discussed in Section 12.5.

During straining, Frank–Read sources generate dislocation loops so that there is a continual increase of their number. Also in moving they will intersect one another. When two non-parallel dislocations cross, then a *jog* or step in one or both of the dislocation lines is formed. A simple example is shown in fig. 11.11 of the jog formed by the intersection of two edge dislocations. When two screw dislocations cross, the jog is an edge dislocation which has the same Burgers vector as the screw dislocation to which it belongs and can move in the same direction as the screw only by shedding vacancies, so that it tends to be sessile. Not only does the creation of jogs involve an increase in strain energy due to the greater length of dislocation line, approximately Gb^3 per jog, but also sessile jogs hinder dislocation movement so that increased stress is necessary to cause further strain. The continued intersecting of dislocation loops eventually builds up a *dislocation network* with anchoring points so close that higher stresses become necessary to activate any more Frank–Read sources.

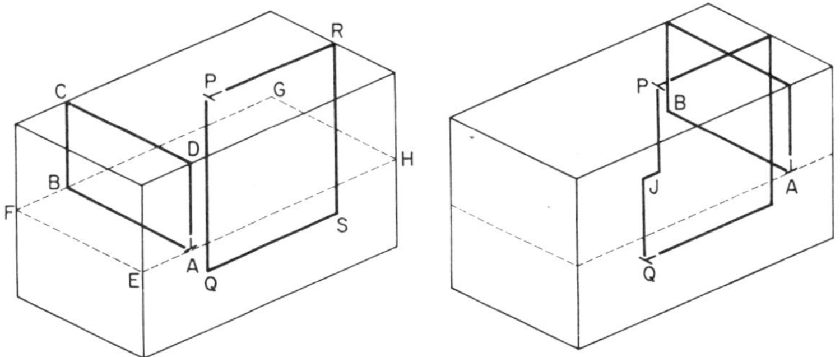

Fig. 11.11.—Half planes of atoms ABCD and PQRS give edge dislocations AB and PQ. AB moving on slip plane EFGH cuts PQ leaving jog at J

Work hardening is used in practice to harden metal sheets and wires by cold rolling and cold drawing, respectively. While this increases the strength considerably, the ductility is greatly reduced, so that the materials are much more brittle.

The effect of work hardening upon recrystallization has been mentioned in Section 5.7. Three stages are observed as the temperature of a cold-worked metal is raised: *recovery*, recrystallization, and grain growth. The first stage—recovery—is the reduction of internal stresses at low temperatures, thought to be due to the movement of dislocations, such as by climb to positions which cause the lattice strains to be reduced. The activation energy for the formation of new nuclei is such that the nucleation rate is appreciable at the recrystallization stage. Grain growth is also a thermally-activated process and proceeds more rapidly at higher temperatures.

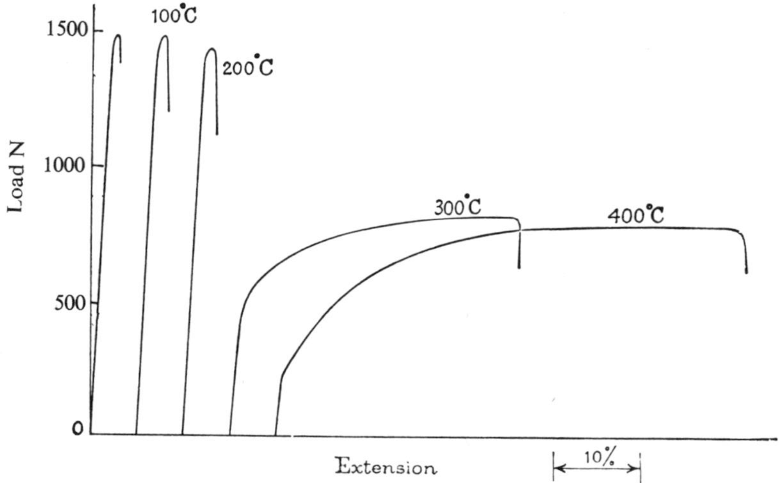

Fig. 11.12.—Load-extension curves for 2-mm diameter copper wire as hard-drawn and after annealing at various temperatures

The effect of annealing temperature upon the load-extension curves of cold-drawn and subsequently annealed copper wires is shown in fig. 11.12. Between 200 °C and 300 °C, in which range recrystallization occurs, there is a considerable drop in strength and a large increase in ductility. Raising the annealing temperature to 400 °C increases the ductility, but above that temperature there is little further change.

11.8. Quench hardening

In steels, the high-temperature form of the iron is the face-centred cubic γ-phase which will dissolve up to 1·7% carbon to form the interstitial solid solution austenite. On slow cooling, two phases form, viz. cementite, which contains nearly all the carbon, and ferrite, which is almost pure α-iron. This change involves diffusion of the carbon atoms and so takes time. If the austenite is quenched, i.e., rapidly cooled by plunging into a cold liquid, the equilibrium changes are suppressed, and the face-centred cubic structure changes by a shear-type transformation to a body-centred tetragonal structure. This is, in effect, a body-centred cubic structure distorted to make one cube edge longer than the other two. The carbon atoms have not had time to diffuse out of the lattice that wants to be ferrite and have distorted the would-be ferrite lattice. As a consequence of this distortion, the lattice will not permit the passage of dislocations, and the material is very hard and non-ductile. The structure is called *martensite*. Various mechanisms have been proposed for the austenite-martensite transformation.*

The variation of the hardness of martensite with the amount of carbon in the steel is shown by the curve marked " martensitic " in fig. 11.8.

There is a *critical cooling rate* which has to be exceeded if 100% martensite is to be formed. At lower rates of cooling, some ferrite and pearlite are also formed.

If the structure is tempered, i.e., reheated to some temperature below the lower critical temperature, the martensite breaks down to the equilibrium phases, i.e., to cementite and ferrite. However, these are not now in the eutectoid structure of pearlite, but the cementite is dispersed as more or less spherical particles in the ferrite. Tempering at a higher temperature reduces the degree of dispersion and so reduces the strength.

The heat treatment of steel is treated at greater length in Chapter 14.

The martensite type of transformation can also occur in other alloys which show a eutectoid reaction, the one of next importance from the engineering standpoint being the heat-treatable aluminium bronzes, i.e., copper-aluminium alloys which contain between 9 and 11% of aluminium. (See the equilibrium diagram in fig. A.3.)

11.9. Precipitation hardening

Dispersion hardening is possible in alloys which have a region of solid solubility in which the solubility decreases with falling temperature as in

* For a review, see E. O. Hall, *Twinning and Diffusionless Transformations in Metals* (Butterworth, 1954).

fig. 11.13. Any alloy with a composition between A and B will be entirely single-phased at the temperature T. On slow cooling, the second phase would start to precipitate when the line AC is crossed. This precipitation is a process of nucleation and growth, and so takes time. On rapid cooling, i.e., quenching, the second phase does not have time to precipitate at the equilibrium temperature, but a supersaturated solid solution is formed, the structure remaining the same as it was at the temperature T. The solid solution will change to the equilibrium structure of two phases as fast as diffusion permits. This may occur at room temperature or a somewhat higher temperature. The second phase precipitates so that it eventually becomes sub-microscopic or microscopic particles dispersed

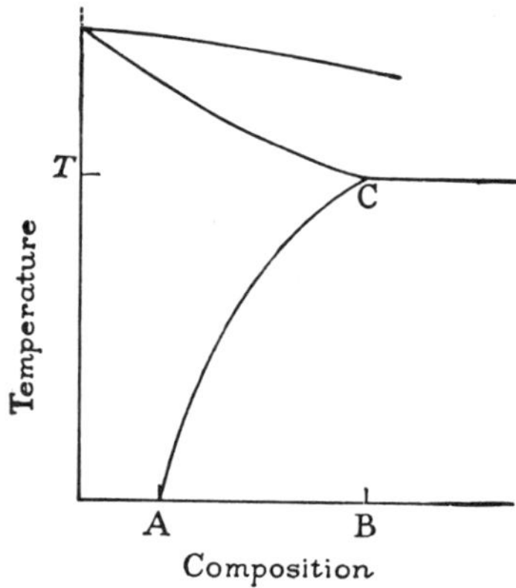

Fig. 11.13.—Form of equilibrium diagram for precipitation hardening alloys

throughout the crystals of the first phase. The mechanism, in the aluminium-copper alloys for example, is that regions very rich in copper atoms develop in certain crystallographic planes of the aluminium crystals. These regions, which are called *Guinier–Preston zones*, are about 2 atoms thick and from 5 to 80 nm in the other dimensions. These zones gradually transform via two intermediate phases to the θ or $CuAl_2$ phase. The Guinier–Preston zones assume different shapes in different alloys. Recent work suggests that the original precipitation occurs at dislocations.

The full heat treatment to produce this kind of hardening is as follows:

1. Solution treat, i.e., heat the alloy to the temperature of maximum solubility and hold at that temperature for a sufficient length of time for all the second phase to dissolve in the solid solution.

2. Quench to produce a supersaturated solution. This is the softest condition and forming to shape of components is frequently done in this condition.

3. Allow precipitate to form at room temperature (*age hardening*) or by raising the temperature (*precipitation hardening*).

At higher temperatures, the precipitate forms more quickly, but coalescence of the precipitate into larger grains also occurs. Also at a sufficiently high temperature the second phase redissolves in the solid solution. A larger grain size of precipitate will cause fewer obstacles to dislocations and so give lower strength. The variation of hardness with time and temperature for a typical alloy is shown in fig. 11.14. Each alloy has its own optimum time and temperature to produce greatest hardness.

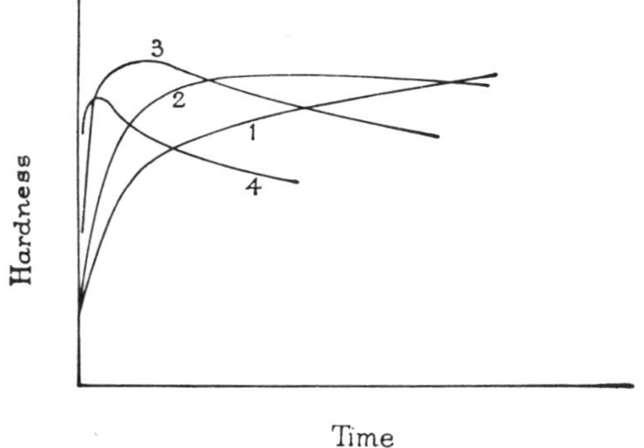

Fig. 11.14.—Hardness of a precipitation hardening alloy due to precipitation treatment. Typical curves obtained for treatment at progressively increasing temperatures 1–4

The most commonly used precipitation-hardening alloys are some of the aluminium alloys, mainly those based on the aluminium-copper system mentioned above (see fig. A.3 for the equilibrium diagram). These can either be age-hardened or precipitation-hardened at temperatures in the region of 100 to 150 °C. Those alloys that will harden at room temperature will remain in the supersaturated condition for a considerable time at only slightly lower temperatures. Hence a refrigerator is a convenient means of storing softened parts in a soft condition until required.

Among many other alloys which can be precipitation-hardened is beryllium bronze, copper alloyed with beryllium. As may be seen from the equilibrium diagram in fig. A.7, those alloys containing about 2% of beryllium can be successfully treated. The solution temperature is about 850 °C and the precipitation treatment is most successful at about 300 °C.

11.10. Dispersed oxide hardening

An oxide or other hard second phase may also be introduced deliberately in a finely divided state into a metal to impede dislocation movement. Many of those systems used in practice are ones which preserve the effectiveness of the dispersion to temperatures approaching the melting point of the parent metal. In this respect they differ from precipitation-hardened alloys which weaken at higher temperatures and so are useful for their creep resistant properties (see Section 15.10).

These two-phase structures are usually made by sintering a finely divided mixture of powdered metal and oxide. Examples are thoria-dispersed nickel which contains a fine dispersion of thoria, ThO_2, in nickel and sintered aluminium powder, SAP, in which aluminium oxide, Al_2O_3, formed as a thin layer ($\sim$ 10 nm thick) on the surfaces of flakes of aluminium metal ($\sim$ 100–200 nm thick) before compacting, agglomerates as particles in the aluminium matrix during the sintering process. The dispersed particle size is usually in the range 10–100 nm and the volume fraction between 1 and 20%.

QUESTIONS

1. Why are the shear hardening curves of aluminium and magnesium so different? [MST]

2. The strength of pure metals may be increased by alloying. Discuss, in terms of the structures produced in equilibrium and non-equilibrium changes, the mechanisms of strengthening in alloys. [MST]

3. Explain the meanings of and the processes involved in (*a*) work hardening, (*b*) quench hardening, (*c*) precipitation hardening, quoting instances in which each method is used in engineering practice. [MST]

4. Discuss the effects on a metal of cold-working and subsequent heating to different temperatures. [S]

5. Describe the variations in the mechanical properties and microstructures of brass with variations in zinc content.

Give two examples each of the use of brass of 70/30 and 60/40 composition and state how the uses and the methods of manufacture adopted are related to the properties of these brasses.

6. Explain the terms *work hardening*, *quench hardening*, and *dispersion hardening*. For the last of these, discuss the effect of the degree of dispersion upon the hardness, using a particular type of alloy as an example in your discussion. [MST]

7. Describe the yield-point phenomenon found in mild steel and give a possible explanation.

Several specimens of mild steel are cold-worked to the same extent. One is immediately placed in boiling water and it shows a higher yield point after 15 min. After what time would the same recovery of yield point be found in a specimen kept at 15 °C after cold-working? For how much longer can the recovery of the yield point be postponed by placing a specimen in a refrigerator at 0 °C? The activation energy for the diffusion of carbon in α-iron is approximately $7{\cdot}5 \times 10^7$ J kmol^{-1}. [MST]

8. What is a *thermally activated process*?

For a precipitation hardening alloy in the "as quenched" condition, the first appearance of a precipitate in the super-saturated solid solution can be detected one hour after quenching when the temperature of the alloy is held at 15 °C. If the temperature after quenching is raised to 100 °C the precipitates can be detected after one minute.

If it is desired to retard the precipitation hardening process so that no precipitates can be detected in under one day, to what temperature must the alloy be cooled after quenching? [MST]

9. Distinguish between *recovery, recrystallization*, and *grain growth* in materials. Discuss the effects of time and temperature upon these phenomena in a work-hardened metal.

A heavily cold-rolled brass sheet must be annealed for 2 min at 400 °C before recrystallization is 50% complete. For how long must the sheet be annealed at 300 °C to produce 50% recrystallization if the activation energy for recrystallization is 166 MJ kmol^{-1}? Take $R = 8{\cdot}3$ kJ kmol^{-1} K^{-1}. [P]

10. What is meant by *recrystallization?* Discuss the factors that affect recrystallization.

The variations of the tensile strength and elongation to fracture of 70 Cu 30 Zn brass as functions of the degree of cold work are shown in fig. 11.15. The degree of cold work is defined as

$$\frac{\text{change in cross-sectional area}}{\text{original area}} \times 100\%$$

A tensile specimen cut from a cold-rolled 70–30 brass plate, 16 mm thick, shows a tensile strength of 550 MPa. It is required to roll a sample of this plate to a final thick-

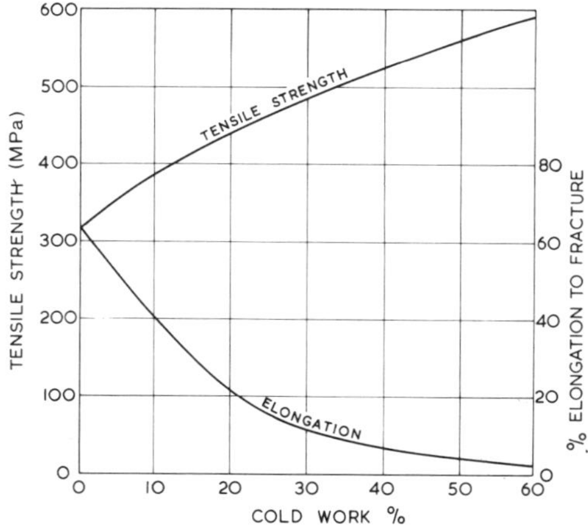

Fig. 11.15.

ness of 3 mm, the rolled sheet then to have a tensile strength of 450 MPa minimum and and elongation to fracture (on the relevant gauge length) of 10% minimum. Specify the steps in the procedure including any necessary heat treatments. Assume that the width of the sheet remains constant during rolling. [E]

11. Discuss the following two strengthening mechanisms of polycrystalline materials:

(i) grain size strengthening

(ii) strain hardening.

A quantitative analysis of the slip band model for yielding in polycrystalline solids leads to the following relationship between yield stress, σ_y, and grain size, d:

$$\sigma_y = \sigma_i + k_y d^{-\frac{1}{2}}$$

where σ_i is the average yield stress of a single grain and k_y is a constant that determines the effectiveness of grain boundaries in raising the yield strength. If the grains in a magnesium oxide tensile specimen are cubical in shape and the yield strength is 80 MPa when 625 grains are present in 1 mm^2 of cross section, what is the strength when 10000 grains appear in 1 mm^2? (Assume σ_i for MgO is 10 MPa.) [P]

12. Distinguish between dispersion strengthening and precipitation strengthening.

If a metal contains strong second-phase particles with an inter-particle spacing of $100b$, where b is the magnitude of the Burgers vector, what will be the ratio of the actual strength to the theoretical strength of the metal? (Assume that the actual strength can be estimated by calculating the stress necessary to bow a dislocation between two adjacent particles).

Estimate the shear strength of a dispersion-strengthened aluminium alloy containing 1% (by weight) Al_2O_3 particles, 10μm in diameter. (Al and Al_2O_3 have densities of 2·7 Mg m^{-3} and 3·96 Mg m^{-3} respectively. The shear modulus, G, of aluminium is 27·6 GPa). [P]

CHAPTER 12

Mechanical Testing of Polycrystalline Materials

12.1. Introduction

The mechanical properties of materials are of prime interest to the engineer and the results of tests to determine these properties are used for various purposes.

Firstly, they may be used in the selection of material for a particular purpose, or for deciding the size or shape of a component of a particular material so that it shall be sufficiently strong to fulfil its purpose.

Secondly, they may be used for quality control. Samples taken from different batches of allegedly the same material are tested and a batch should be rejected if the test results fall outside acceptable limits. These limits are quoted in a Specification, possibly a private agreement between supplier and user, or, more generally, in a Specification of nation-wide or international application. British Standard Specifications, for example, cover the methods of sampling and testing as well as the requirements from the test results for almost all materials in common engineering use.

The most widely quoted mechanical properties are those determined by a tensile test, which is discussed in this chapter together with other tests that give related results. However, the tensile test does not tell the whole story and further information is needed to permit satisfactory design under conditions where complex stress systems will be encountered or the component will be subject to impact, repeated or long-term loading. These factors are considered more fully in Chapters 13–15.

12.2. Tensile testing machines

Machines for performing tensile tests comprise means for straining (i.e., stretching) a specimen and simultaneously measuring the load required to perform the straining. The simplest are of the single-lever type shown in fig. 12.1.

One end of the specimen is attached to the straining head which is moved by a screw mechanism with a hand or electric drive, or by hydraulic pressure. The other end is attached to a lever at a point near the fulcrum, and a weight is moved along the lever to apply a load to the specimen. The load may be set or increased at any desired rate, and the straining

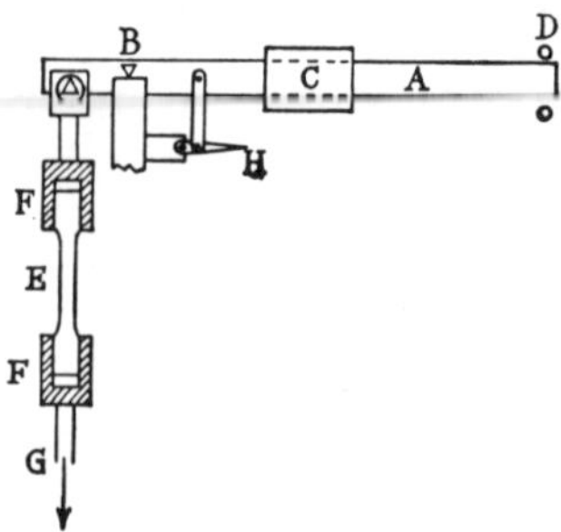

Fig. 12.1.—Single-lever testing machine
A—beam
B—fulcrum
C—moving weight
D—stops to limit movement of beam
E—specimen
F—specimen grips
G—straining head
H—pointer to indicate position of beam

head then moved as the specimen extends to keep the lever in a position of balance. With this type of machine it is not easy to follow rapid decreases of load during extension.

In other types of machines, the straining head is also moved by mechanical or hydraulic means, but the load is continuously indicated, either by measurement of the fluid pressure in some hydraulic machines or by a weighing device. This is a stiff spring connected to the specimen either directly in machines of small capacity or through a system of levers in larger machines, the deflection of the spring being indicated by a pointer and dial, or other means. Any decrease of load with increase of strain is immediately indicated.

In the most modern types of testing machine the weighing device is an extremely stiff load cell which deforms elastically. The strain which is proportional to the load is measured electrically, the output being fed directly to an indicator or recorder.

12.3. Tensile specimens

Specimens for tensile tests are usually machined to a circular cross-section, but specimens from plates or sheets may be of rectangular section. Each must have a portion of uniform cross-section on which extension measurements can be made. To ensure that any fracture will occur in this uniform portion, the ends are usually machined to a larger cross-section with smooth blending curves into the central uniform part. The specimens are pulled by serrated grips at the ends, or by supporting at the shoulders of the large ends, or by threaded ends, the form of the specimen for each of these methods being shown in fig. 12.2.

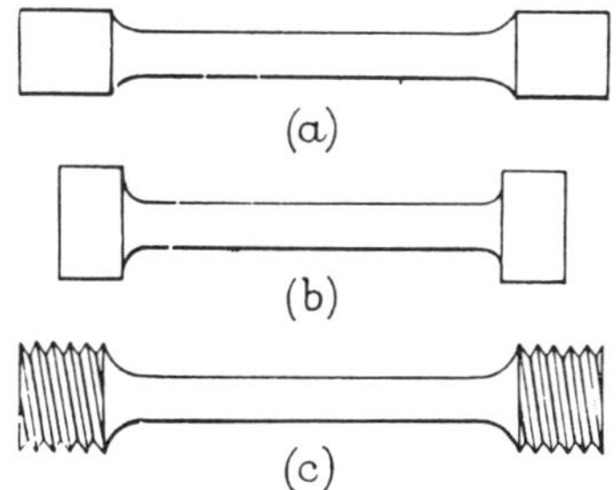

Fig. 12.2.—Types of tensile specimen
(*a*) for serrated grips
(*b*) for shoulder grips
(*c*) for threaded grips

The extension may be taken as the movement of the grips, but the exact length in which this extension occurs is difficult to define because of the changing cross-section at the ends. For more accurate work, extension measurements are made on a definite length of the uniform portion by an extensometer of suitable accuracy. This length is known as the *gauge length.*

12.4. Behaviour of materials in tension

When a specimen is pulled in tension, it can behave in one of three ways. In each it first stretches in an elastic manner, in the case of metals obeying Hooke's law, but for some materials such as rubber the elasticity is non-linear. In the first way it then deforms plastically and stretches, so

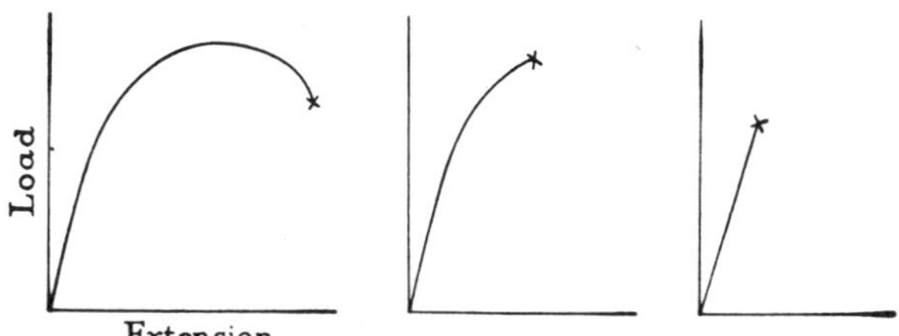

Fig. 12.3.—Possible types of behaviour of materials in a tensile test
(*a*) ductile (i.e., forms a neck), (*b*) and (*c*) non-ductile, fracture occurring in (*c*) without previous plastic deformation

that it does not return to its original length on unloading. Eventually a neck forms, and further deformation occurs only in the region of the neck, with a decrease in the applied load, until fracture finally occurs at the neck. In the second way, plastic deformation starts, but the specimen breaks without forming a neck. In the third way, the specimen breaks before the onset of plasticity. The form of the load–extension curve for each case is shown in fig. 12.3. The first type of behaviour is known as *ductile extension*, while the others are known as *non-ductile*, the last-named often being called *brittle*. The term brittle is, however, also used in another context as explained in Chapter 13.

12.5. Load-extension curve for a ductile material

The typical shape of the load–extension curve of a ductile metal is shown in fig. 12.4. The various points marked on the curve are as follows:

A is the limit of the initial straight line OA, i.e., it is the *limit of proportionality*. Beyond this point Hooke's law is no longer obeyed. The slope of this line gives *Young's modulus* for the material (see Section 12.7).

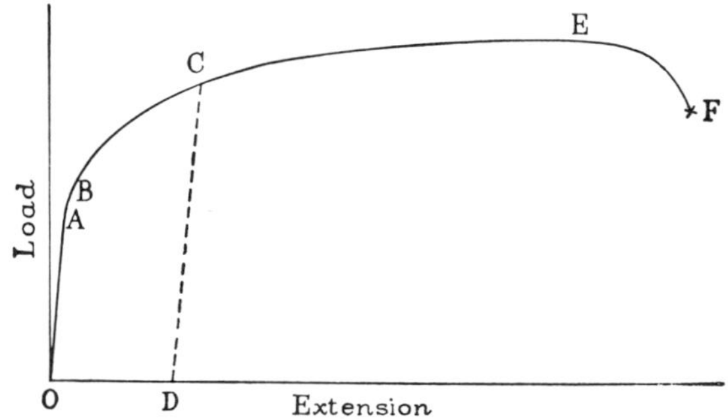

Fig. 12·4.—Load–extension curve for a typical ductile metal

B is the limit to which the specimen can be strained and still return to the original length on unloading. It is the limit of elasticity or *elastic limit* and is not necessarily coincident with A.

C is a point in the plastic range. If the specimen is loaded to C and then unloaded, it will behave in a linear manner (line CD) and will show a permanent set OD. On reloading, the line DC is retraced and then the specimen yields plastically again on being taken beyond C. The load at C divided by the original area is known as the *yield stress* for the material in the condition D.

E is the maximum load. On a lever-type machine, where the load is controlled, the specimen would continue deforming to fracture. On a strain-controlled machine, the fall-off of load EF can be observed.

F is the point at which the separation of the test piece into two parts occurs.

The proportional limit and the elastic limit can be determined only to the limit of accuracy set by the extensometer used. In practice it is usual to quote the proof stress (except in the case of mild steel and other materials with a similar behaviour which will be considered below). The *proof stress* is defined as the stress (i.e., load/area) which produces, while the load is still applied, a non-proportional extension equal to a specified percentage of the extensometer gauge length. A commonly used percentage is 0·1 %, the corresponding stress being referred to as the 0·1 % proof stress or 0·1 % P.S. The accuracy of available extensometers is such that proof stresses for extensions of this magnitude can be determined with sufficient precision.

Another quantity sometimes quoted is the *permanent set stress*, being that stress which, after removal of the load, has produced a permanent extension which is a specific percentage of the gauge length. If an unloading line, such as CD, is linear and parallel to the initial elastic line OA, then the proof stress and permanent stress for the same percentage extension will be equal. In practice, there is usually a slight difference between these quantities.

The maximum load divided by the original cross-sectional area is called the *tensile strength*. It has also been known as the *ultimate tensile stress* (U.T.S.), the *maximum stress* (M.S.), or *tenacity*.

When a material is deformed plastically, it *work hardens*, that is, the stress has to be increased to give further deformation. If a material with a stress–strain curve of the form shown in fig. 12.4 is deformed to C and unloaded, it returns to point D. On reloading, it deforms elastically to C before any further strain occurs. The point C is known as the *yield stress* for the material in condition D. Most annealed materials (other than mild steel) show practically no elastic range, and therefore cannot be said to have true yield stresses.

12.6. Yield-point phenomena

In mild steel and some other materials there is another phenomenon. At the elastic limit there is a sudden yield and fall-off of load. The material continues to deform at a lower load until work hardening sets in. The load–extension curve takes the form shown in fig. 12.5.

A is the *upper yield point*.

B is the *lower yield point*.

The stresses at these points are the *upper yield-point stress* and *lower yield-point stress*, respectively.

CD is the work-hardening curve.

The apparent value of the upper yield-point stress is very sensitive to axial loading and to stress concentrations at the grips, and any departure from perfectly uniform stress across the test specimen causes a reduction of the load at which yielding commences.

When the load drops to B in fig. 12.5, regions of plastically-deformed material form around stress concentrations such as fillets and grips and, on further straining, these regions extend until the whole specimen has yielded at C. Work hardening then commences.

If the surface is sufficiently smooth, it is possible to observe the boundaries of the deformed regions travel along the specimen. These are known as *Luders lines.*

An explanation of the cause of a sharp yield point is given in Section 11.4.

For mild steel the value of the lower yield-point stress is about half of the tensile strength.

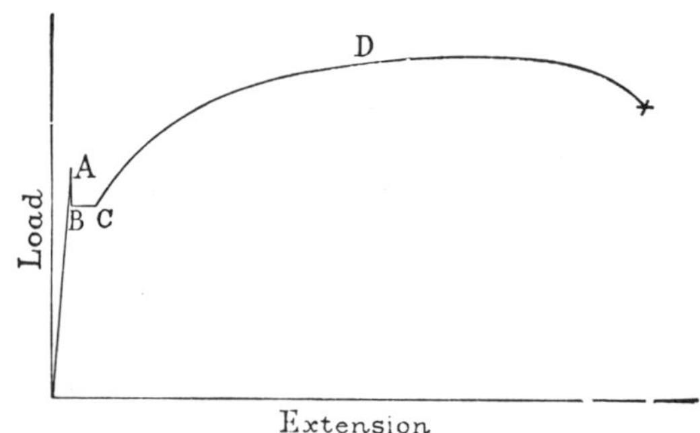

Fig. 12.5.—Typical load–extension curve for a low carbon steel

The effect of a yield-point phenomenon can be demonstrated by a simple experiment. A piece of annealed copper wire and one of annealed mild steel are in turn bent around a finger. The copper wire bends into a smooth curve because, as each region deforms, it work hardens and requires a greater bending moment to cause further deformation than the neighbouring section. A region of the steel wire starts to deform when the upper yield-point stress is reached. The stress for further yielding being less, bending will continue at that point rather than at immediately neighbouring sections, so that the wire takes a polygonal form. Two such

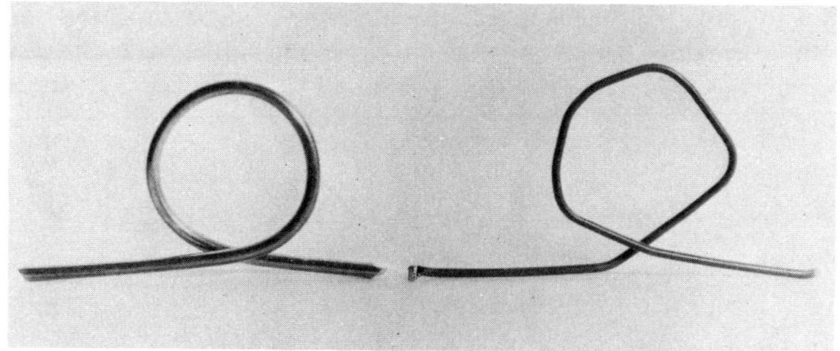

Fig. 12.6.—Copper and mild-steel wires bent around a finger

wires are shown in fig. 12.6. Each bend in the steel wire forms with a distinct jerk.

It should be noted that the yield point or elastic limit marks the onset of plasticity, and the tensile strength is the stress at which necking commences. The tensile strength in a ductile material is not associated with fracture. Hence all these stresses are plastic properties.

12.7. The influence of test-piece dimensions

If tensile tests are made upon different-sized specimens of the same material, the load-extension curves will be different. Figure 12.7 shows

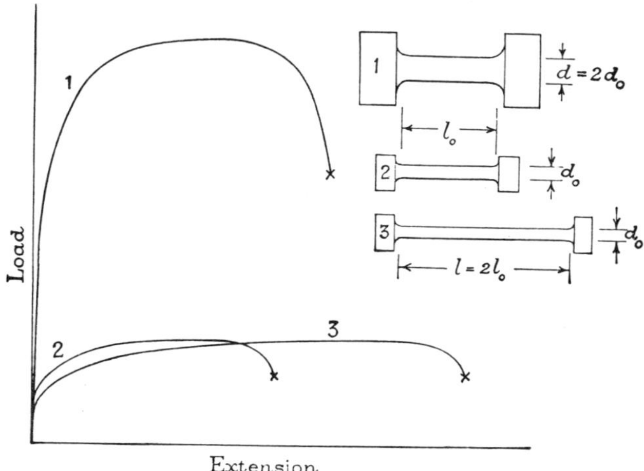

Fig. 12.7.—Load–extension curves for three specimens of the same material but of different dimensions

curves for three specimens with different values of diameter d and gauge length l. The three curves are different. In an attempt to make the curves independent of the test-piece size, it is usual to plót a curve of stress σ against the strain ε. These quantities are defined as

$$\sigma = \frac{P}{A_0}$$

and

$$\varepsilon = \frac{l-l_0}{l_0}$$

where P = load, A_0 = original cross-sectional area, l_0 = original length and l = length at load P. The stress–strain curves derived from the three load-extension curves of fig. 12.7 will be identical up to the point of maximum load. Beyond that point they will differ as shown in fig. 12.8.

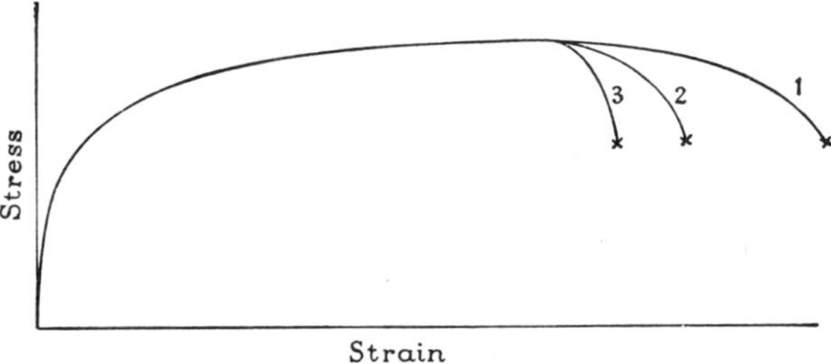

Fig. 12.8.—Stress–strain curves derived from curves in fig. 12.7

For the elastic range, Young's modulus is $E = \sigma/\varepsilon$.

The total extension to fracture is of such magnitude for ductile metals that it can be determined with sufficient accuracy on gauge lengths of 50 to 200 mm by means of a pair of dividers and a scale. This extension to fracture divided by the original length is a commonly quoted property known as the *elongation after fracture*. It is obvious from fig. 12.7 that the elongation after fracture is dependent upon the dimensions of the specimen.

The elongation is made up of two parts: first, that due to the uniform plastic extension of the whole specimen and, second, the local extension in the region of the neck.

Consider a specimen on which six uniformly-spaced marks A, B, C, D, E, F have been placed as shown in fig. 12.9. Suppose after testing to fracture, the two portions are fitted together and that the marks are now at A′, B′, C′, D′, E′, F′ as shown. Also suppose the measured lengths are as shown. Regions AB, BC, DE, and EF have all extended uniformly up to the maximum load and have an elongation of 20%.

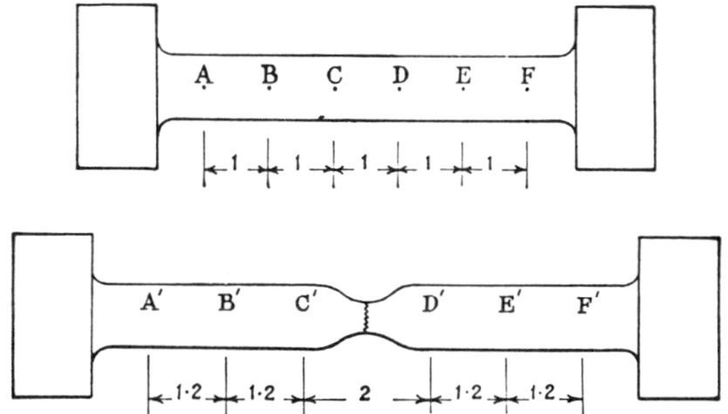

Fig. 12.9.—Specimen with equally spaced gauge marks before and after fracture

Region CD contains the neck, and has a total elongation of 1 (i.e., 100%) of which 0·2 was due to the uniform extension, and hence the other 0·8 must have been due to the neck.

If the gauge length is taken as 1, then the elongation would be quoted as 100%. If, however, BE were taken as the gauge length, the elongation would be

$$\frac{4{\cdot}4-3}{3} = 47\%$$

and if AF were the gauge length, the elongation would be

$$\frac{6{\cdot}8-5}{5} = 36\%$$

It is therefore obvious that the elongation is not independent of the gauge length.

If two specimens of different diameters, d_1 and d_2, but of identical material, are tested, the necks are of geometrically similar shape. This is *Barba's similarity law*. Therefore to compare results from different sized specimens, the gauge length taken must bear a fixed proportion to the diameter of the specimen.

In British Standard practice,* the gauge length is chosen so that

$$l_0 = 5{\cdot}65\sqrt{A_0} = 5{\cdot}0d$$

Hence for a gauge length of 50 mm, the specimen diameter should be 10 mm.

* B.S. 18 *Methods for Tensile Testing of Materials*, Part 2: 1971 *Steel (General)*, British Standard Specifications, published by British Standards Institution.

The elongation is a measure of ductility of the specimen. Another measure of ductility is the *reduction of area after fracture* which is the reduction in the area of cross-section at the neck expressed as a percentage of the original area. Thus

$$\text{reduction of area} = \frac{\frac{1}{4}\pi d_0^2 - \frac{1}{4}\pi d_n^2}{\frac{1}{4}\pi d_0^2} \times 100\,\%$$

$$= \left[1 - \left(\frac{d_n}{d_0}\right)^2\right] \times 100\,\%$$

where d_0 and d_n are the initial diameter and the diameter at the neck, respectively.

12.8. True stress

When plastic deformation occurs in a tensile test there is a decrease in the area of cross-section of the specimen. The *true stress* is therefore not the value already considered, i.e., P/A_0, which we now term the *nominal stress*, but is $\sigma_t = P/A$, where A is the actual area which becomes progressively smaller as deformation proceeds. The effect that the change of definition of stress has on the stress-strain curve is shown in fig. 12.10.

Under elastic loading, the cross-section diminishes, the diametral strain being $-\nu$ times the longitudinal strain, where ν is Poisson's ratio. If ν is less than 0·5, as is usual for metals, then there is an increase in volume. Plastic deformation causes a negligible change in volume and if

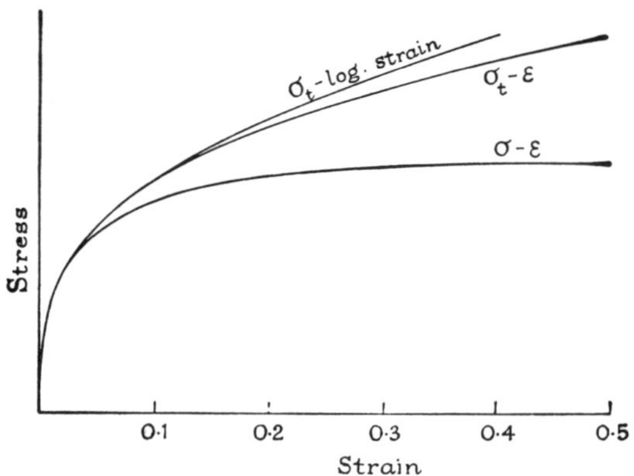

Fig. 12.10.—Effect of different definitions of stress and strain upon shape of stress–strain curve

the plastic strain is large compared with the elastic strain, any elastic volume change can usually be neglected. Hence

$$Al = A_0 l_0$$

or

$$A = \frac{A_0 l_0}{l} = \frac{A_0}{1+\varepsilon}$$

The true stress is then

$$\sigma_t = \frac{P}{A} = \frac{(1+\varepsilon)P}{A_0}$$

$$= (1+\varepsilon)\sigma$$

Now the true stress will be related to strain by some relationship which can be expressed as $\sigma_t = f(\varepsilon)$. Then

$$P = \frac{A_0 \sigma_t}{(1+\varepsilon)} = \frac{A_0 f(\varepsilon)}{(1+\varepsilon)}$$

from which

$$\frac{dP}{d\varepsilon} = \frac{A_0}{(1+\varepsilon)^2}\left[(1+\varepsilon)\frac{d\sigma_t}{d\varepsilon} - \sigma_t\right]$$

The maximum load P_{max} will occur at a strain for which $dP/d\varepsilon = 0$, i.e.,

$$(1+\varepsilon)\frac{d\sigma_t}{d\varepsilon} = \sigma_t$$

or

$$\frac{d\sigma_t}{d\varepsilon} = \frac{\sigma_t}{1+\varepsilon}$$

A tangent to the curve $\sigma_t = f(\varepsilon)$ from the point $\sigma_t = 0$, $\varepsilon = -1$, will touch the curve at a point Q which gives the value of true stress and strain corresponding to P_{max}. This may be seen from fig. 12.11 since

$$\tan\alpha = \frac{\text{QB}}{\text{AB}} = \frac{\sigma_t}{1+\varepsilon} = \frac{d\sigma_t}{d\varepsilon}$$

The point R at which AQ cuts the stress axis gives

$$\text{OR} = \frac{\sigma_t}{1+\varepsilon} = \sigma_{max}$$

i.e., OR is the value of the maximum nominal stress, i.e., the tensile strength. This graphical method for determining tensile strength is known as Considèré's construction.

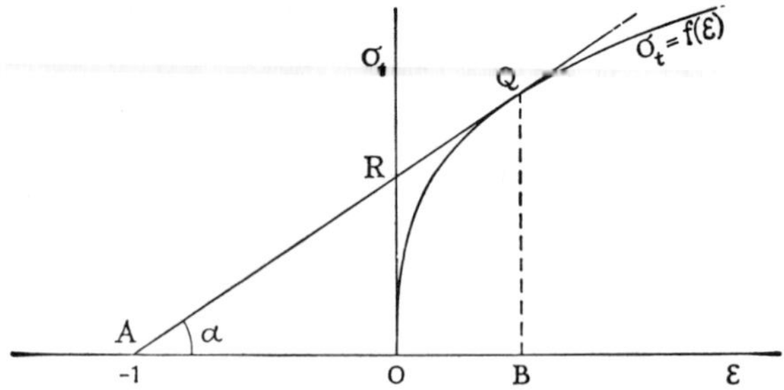

Fig. 12.11.—Considère's construction

In many practical cases, it is found that the true stress–strain curve may be closely represented by a law of the form*

$$\sigma_t = a\varepsilon^b$$

For a material which has such a stress–strain relationship,

$$\frac{d\sigma_t}{d\varepsilon} = ba\varepsilon^{b-1}$$

Therefore the value of strain corresponding to the tensile strength is given by

$$ba\varepsilon^{b-1} = \frac{a\varepsilon^b}{1+\varepsilon}$$

Therefore

$$\varepsilon = b(1+\varepsilon)$$

which gives

$$\varepsilon = \frac{b}{1-b}$$

Then

$$\text{tensile strength} = \sigma_{\max} = \frac{\sigma_t}{1+\varepsilon}$$

$$= \frac{a\left(\frac{b}{1-b}\right)^b}{1+\left(\frac{b}{1-b}\right)}$$

$$= a(1-b)\left(\frac{b}{1-b}\right)^b$$

or

$$\text{tensile strength} = ab^b(1-b)^{1-b}$$

* A. Nádai, *Plasticity* (McGraw-Hill, 1931), Chapter 15.

12.9. Logarithmic strain

When an increment of extension δl occurs during straining, it is doing so on an actual length l and not on the original length l_0. Hence strain may be defined as $\Sigma(\delta l/l)$. In the infinitesimal form this becomes

$$\int_{l_0}^{l} \frac{dl}{l} = \log_e \frac{l}{l_0}$$

$$= \log_e (1+\varepsilon)$$

In this form the strain is referred to as the *logarithmic strain* or *true strain* ϵ.

The effect of using this form of strain on the shape of the stress–strain curve is shown in fig. 12.10, and it can be seen that the difference between ε and ϵ is very small for strains up to 0·1.

The true stress–logarithmic strain curve can also be represented with sufficient accuracy for many metals by a power law of the form

$$\sigma_t = k\epsilon^n$$

where n is referred to as the *strain hardening exponent* and k is the *strength coefficient.* If true stress is plotted against logarithmic strain on a log–log basis, the line should be straight and of slope n, intersecting the $\epsilon = 1$ line at the value of k. Necking commences when $\epsilon = n$.

12.10. Ideal plastic material

An ideal plastic material is defined as one that behaves in a linear elastic manner until the yield stress is reached, and then any further deformation continues at this constant yield stress, as shown in fig. 12.12.

A work-hardened material and also mild steel in the initial stages of deformation approximate to this behaviour, and are often treated as such in considering the plastic behaviour of structures.

12.11. Other related tests

The tensile test is expensive to perform because a specimen with suitably-shaped enlarged ends and a reduced parallel central portion has to be prepared, involving extensive machining. Tests that require less preparation of specimens are compression and hardness tests. In the following sections these tests and their relationship to the tensile test will be considered.

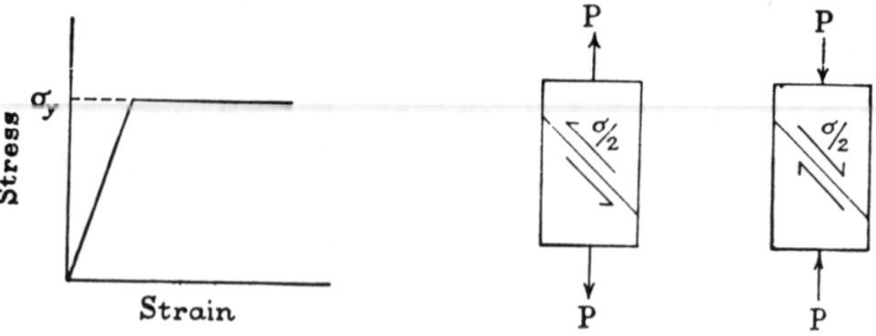

Fig. 12.12.—Stress–strain curve for an ideal plastic material

Fig. 12.13.—Shear stress due to uniaxial tensile and compressive loading

12.12. Compression tests

Plastic deformation is a shear process and occurs under uniaxial stressing when the shear stress reaches a definite value. Now for a uniaxial stress system the applied stresses in tension and compression would be as shown in fig. 12.13. The maximum shear stress is on a plane inclined

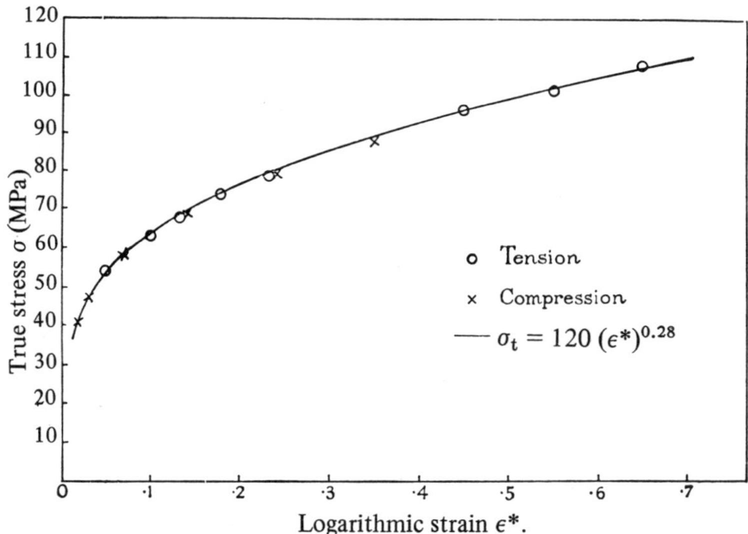

Fig. 12.14.—True stress–log. strain curve for partially annealed aluminium. Tension and compression results lie on same curve—empirically fitted by $\sigma_t = 120(\epsilon^*)^{0\cdot28}$ MPa. [after Tabor]

at 45° to the direction of the direct stress and will have a value of $\frac{1}{2}\sigma$ where σ is the value of the direct stress. Hence we would expect the yield stress for a material in any given condition to be the same whether the specimen is tested in tension or compression. It is found by experiments

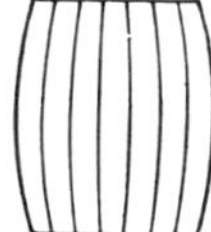

Fig. 12.15.—Barrel shape acquired by cylindrical specimen during compression

on pure metals that if two specimens of identical material are tested, one in tension and one in compression, the true stress–logarithmic strain curves will be practically identical as shown in fig. 12.14.

This depends on the compression test being a pure compression test, i.e., the stress is uniaxial. In practice, since the material wishes to expand laterally, it will be restricted at the compression anvils due to friction, and the specimen will assume a barrel shape as in fig. 12.15. Also, due to friction, a greater load is necessary to cause yielding. If the friction is kept to a low value by the use of suitable lubricants, barrelling is much less pronounced, and the agreement with the tensile-test results is improved.

If the specimens have been cold-worked prior to testing, the results for tension and compression may not agree. For example, if a specimen is

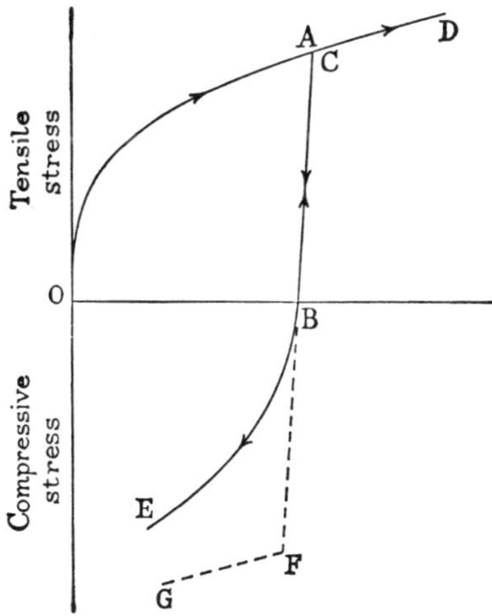

Fig. 12.16.—Effect of reversal of stress

OA—initial loading in tension
AB—recovery on unloading
BCD—reloading in tension
BFG—image of BCD
BE—reloading in compression

pretensioned, its yield point for further tension is raised, but that for compression is lowered. The effect, which is shown in fig. 12.16, is known as the *Bauschinger effect*.

12.13. Hardness tests

Hardness has been defined and measured in a variety of ways. The principal methods fall into the following three classes:

Resistance to indentation by a particular shape of indenter.
Resistance to scratching.
The rebound of a steel ball from the surface.

The first of these—static indentation hardness—is the only class that will be considered in this book, and three tests only of that class.

12.14. Brinell hardness

A spherical hard steel ball is pressed into the surface under a steady load, the load maintained for a few seconds for a steady state to be reached, and the indenter removed. The diameter of the permanent impression is measured with a suitable measuring microscope, either by a micrometer eyepiece or by projection on to a graduated screen. The *Brinell Hardness Number* (HB) is given as the load applied divided by the area of the curved surface of the identation, that is,

$$\text{HB} = \frac{2W}{\pi D^2[1-\sqrt{\{1-(d/D)^2\}}]}$$

where W = applied load, d = diameter of impression and D = diameter of ball.

Tables for converting the measured values of W, d, and D to hardness are given in B.S. 240.* The hardness, which has the dimensions of stress, is always expressed in kgf mm^{-2}.

In most cases the HB is not constant for a given specimen under test, but depends upon the load and size of ball used. It is found that for geometrically similar indentations in a homogeneous medium, whatever their size, the hardness number is constant. Thus if d/D for one indentation is the same as d/D for another indentation with a ball of different diameter and an appropriately different load, the same HB results. Hence it

* B.S. 240: *Method for Brinell Hardness Test*, Part 1: 1962, *Testing of Metals*.
According to this Standard, results should be quoted in the form

226 HB 10/3000

where the standard abbreviation HB is preceded by the hardness number and followed by the ball diameter in mm and the load in kgf.

is always necessary to state the values of D and W used when quoting an HB. It is usual to choose a value of W so that d/D lies between 0·25 and 0·5.

Balls of tungsten carbide are often used in place of steel balls when materials of high hardness are being tested, as these are much harder and deform less.

12.15. Diamond pyramid hardness

This method was introduced by Smith and Sandland in 1922 and developed by Vickers-Armstrong Ltd. For this reason it is often referred to as the *Vickers hardness*

The indenter is a diamond in the form of a square pyramid, the angle between opposite faces being 136°. This angle was chosen by analogy with the Brinell test in which the average d/D ratio is 0·375. When tangents are drawn from the edges of an impression with this d/D ratio, as in fig. 12.17, the included angle is 136°.

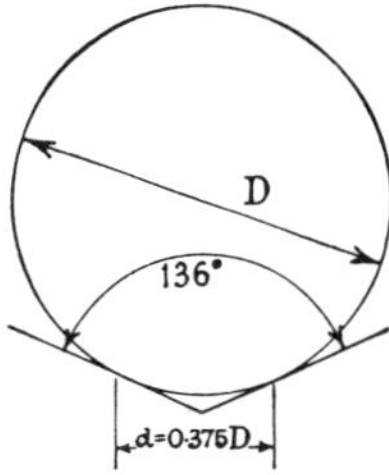

Fig. 12.17.—Relation of diamond indenter to Brinell indentation

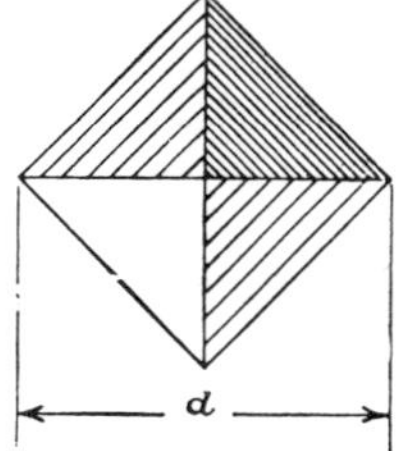

Fig. 12.18.—Plan view of diamond impression

The impression is made by applying a suitable load, and the length d of its diagonal (fig. 12.18) is measured. The hardness HV is defined as the load W divided by the area of the sloping sides of the impression:

$$\mathrm{HV} = \frac{2W}{d^2}\cos 22^\circ$$

$$= 1{\cdot}854\,\frac{W}{d^2}$$

The loads used may be from 1 to 120 kgf, depending upon the hardness of the material under test. For most work the load is chosen to make the value of d lie in the range 0·5–1 mm.

Tables for converting d to HV are given in B.S. 427.* The HV is always quoted in kgf mm^{-2}.

Since all impressions are geometrically similar, the HV is independent of the load used, provided that the material is homogeneous in its mechanical properties as the depth below the surface varies.

At low hardnesses, the HB and HV for any one material are equal.

12.16. Rockwell hardness

This test is based on the measurement of the depth of penetration of the indenter. A load of 10 kgf is first applied to the indenter, and the depth of penetration then reached is taken as the zero for further measurements. A further load (called the major load) is then applied and removed as in fig. 12.19, leaving only the minor load. The depth, d mm, of the indenter relative to the zero position (i.e., that before the application of the major load) is recorded on a suitable dial gauge.

Different scales of hardness, using different shapes of indenter and different major loads, are available.† The commonest are known as Rockwell B (HRB) and Rockwell C (HRC).

For softer materials, a 1·59-mm diameter steel ball is used, the major load is 90 kgf (100 kgf total load) and the hardness is

$$\mathrm{HRB} = 130 - \frac{d}{0{\cdot}002}$$

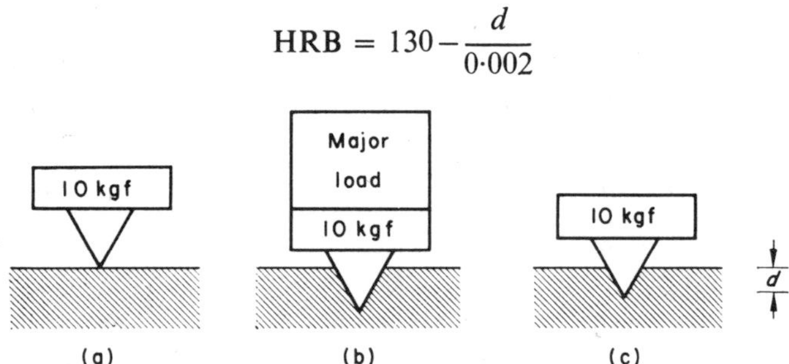

Fig. 12.19.—Stages of Rockwell hardness test; d is difference in indenter height between (a) and (c)

* B.S. 427: *Method for Vickers Hardness Test*, Part I: 1961, *Testing of Metals*. Results should be quoted in the form

648 HV 30

where the standard abbreviation is preceded by the hardness number and followed by the load in kgf.

† A complete list is given in B.S. 891: *Method for Rockwell Hardness Test*, Part I: 1962, *Testing of Metals*. Results should be quoted in the form

60 HRA

where the standard abbreviation HR is preceded by the hardness number and followed by the appropriate letter.

For harder materials, a conical-shaped diamond of 120° apex angle is used, the major load is 140 kgf (total load 150 kgf), and the hardness is

$$\text{HRC} = 100 - \frac{d}{0{\cdot}002}$$

On the usual forms of hardness tester, the dial gauge is calibrated in divisions corresponding to intervals of 0·002 mm and numbered suitably for both B and C scales.

The conversion of Rockwell hardness values to HV and HB can be done by empirical curves. Tabular values from these curves are given in B.S. 860.*

12.17. Relationship of HV to yield stress

Since the deformation under the indenter is plastic, it should be possible to relate the hardness to the plastic properties of the material. The plastic flow is a three-dimensional problem and not amenable to simple treatment, particularly when work-hardening effects have to be considered. It has been shown that for ideal plastic materials (see Section 12.10)

$$\text{HV} \approx 2{\cdot}9\sigma_y$$

where σ_y is the yield stress.†

Some experimental values for work-hardened materials which approximate to an ideal plastic material are given in Table 12.1.

TABLE 12.1—HARDNESS AND YIELD STRESS OF SOME WORK-HARDENED MATERIALS

Material	σ_y (kgf mm^{-2})	HV	HV/σ_y
Aluminium	16·2	47·5	2·93
Copper	40·4	120	2·97
70/30 Brass	51·8	144	2·78
Mild steel	54·4	166	3·06

In a work-hardening material there will be different amounts of work hardening at different points under the indenter, so that the yield stress is not uniform. If the average amount of strain under the indenter were known, it would be possible to say that the average yield stress would be the yield stress corresponding to that average value of strain. It is found

* B.S. 860: 1967. *Approximate Comparison of Hardness Scales.*

† D. Tabor, *The Hardness of Metals* (Oxford University Press, 1951).

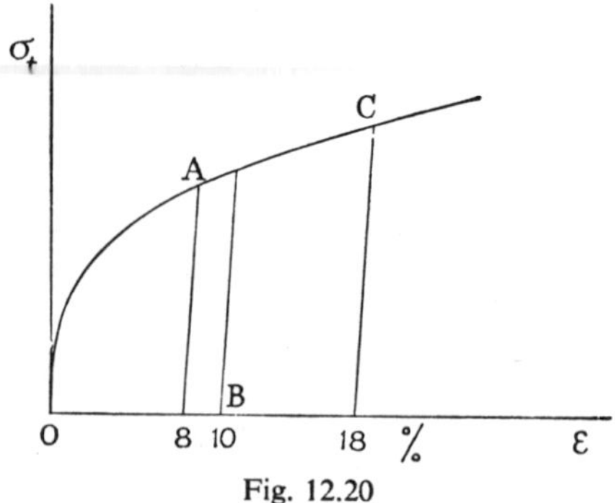

Fig. 12.20

by experiment that the average value of strain to be considered is about 8%, so we may say that the indentation takes the material on average to a state corresponding to 0·08 more strain than it had previously undergone.

Thus for virgin material O (fig. 12.20) the average condition under the indenter will be A which corresponds to a strain of 0·08. For material given 10% strain originally, B, the average condition under the indenter will be C, which corresponds to a strain of 0·18, etc. The stresses corresponding to A and C are the values of the representative yield stress Y_r such that

$$\mathrm{HV} \approx 2{\cdot}9\ Y_r$$

The results of some practical tests on an annealed copper specimen which verify this are given in fig. 12.21.

12.18. Relationship between tensile strength and HV

For a material with true stress–linear strain curve given by

$$\sigma_t = a\varepsilon^b$$

it has been shown (p. 196) that

$$\sigma_{\mathrm{max}} = ab^b(1-b)^{1-b}$$

Now the diamond indenter produces an average strain of 8%, so that the representative yield stress is

$$Y_r = a(0{\cdot}08)^b$$

and

$$\mathrm{HV} = 2{\cdot}9a(0{\cdot}08)^b$$

Therefore

$$a = \frac{\mathrm{HV}}{2{\cdot}9}(12{\cdot}5)^b$$

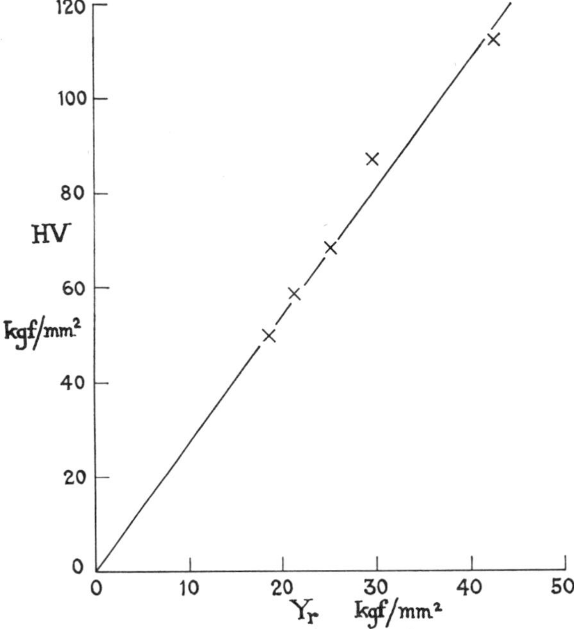

Fig. 12.21.—Hardness and representative yield stress under indenter for various stages of work hardening an annealed copper specimen

Substitution of this value in the expression for the tensile strength gives

$$\sigma_{max} = \frac{\text{HV}}{2{\cdot}9}(12{\cdot}5b)^b(1-b)^{1-b}$$

In a class of materials which have stress–strain curves of similar shape, i.e., b is the same but a differs, there will be a constant relationship between tensile strength and HV.

For annealed steels $b \approx 0{\cdot}2$, so that

$$\frac{(12{\cdot}5b)^b(1-b)^{1-b}}{2{\cdot}9} = \frac{(12{\cdot}5 \times 0{\cdot}2)^{0{\cdot}2}(0{\cdot}8)^{0{\cdot}8}}{2{\cdot}9}$$

$$= 0{\cdot}346$$

i.e.,

$$\text{tensile strength (in kgf mm}^{-2}) = 0{\cdot}346 \text{ HV}$$

But

$$1 \text{ kgf mm}^{-2} = 9{\cdot}80665 \text{ MPa}$$

so that

$$\text{tensile strength (in MPa)} = 0.346 \times 9{\cdot}80665 \text{ HV}$$
$$= 3{\cdot}4 \text{ HV}$$

Such a result is useful in estimating the strength of a steel from a single hardness measurement.

QUESTIONS

1. Sketch curves connecting the nominal stress and linear strain obtained from tensile tests of (*a*) a typical ductile material, (*b*) a typical non-ductile material, (*c*) 0·2% carbon steel.

The following figures were obtained in a tensile test of a test piece with a 50-mm gauge length and a cross-sectional area of 160 mm^2.

Extension (mm)	0·05	0·10	0·15	0·20	0·25	0·30	1·25	2·5	3·75	5·0	6·25	7·5
Load (kN)	12	25	32	36	40	42	63	80	93	100	101	90

The elongation and reduction of area were 16% and 64%, respectively.

Calculate the maximum allowable working stress if this is to equal (*a*) 0·25 × tensile strength, (*b*) 0·6 × 0·1% proof stress.

What would have been the elongation and reduction of area if a 150-mm gauge length had been used? [P]

2. British Standard Specification B.S. 18: 1962 states that for a tensile test on a ductile metal the specimen gauge length l must be related to the specimen cross-sectional area A by the expression $l = 5{\cdot}65\ \sqrt{A}$. Explain briefly why it is necessary to specify this relationship.

In the following table all three tensile specimens were turned from the same ductile metal. Calculate the percentage elongation of specimen No. 3.

Specimen No.	Gauge length (mm)	Diameter (mm)	Percentage elongation
1	25	2·5	22
2	50	6·25	25
3	75	12·5	—

[P]

3. The following observations were made during a tensile test on a 15-mm diameter specimen of pure aluminium:

Load (kN)	Extension on 50-mm gauge length (mm)
4	0·02
6	0·125
8	0·625
10	1·5
12	4·75

By plotting on log–log graph paper, or otherwise, determine a mathematical relationship between true stress and linear strain. Hence determine the tensile strength for the material. [MST]

4. Show that the tensile strength of a ductile material is determined by the plastic properties of the material and not by rupture strength.

Calculate the tensile strength for a material whose true yield stress σ_t and linear strain ε are related by $\sigma_t = 27\ \varepsilon^{0{\cdot}25}$. [P]

5. Calculate the tensile strength of a brittle material for which $a = 40$, $b = 0{\cdot}25$, and the elongation to fracture on a 50 mm gauge length is 25%. [MST]

6. A frictionless uniaxial compression test on an annealed aluminium cylinder initially 17 mm diameter and 25 mm long gave the following results

Load (kN)	Reduction in height (mm)
11	0·1
16·5	0·7
22	1·9
27·5	3·6
33	5·1
38·5	6·6
44	8·0

Plot the curve of true stress against logarithmic strain.

Determine the tensile strength and the value of the linear strain at which necking would commence in a tensile test on the material. [MST]

7. The tensile stress–strain curve of a metal may be represented by $\sigma = a\,\epsilon^b$, where σ = true stress, ϵ = true (logarithmic) strain, and a and b are constants. Show that the condition for a neck to form arises when the slope of the curve equals the true stress at that point and hence, or otherwise, that necking occurs when $\epsilon = b$. [MST]

8. As a result of slack quenching the tensile properties of a 50-mm diameter steel bar are found to vary over the cross-section. For the centre 25-mm diameter portion the tensile properties may be represented by the equation

$$\sigma_t = 90\ \varepsilon^{0.2}$$

where σ_t is the true stress in MPa and ε is the linear strain, while for the remainder of the cross-section the corresponding equation is

$$\sigma_t = 130\ \varepsilon^{0.2}$$

Calculate the maximum tensile load the bar will sustain before fracture. [MST]

9. Why are hardness tests frequently used in place of tensile tests to determine the mechanical strength of a metal?

It has been shown that HV = $2{\cdot}9\,Y_r$, where HV is the diamond pyramid hardness of a metal in any particular state of strain and Y_r is the yield measured at a strain 0·08 more than that at which the hardness is measured. A certain metal has a stress–strain curve given by $\sigma_t = a\varepsilon^{0.32}$, where σ_t is the true tensile stress, ε is the linear tensile strain, and a is a constant. Derive an expression for the tensile strength of a specimen of the metal in terms of the measured value of HV for the specimen. [MST]

10. Define *hardness*. What advantage is gained by the use of a pyramidal rather than a spherical indenter in a hardness test on metals?

A Vickers' Diamond Hardness test is made on a certain steel with a load of 30 kgf and a square pyramidal indenter of included angle 136°. The impression measures 0·654 mm across the diagonals. Determine the Vickers Hardness Number.

What maximum load should be used to investigate the hardness of a particular grain in this material if the grain diameter is 0·050 mm and its hardness is not expected to be less than that of the bulk specimen? [MST]

11. What general physical principle is used when Brinell indentations made with spheres of different diameters are compared?

How is the hardness of a work-hardened metal related to its yield stress, and why?

The true stress–strain relationship for a metal with Young's modulus $E = 1{\cdot}3 \times 10^5$ MPa can be approximated by $\sigma_t = 350\ \varepsilon^{0.4}$ MPa. Find the ratio of the plastic strain to the elastic strain at maximum tensile load. [P]

12. A tensile test on a particular metal gave a true stress–linear strain curve represented by

$$\sigma_t = 420\ \varepsilon^{0.33}\ \text{MPa}$$

If a specimen, 25 mm diameter, of this metal in the same initial condition is subjected to a compressive linear strain of 20 %, what load would be necessary?

Estimate the diamond pyramid hardness of the specimen after compression. [P]

CHAPTER 13

Plastic Flow and Fracture

13.1. Introduction

One of the important considerations in the design of an engineering component or structure is that it shall not deform excessively or break under the loads or forces to which it will be subjected during its working life, that is to say, its mechanical strength shall be sufficient. For any particular design, the stresses at all points can be determined, by calculation or some experimental method of stress analysis, for the loads and forces to which that component will be subjected. The stresses must obviously not exceed the permissible stress for the material chosen and the factors governing the choice may vary. In some cases, limitations of space or weight may dictate the size and hence the stresses, the material being chosen accordingly; or the material may be chosen with some other factor in mind, such as cost or corrosion resistance, and the size of the component made to fit.

The maximum allowable stress in the component is usually taken as some fraction f of the minimum stress that will cause plastic deformation in a ductile material or fracture in a brittle material. $1/f$ is known as the *factor of safety*.

13.2. Plastic flow under multi-axial stresses

In a tensile or compressive test the specimen is subjected to stress in one direction only—uniaxial stress. In many engineering components the shape is such that at a point there are stresses in more than one direction—a *multi-axial stress* system. At any point a stress system can be defined by normal stresses on three mutually perpendicular planes, the directions of these planes being chosen so that the shear stresses on them are zero. These normal stresses are known as the *principal stresses.* The normal and shear stresses on a plane in any other direction through the point can be calculated by formulae which are given in many textbooks on strength of materials.

The conditions for the onset of plastic deformation under multi-axial stressing has been the subject of extensive investigations. For a material with constant yield stress in uniaxial tension, i.e., no strain hardening, the most widely-accepted criterion is that the von Mises plasticity function

$$[(\sigma_1-\sigma_2)^2+(\sigma_2-\sigma_3)^2+(\sigma_3-\sigma_1)^2]$$

reaches a value of $2\sigma_y^2$, where σ_y is the yield stress in simple tension and $\sigma_1, \sigma_2, \sigma_3$ are the three principal stresses.

For the case when $\sigma_3 = 0$, the values of σ_1 and σ_2 for which yielding will just occur lie on an ellipse as shown in fig. 13.1. Addition of a hydrostatic stress σ_v to the three principal stresses does not alter any of the terms $(\sigma_1-\sigma_2)^2$, etc., in the von Mises function. Hence for any value of σ_3 that is not zero, the boundary curve for yielding is an ellipse of the same shape but with its centre displaced along the line $\sigma_1 = \sigma_2 = \sigma_3$, the σ_3-axis being perpendicular to both the σ_1 and the σ_2-axes. The surface generated by the ellipse moving along this line is a circular cylinder of radius $\sqrt{(\frac{2}{3})} \cdot \sigma_y$.

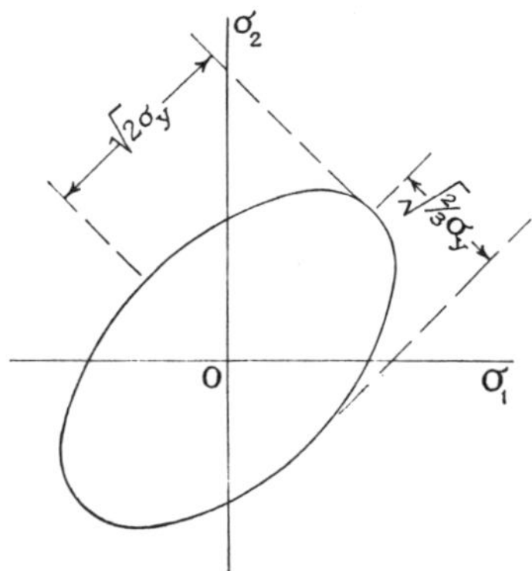

Fig. 13.1.—Boundary of yield under plane stress system according to von Mises plasticity function

13.3. Ductile fracture

It was shown in Chapter 12 that the maximum stress that a material can stand in simple tension, i.e., the tensile strength, is a plastic property dependent only upon the plastic deformation law of the material. At that stress, necking commences and the cross-section decreases until the specimen separates into two parts. The fracture is described as a cup-and-cone fracture, having a flat central part surrounded by a sloping edge or lip which may be entirely on one part or partly on each. A typical break is shown in fig. 13.2.

If the two parts are fitted together and sectioned as in fig. 13.3 it will be seen that the cup portion had failed before the cone, and that in the cup the metal has drawn out into separate fibres, each of which finally fractures. It has been shown* that, in the last stages of deformation at the neck, holes develop in the material at impurities and open up so that there is an internal as well as an external reduction of area, and the net area just before separation of the two portions is probably very close to

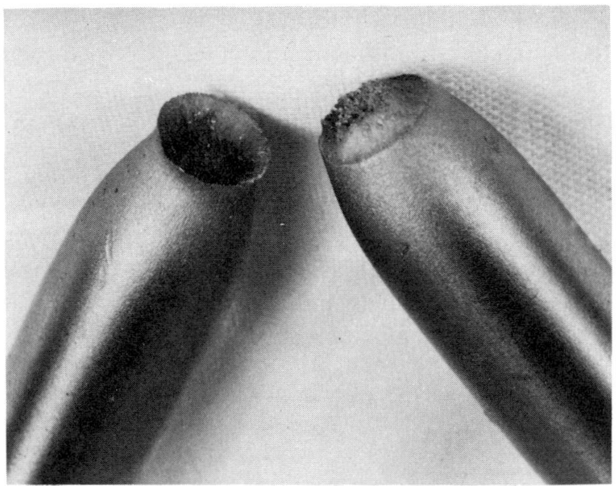

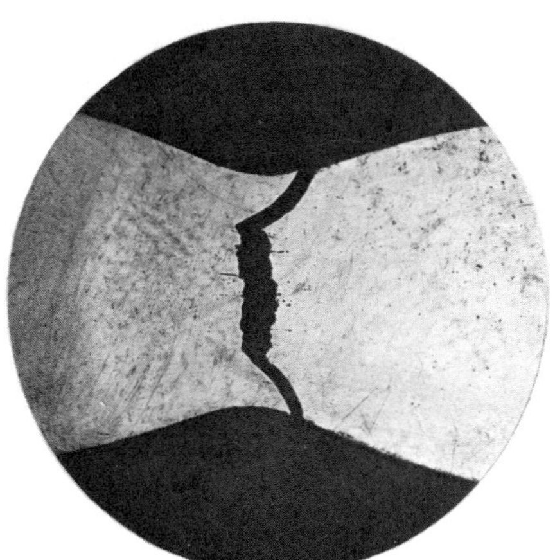

Fig. 13.3.—Section of cup-and-cone fracture (× 3)

* C. F. Tipper, *The Fracture of Mild Steel Plate*, Admiralty Ship Welding Committee, Report No. R3. (H.M.S.O., London, 1948).
K. E. Puttick, *Phil. Mag.*, Vol. 4, 1959, p. 964.

zero. Bridgman* has shown that for tensile tests conducted under externally-applied hydrostatic pressure the reduction of area at fracture is increased, mainly by reduction of the area of the cup. Under external pressure, there can be a shear stress causing deformation while the principal stresses are still compressive, so that there would be no tendency for holes to form at any impurities.

Also as purer and purer material is used, it is found that the observed reduction of area becomes greater, presumably because there is less opportunity for internal holes to be formed.

Hence ductile fracture is not really a fracture, but plastic deformation to the end.

13.4. Actual failures of engineering components

Components designed with the maximum stress criterion described in Section 13.1 modified as in Section 13.2 in cases of multi-axial stressing would be expected not to fail under the applied loads and forces. If these loads and forces were exceeded, plastic deformation would ensue, leading in cases of sufficient overload to ductile fracture.

Many components designed solely on the criterion that the stresses due to the applied loads shall not be sufficient to cause plastic deformation, nevertheless are found to fail in service. The causes of such failures are numerous, and their significance must be appreciated by designers and allowed for in appropriate cases. Among the more important are:

brittleness or lack of toughness
fatigue
creep
corrosion

The first three of these are considered in some detail in the rest of this and the following two Chapters, together with testing methods used for assessing their relevant effects in various cases. Corrosion is discussed in Chapter 18.

It has been shown in the last chapter that the tensile, compression, and static indentation hardness tests can all be related to one another via the fundamental plastic law of the material. Thus from the results of a complete tensile test it would be possible to predict the results of a compression and a hardness test. No further fundamental information would therefore be gained by performing these two tests on material for which the results of a tensile test are already available.

The three types of mechanical tests now to be described bear no straightforward relationship to the tensile test and the results cannot be forecast from the tensile-test results.

* P. W. Bridgman, *Studies in Large Plastic Flow and Fracture* (McGraw-Hill, 1952).

13.5. Theoretical cohesive strength of a solid

As mentioned in Section 12.4, brittleness is exhibited where a tensile specimen fractures with little or no plastic deformation and without forming a neck. If dislocation movement is completely inhibited so that there can be no plastic flow, then the material will fail by actual separation of layers of atoms. The fracture strength of a perfect solid material which fails in this manner can be estimated theoretically in the following way.

The variation of interatomic forces with interatomic spacing was discussed in Section 4.13. The curve of total interatomic forces per unit area, i.e., the stress σ, against the change x in the inter-atomic spacing from the equilibrium value a_0 will be of the same shape and is shown in fig. 13.4. The work which has to be done to separate two surfaces of unit area will be equal to the area under the curve. It is also equal to the surface energy of the two new surfaces so produced.

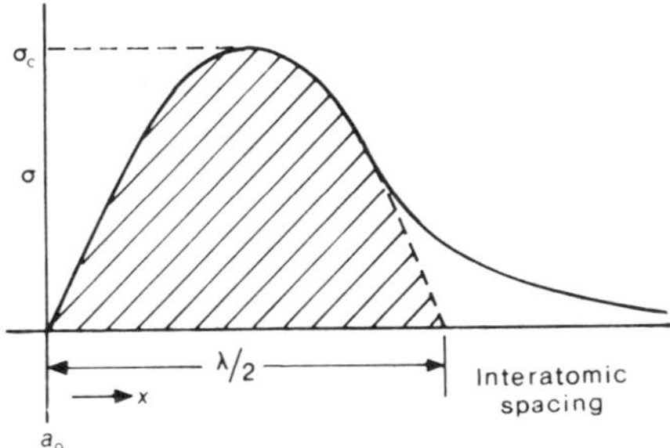

Fig. 13.4.— Variation of tensile stress with interatomic spacing. Shaded area is sine curve approximation.

The curve can be approximated by a sine curve

$$\sigma = \sigma_c \sin \frac{2\pi x}{\lambda} \tag{13.1}$$

where σ_c is the theoretical cohesive strength of the solid. The values of σ_c and λ are chosen (i) to make the slope at $x = 0$ satisfy the linear relationship for Young's modulus and (ii) to make the area under the curve equal $2\gamma_s$, where γ_s is the surface energy per unit area.

The strain is $\varepsilon = x/a_0$ so that the first condition becomes

$$\frac{d\sigma}{d\varepsilon} = a_0 \frac{d\sigma}{dx} = E$$

when x/a_0 is small. But from equation (13.1)

$$\frac{d\sigma}{dx} = \frac{2\pi}{\lambda}\sigma_c \cos\frac{2\pi x}{\lambda}$$

which has the value $2\pi\sigma_c/\lambda$ when $x = 0$.

Therefore

$$2\pi a_0 \sigma_c = E\lambda \tag{13.2}$$

The area under the sine curve is

$$\int_0^{\lambda/2} \sigma_c \sin\frac{2\pi x}{\lambda}\,dx = \frac{\sigma_c \lambda}{\pi} = 2\gamma_s \tag{13.3}$$

Hence from equations (13.2) and (13.3)

$$a_0 \sigma_c^{\,2} = E\gamma_s$$

so that

$$\sigma_c = \sqrt{\left(\frac{E\gamma_s}{a_0}\right)} \tag{13.4}$$

For a glass, typical values are $E \approx 40$ GPa, $\gamma_s \approx 1$ J m^{-2}, and $a_0 \approx 2 \times 10^{-10}$m, so that the theoretical strength is about 14 GPa, i.e., about $E/3$. For most materials, including metals, the theoretical strength calculated from equation (13.4) is approximately $E/6$, though a more refined calculation will give about $E/10$.

13.6. Actual cohesive strength

The actual strengths of brittle materials are usually of a much lower order of magnitude, but if special care is taken to produce nearly perfect materials, then strengths that approach the theoretical values may be realized. For example, freshly drawn glass fibres can withstand a very high tensile stress (about 3 GPa) without fracture, but if a scratch is made on the surface (even a feather drawn lightly across the surface will produce a sufficiently deep scratch) the glass fractures under much lower stresses. Also it is observed that the strength of the freshly drawn fibres decreases with time, which is due to the corrosive effect of the atmosphere on glass, producing minor surface flaws.

The observation that the presence of crack-like flaws in real solids can account for their low strengths relative to ideal solids was made by Larmor* in 1892.

* J. Larmor, *Phil. Mag*, Vol. 33, 1892, p. 70.

A crack concentrates the applied stress at its tip. A stress concentration factor is defined as the ratio of the maximum stress due to a hole or notch to the mean stress that would exist in the absence of the hole or notch. In an elastic medium that behaves according to Hooke's law, the stress concentration factor K_t due to an elliptical hole (fig. 13.5*a*), the major axis of which is perpendicular to the mean stress direction, is given by

$$K_t = 1 + 2\sqrt{\left(\frac{c}{\rho}\right)}$$

where $2c$ is the length of the major axis and ρ is the radius of curvature of the hole at the end of the major axis—the point where the stress is greatest.

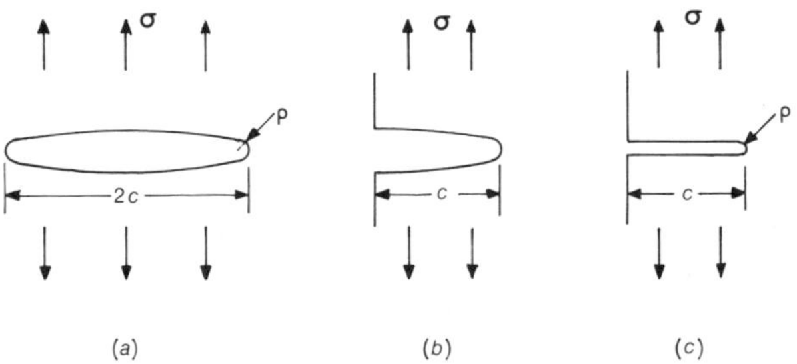

Fig. 13.5.—Stress concentration due to (*a*) elliptical hole, (*b*) semi-elliptical edge notch, (*c*) sharp edge crack in a unidirectional stress field

A similar result applies for a semi-elliptical surface crack of depth c (fig. 13.5*b*). A sharp crack (fig. 13.5*c*) is approximately an ellipse in which $c \gg \rho$, and so

$$K_t \approx 2\sqrt{\left(\frac{c}{\rho}\right)}$$

For a relatively small crack in a large sheet of material the mean stress that would exist in the material if the crack were absent is equal to the macroscopic stress σ away from the crack. The stress at the crack tip will be

$$\sigma_{tip} = K_t\sigma = 2\sigma\sqrt{\left(\frac{c}{\rho}\right)} \qquad (13.5)$$

When this stress reaches the cohesive stress (given by equation (13.4) the breaking of interatomic bonds can occur at the crack tip. Hence the strength in the presence of a crack would be expected to be

$$\sigma = \frac{1}{2}\sqrt{\left(\frac{\rho}{c}\frac{E\gamma_s}{a_0}\right)} \tag{13.6}$$

It is however extremely difficult to define the distance to which the atoms must separate for the crack to extend and so to attach a value to ρ. However it will be of the order of a_0, as will be shown later. Also if some plastic flow could occur so that extension of the crack were to cause a blunting of the tip, ρ would increase, and for a constant σ the crack might not continue to grow.

13.7. Griffith equation

Griffith* attacked the problem of fracture from a different standpoint by applying the "theorem of minimum energy". Under a given loading, fracture of a body can continue only if by so doing there is a decrease of potential energy of the system. There are three factors that contribute to the total energy of a stressed body containing a crack.

(i) When the body deforms due to the action of external forces, work W is done on the body so that the potential energy of the forces changes by $- W$.

(ii) The stressed body contains elastic strain energy, equal to $\sigma^2/2E$ per unit volume where there is a uniaxial stress of magnitude σ.

(iii) As crack surfaces are formed, work is done against the cohesive forces which act between the atoms, as discussed in Section 13.5, and this appears as surface potential energy.

Suppose that for a small increase of crack size, these energies change by $- \delta W$, δF_{el} and δF_{tip}, respectively. Then the crack will propagate if and only if the sum of these three terms is equal to or less than zero, i.e.,

$$\delta F_{el} + \delta F_{tip} - \delta W \leqslant O \tag{13.7}$$

We will apply this to the case of a large plate of thickness t of a material which behaves in a linear elastic manner and which is subject to uniaxial stress. Let it contain a small crack of length $2c$ which is perpendicular to the stress direction (fig. 13.6*a*). Let the stressing be due to a load P applied along one edge which causes a displacement u of the edge.

With a crack of constant length $2c$ present, the load–displacement curve would be a straight line OA as shown in fig. 13.6*b*. With a crack of

* A. A. Griffith, *Phil. Trans. Roy. Soc.* A, Vol. 221, 1921, p. 163.
That paper contained some errors which were corrected in *Proc. 1st Int. Congress Applied Mech.*, Delft 1924.

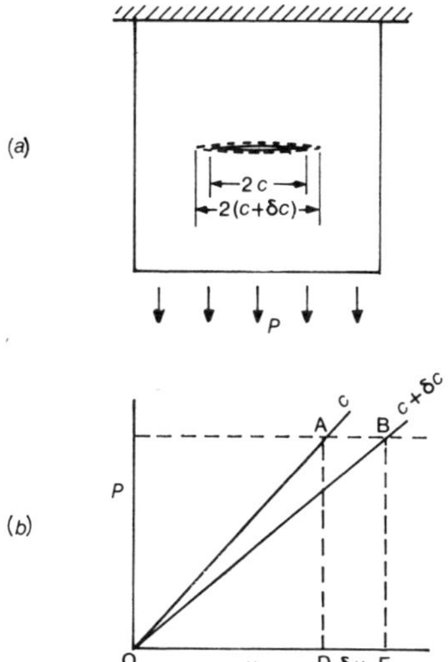

Fig. 13.6.—(*a*) Plate with crack growing under constant load.
(*b*) Variation of compliance with crack length

length $2(c + \delta c)$, the load–displacement curve would be a line OB of different slope. The slope $\delta P/\delta u$ of the line is known as the stiffness of the sheet and its reciprocal $\delta u/\delta P$ is known as the *compliance.*

Growth of the crack from length $2c$ to length $2(c + \delta c)$ would cause an increase in the compliance of the plate and, if it takes place while the load P remains constant, the loaded edge will displace δu in the direction of P. The work done by the load is

$$\delta W = P\delta u$$

The stored elastic strain energies at load P when the crack lengths are $2c$ and $2(c + \delta c)$ are the areas OAD and OBE under the load–displacement curves. The increase in elastic strain energy is therefore

$$\delta F_{el} = \text{area ABED} - \text{area ABO}$$

$$= \frac{1}{2} P\,\delta u$$

i.e.,

$$\delta W = 2\delta F_{el}$$

Griffith showed this to be a general result for a body stressed in any manner.

For the crack to propagate, we get, from equation (13.7),

$$2\,\delta F_{el} \geqslant \delta F_{el} + \delta F_{tip}$$

or

$$\delta F_{el} \geqslant \delta F_{tip}$$

Using a solution for the stress distribution around an elliptical hole derived earlier by Inglis, Griffith showed that in a plate of thickness t the increase in elastic strain energy due to the hole was given by

$$F_{el} = \frac{\pi c^2 t \sigma^2}{E}$$

The energy of the crack surfaces is $4\,ct\gamma_s$. Hence the crack extension condition becomes

$$\frac{d}{dc}\left(-\frac{\pi c^2 t \sigma^2}{E} + 4ct\gamma_s\right) \leqslant 0$$

i.e.,

$$-\frac{2\pi ct\sigma^2}{E} + 4t\gamma_s \leqslant 0$$

giving

$$c \geqslant \frac{2E\,\gamma_s}{\pi\sigma^2}$$

This is the critical length of crack, i.e., any cracks of length equal to or greater than that value will propagate spontaneously at that stress level.

Thus if there is an internal crack of length $2c$ in an infinite body or a surface crack of depth c in a semi-infinite body in a brittle material, the minimum stress necessary to cause it to propagate is

$$\sigma = \sqrt{\left(\frac{2E\gamma_s}{\pi c}\right)} \tag{13.8}$$

A further case to consider is a similar plate subject to a fixed displacement as in fig. 13.7*a*. The load–displacement curves for the two crack lengths will be as before, but because u remains constant while the crack increases in length, the load will reduce by δP, from A to F on fig. 13.7*b*. No work is done by the applied load. The energy change at the crack tip is as before

$$\delta F_{tip} = 4\,\gamma_s t \delta c$$

There is a decrease in stored elastic strain energy equal to area OAF. For a very small change in crack length

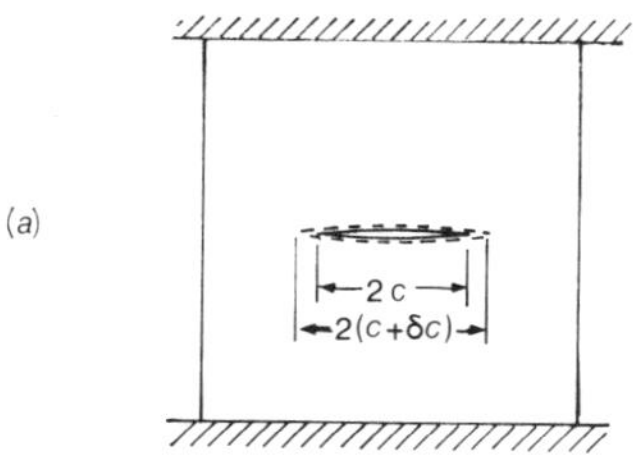

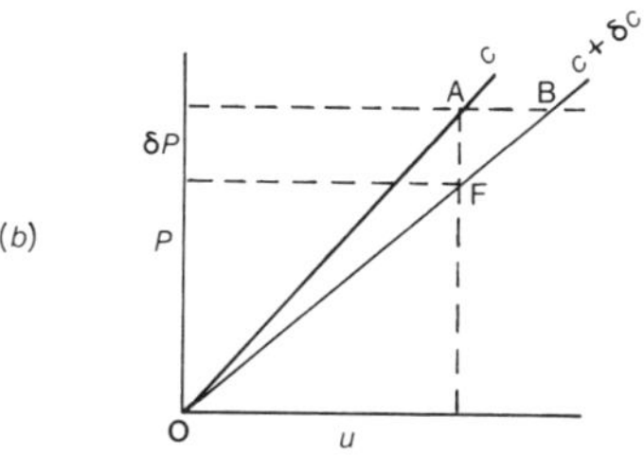

Fig. 13.7.—(*a*) Plate with growing crack under fixed displacement. (*b*) Variation of compliance with crack length

$$\text{area OAF} \approx \text{area OAB}$$

$$= \frac{2\pi c t \sigma^2}{E} \delta c$$

Hence, using equation (13.7),

$$-\frac{2\pi c t \sigma^2 \delta c}{E} + 4\,\gamma_s t \delta c \leqslant 0$$

giving

$$\sigma = \sqrt{\left(\frac{2E\gamma_s}{\pi c}\right)}$$

as before.

The two variables are σ and c, while E and γ_s are properties of the material. We can re-arrange the equation to separate them.

$$\sigma\sqrt{(\pi c)} = \sqrt{(2E\gamma_s)}$$

Note that the longer the crack, the lower is the stress necessary to cause fracture.

Also it may be seen that Griffith's equation, equation (13.8), is of the same form as the relationship derived earlier, equation (13.6), and is identical if

$$\frac{\rho}{4a_0} = \frac{2}{\pi}$$

i.e.,

$$\rho \approx 2{\cdot}5a_0$$

That is, it would appear that a crack has to open to about $5a_0$ before the surfaces can be regarded as separate.

Griffith tested his theory by experiments on thin-walled glass spheres and cylinders containing cracks of known dimensions. He pressurized these internally until they fractured by spread of the cracks and, from the pressure reached in each case, he calculated the value of $\sigma\sqrt{(c)}$. This quantity was reasonably constant, confirming the general nature of the theory, though the value was about 80% greater than that expected from values of E and γ_s.

For materials which also undergo some plastic deformation before fracture, the Griffith equation is not strictly applicable, although a modification of it has wide application (see Section 13.20).

13.8. Toughness

The term *toughness*, which is not always used in a strictly defined sense, is usually taken in a metallurgical context to mean the converse of brittleness. In a quantitative manner it can be quoted as the energy consumed or work done in rupturing a specimen. Obviously the geometry of the specimen is one of the governing factors, and so a standard size and shape of specimen must be used for comparative purposes. Since work is the product of force and distance, the toughness is dependent upon both yield stress and ductility. If the test is not a simple tensile test on a parallel-sided specimen, the yield stress under conditions of multi-axial stress is important.

The toughness as determined from a test on a specimen which contains a machined notch, that is, the energy consumed in rupturing the specimen, should be more accurately described as the *notch toughness*.

It might be expected that it would be possible to calculate the result of any particular test by working from the tensile-test results, since by knowing the strain at any point we can get the stress. But it is found that this is not so. Two different specimens (even two different plain carbon steels) may give identical stress–strain curves in tension, but behave quite differently in a particular type of test known as a notch impact test (see Section 13.10). For example, the following results for two steels may be compared:

	Yield point	Tensile strength	Elongation %	Reduction of area %	Toughness
(*a*)	753 MPa	857 MPa	28·6	64·0	106 J
(*b*)	709	840	26·5	63·7	12

It will be observed that, although the tensile results are almost identical, the work necessary to cause fracture of one specimen is nearly nine times greater than the work to cause fracture of the other specimen.

Specimen (*a*), which would be classed as a tough specimen, would have deformed a lot, eventually tearing at the root of the notch and so absorbing a lot of energy. Specimen (*b*) would have deformed very little before

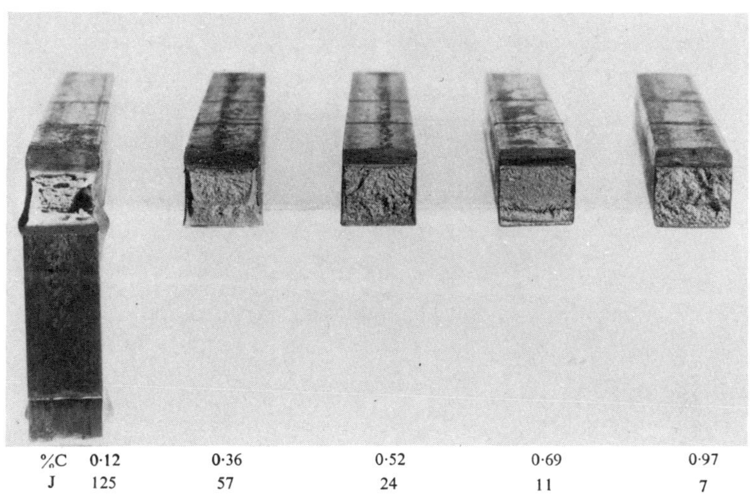

%C	0·12	0·36	0·52	0·69	0·97
J	125	57	24	11	7

Fig. 13.8.—Fractured Izod specimens. Plain carbon steels

snapping in a brittle manner, so absorbing little energy. When iron and steel fail in a brittle manner, they do so by splitting or *cleavage* along certain planes in the crystals and produce a bright *crystalline* appearance, easily distinguishable from the *fibrous* appearance of a ductile fracture. Some fracture surfaces are seen in fig. 13.8. In the tensile test both steels (*a*) and (*b*) quoted above would have shown a ductile fracture. The change from ductile to brittle fracture in the notch impact test on steel (*b*) is brought about by the effects of the notch as described in the next section.

13.9. The influence of the notch

As the load is applied to the specimen, a complex stress system builds up at the root of the notch. There is a stress concentration at the root so that the yield stress is first reached there. Because volume changes are negligible in plastic deformation, yield in tension can occur only if accompanied by a contraction in one or both perpendicular directions. The surrounding material is at a stress lower than yield and opposes any such contraction, so that tensile stresses are set up in these directions, i.e., a state of triaxial tensile stress exists (fig. 13.9). As may be seen from

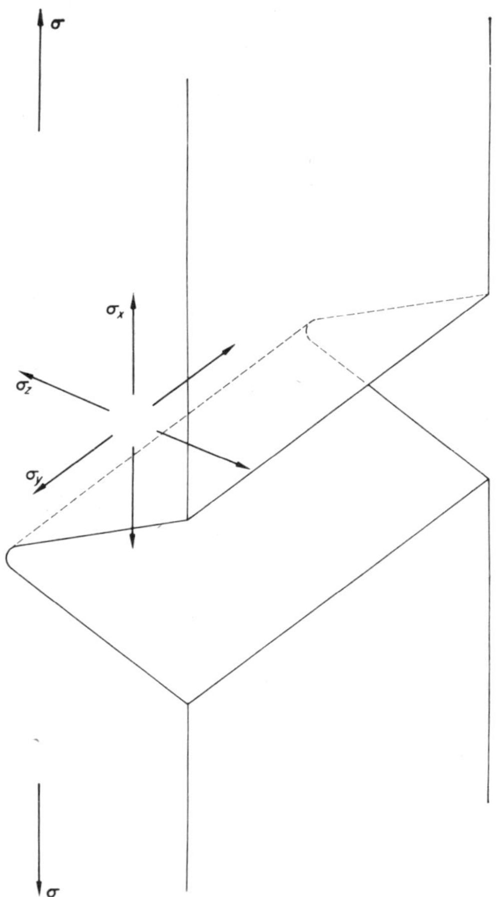

Fig. 13.9.—Triaxial stress condition at notch due to average tensile stress σ. Stress concentration effect maximum principal stress $\sigma_x > \sigma$. Restraining effect of less highly stressed neighbouring regions causes principal stresses σ_y and σ_z

Section 13.2, a higher maximum tensile stress is then necessary to cause yielding. When this is reached by further stressing of the test piece, plastic deformation will occur locally to the notch root and eventually be sufficient for ductile fracture to commence. The leading edge of this crack will be of atomic width and produce a greater stress concentration than the original machine-made notch. The maximum principal stress at the crack tip which is necessary to continue plastic deformation will bear a definite ratio (said to be of the order of 3) to the yield point in simple tension. If at any time this maximum principal stress exceeds that necessary to overcome the cohesive forces in the material, then the crack could spread in a brittle manner.

Any factor which raises the yield point in a simple tension test, namely, low temperature or a high rate of strain, could therefore change the mode of fracture from ductile to brittle.

13.10. Notch impact tests

The number of tests for determining notch toughness is very large.* The common ones in regular use are the Izod and the Charpy tests.†

The Izod specimen is a bar of square cross-section across which is machined a notch of standard size. The dimensions of the specimen and of the notch are shown in fig. 13.10. The Izod specimen is usually made with three notches, so that three measurements of toughness can be obtained from one specimen.

The bar is clamped just below a notch and hit at a point 22 mm from the notch by a swinging pendulum, i.e., it is loaded as a cantilever. The pendulum is released from a height such that at the bottom of its swing, when it hits the specimen, its kinetic energy is 163 J. After hitting the specimen, the pendulum continues to swing to a height which gives a measure of the work done in bending or breaking the specimen. The greater the work done, the less high does the pendulum swing. The angle of swing or the energy absorbed is read off a calibrated scale.

Two of the standard Charpy specimens are shown in fig. 13.11. The Vee notch specimen is the one more commonly used in Great Britain. The specimen is fractured by a swinging pendulum as for the Izod test, but the specimen is loaded as a beam instead of as a cantilever.

* C. F. Tipper, " The Brittle Fracture of Metals at Atmospheric and Sub-Zero Temperatures," *Met. Rev.*, Vol. 2, 1957, p. 195.

† B.S. 131. *Methods for Notched Bar tests.*
Part 1: 1961. *The Izod Impact Test on Metals.*
Part 2: 1972. *The Charpy V-notch Impact Test.*
Part 3: 1972. *The Charpy U-notch Impact Test on Metals.*
Part 5: 1965. *Determination of Crystallinity for Metals.*

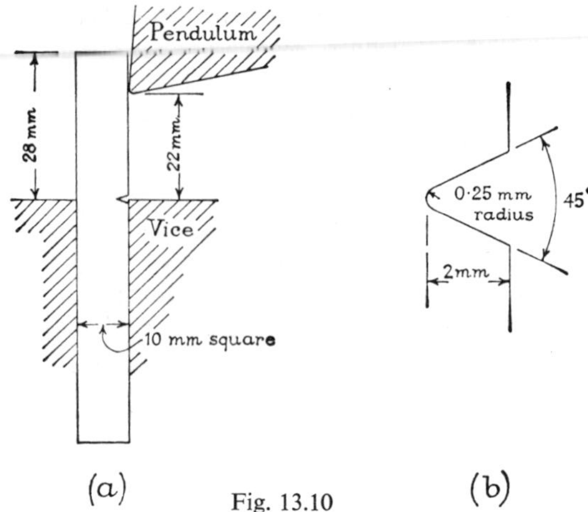

Fig. 13.10

(*a*) Izod specimen and mounting. (*b*) Form of standard Vee notch

13.11. Factors of Izod and Charpy tests

The notch impact bend tests (to give them their fully descriptive name) have three factors not present in a tensile test, each of which may have an effect upon the energy value. These are the notch, the bending, and the impact or high rate of strain. It is important to consider the relative importance of each.

It would be possible to do the tests on un-notched specimens. The impact test for cast iron is in fact performed on an unnotched bar, but this is a material which is basically brittle. The notch, as already explained,

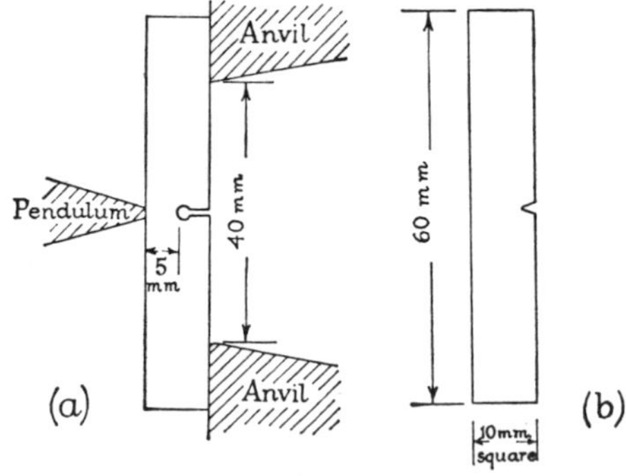

Fig. 13.11

(*a*) Charpy specimen with keyhole notch, showing mounting and striking arrangement.

(*b*) Charpy specimen with standard Vee notch

has the effect of introducing multi-axial stresses, and in some cases causing brittle fracture. Hence the notch plays a very important part.

The bending is merely a geometrical factor, being the manner by which the tensile stress is introduced, and plays a very small part. A notch bend test does not discriminate steels in a different manner from a notch tensile test.

An increased rate of straining is found to raise the yield stress. In certain cases this effect may be sufficient to change the mode of fracture, when there will be a large decrease in the energy value. If the mode is not changed, then, since the work to fracture is a function of stress and strain, the higher speed of loading will raise the stresses and hence the energy. In a ductile fracture, where there is much plastic deformation, the raised yield stress may contribute as much as 40% of the total energy in an impact test. In a brittle fracture, where there is very little plastic flow, the contribution is much less.

By straining notched specimens slowly in a suitable machine, the work to fracture can be evaluated from the area under the load–deformation curve and compared with the value for impact loading. Some actual tests on two steels gave the following results:

	Slow loading (J)	Impact loading (J)
Low-carbon steel	30	51
High-carbon steel	3·1	3·7

Thus the notch is the most important feature of the notch bent impact test.

13.12. Uses of notch tests

The Izod test was first devised for checking the suitability of different heat treatments applied to steels, but it has come into much greater prominence in recent years in the study of the low-temperature brittleness of mild steel. The effect of heat treatment upon the Izod value will be considered in Chapter 16.

The energy values given by a notch bend test cannot be used directly for design purposes. Hence the main purpose of the test is for sorting materials into an order of merit, either to choose the best for a particular application or to compare with others of proved value.

13.13. Notch brittleness of mild steel

If we perform Izod tests on a steel over a range of temperatures, it is found that there is a sudden change from high-energy and fully-fibrous fractures to low-energy and fully-crystalline fractures. The drop in energy

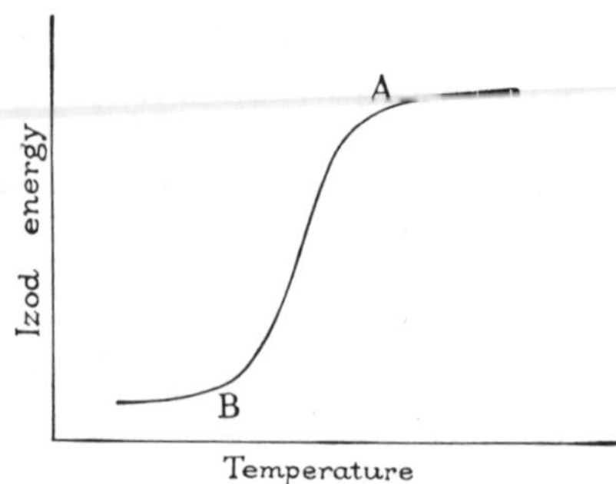

Fig. 13.12.—Typical curve for variation of Izod value with temperature for a mild steel

is rapid over a small range of temperatures as may be seen from the typical curve in fig. 13.12. Points A and B which mark the limits of the transition range may be only 20 to 30 K apart. A normal unnotched tensile test would, however, show only a small and gradual change over a 100 K range, the yield point being raised as the temperature falls; but this increase, like the effect of the notch and increased rates of strain, promotes brittleness.

The form of curve so obtained is typical of all body-centred cubic and hexagonal close-packed metals, but the phenomenon is not shown by face-centred cubic metals.

13.14. Transition temperatures

The transition temperature which marks the region in which the energy value and type of fracture undergo their great change may be defined in one of several ways, and different ways may give somewhat different values of the temperature. Hence it is important to note the basis of definition for any quoted values of transition temperature. The most common bases of definition are:

The point of greatest slope of the energy–temperature curve.

The temperature at which the fracture appearance shows a definite percentage of crystallinity, e.g., 50%.

The temperature at which the energy has a definite value, e.g., 20 J.

13.15. Tests on large plates

When large plates of steel are tested in tension and notches are deliberately placed in them, it is found that the plate tears if the temperature is above a certain value, but breaks in a brittle manner if the temperature is below that value. The particular value, which varies from steel to steel, is found to correlate closely with the transition temperature as defined in either of the first two ways listed above. The actual energy value is of much less significance; what is a low-energy value for one steel may be a high-energy value for another steel as may be seen from the curves in fig. 13.13. The important factor is that the transition temperature in low-carbon steels is near the ambient temperature.

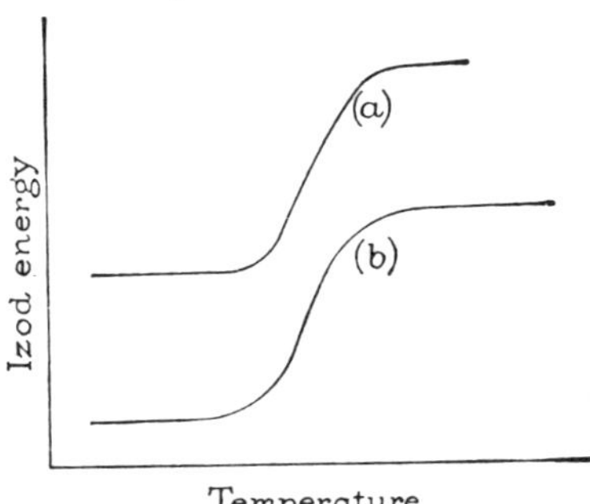

Fig. 13.13.—Comparison of transition curves for two hypothetical steels. The energy value of steel (*b*) for a fully ductile fracture is equal to that for an almost fully brittle fracture of steel (*a*)

13.16. Casualties due to brittle fracture

Abnormal conditions not contemplated in design do occur and have brought about failures due to brittle fractures. Stress concentrations, rough edges, defective welds, etc., can all lead to small notches or cracks which, if the steel is below the transition temperature, may start a brittle fracture. The mean stress need not necessarily be high, because the stress concentration can raise it locally, and also, as has been shown by Wells* and Mylonas,† residual stresses due to welding or plastic deformation can add to small applied stresses to give local stresses equal to the yield stress. In a welded structure the fracture can pass through welds into neighbouring plates, and if these are likewise below their transition temperature the result may be a failure of large extent.

Many brittle fractures have occurred in ships, several welded vessels having broken completely in two. Other large-scale fractures have occurred in large oil-storage tanks, in welded pipelines and in welded bridges.

Poor design and poor workmanship have contributed to these failures, but even when these factors have been largely remedied, casualties have occurred. If the steels used had had a much lower transition temperature,

* A. A. Wells, *Trans. Inst. Naval Arch.*, Vol. 98, 1956, p. 296.

† C. Mylonas, D. C. Drucker, and J. D. Brunton, *The Welding Journal*, Research Suppl., Vol. 23, 1958, p. 473S.

these failures would not have occurred. In a large number of cases which have been thoroughly investigated, it was found that the steel temperature at the time of the fracture was below the transition temperature, as shown by an Izod or Charpy test or a notch tensile test.*

13.17. The effect of metallurgical factors

The transition temperature of a steel is found to be related to the composition—each element, even when present in small quantities, having an effect—and to the grain size. As both of these factors are dependent upon the practices adopted in steelmaking, a brief outline of the processes is necessary for an understanding of these effects.

Steel is made from pig iron (see p. 112) by refining, which is principally the removal of the unwanted carbon and other elements by oxidation. Most mild steel was made by one of two processes, either (*a*) the Bessemer process, in which air was blown through the molten pig iron to oxidize the unwanted elements, the heat of reaction being sufficient not only to keep the steel molten, but also to raise its temperature; or (*b*) the open-hearth process, in which the molten metal was contained in a shallow hearth and heated by flames burning across the top, iron oxide being added to remove the carbon, etc. These methods are now largely ousted by the Basic Oxygen process in which oxygen is blown on to the molten steel which is contained in a convertor. In each process any necessary additions of alloying elements are made at an appropriate stage.

The molten steel in the final stages of refining contains a considerable amount of iron oxide in solution. This may be removed by the addition of certain elements (usually manganese, silicon, or aluminium) to the metal just before pouring into the moulds. The process is known as *killing*. Unkilled steel is known as *rimming* steel. The major part of the steel used for structural purposes is *semi-killed*, i.e., de-oxidation is not carried to completion.

Finally, the steel ingots are hot-rolled to plates or sections, while the steel is at temperatures well above the A_3-temperature. During rolling, austenite grains are broken down. If rolling is completed at temperatures well above the A_3-temperature, grain growth may subsequently occur while cooling before the austenite transforms. Thick plates usually have a higher finishing temperature, unless special steps are taken to cool the steel between passes, and also cool more slowly to the A_{r3}-temperature, permitting more grain growth.

Bessemer steels generally contained a higher proportion of nitrogen, which gave higher transition temperatures in notch tests, and these steels

* J. F. Baker and C. F. Tipper, "The Value of the Notch Tensile Test," *Proc. Inst. Mech. Engrs.*, Vol. 170, 1956, p. 65.

have been forbidden for structural and shipbuilding purposes in the United Kingdom for many years. The adoption of the basic oxygen process has removed this objection.

The grain size of the steel is affected by the elements present, especially those used for de-oxidizing, aluminium in particular promoting fine grain size (see p. 70). Fully-killed and fine-grained steels have lower transition temperatures and are to be preferred where there might be a danger of brittle fracture.

Increased manganese content lowers the transition temperature and increased carbon raises it. The practice adopted has been to keep the Mn/C ratio high, thereby compensating for loss of strength due to less carbon by increase of strength due to added manganese. The ratio must not be too high, however, or trouble arises in welding.

Strain ageing (see p. 175), cold working, and any other process which normally raises the yield point of the steel and also reduces its ductility have the effect of raising the transition temperature. Fully-killed fine-grain steels show little or no strain ageing.

The high transition temperature of the steel used in a lot of the earlier welded ships was due to a low manganese content. Specifications for ship steel, which required that the steel should satisfy a tensile test, were in the course of time modified to demand a composition which would ensure a sufficiently low transition temperature. The need has since arisen for thicker plates, thicknesses of 40 to 50 mm being needed instead of the customary maximum thickness of about 25 mm. Hence the transition temperature of the plates has tended to rise again due to the grain growth mentioned above which has an embrittling effect. This can be overcome either by the use of expensive alloying elements, or by modifying rolling practice so that the plates have lower finishing temperatures, or by subjecting the rolled plates to a normalizing treatment (see p. 283) to reduce the grain size.

13.18. Fracture mechanics

Griffith's equation for the critical stress at which a crack will extend in a brittle material was derived in Section 13.7. In that treatment, the only effects of the stress that were considered were elastic deformation of a solid and local deformation to the extent of breaking the atomic bonds. Among structural materials, metals and their alloys exhibit a more varied behaviour under applied stress, in particular they are able to deform plastically. Such plastic flow may relieve the stress concentration at the end of a crack so that brittle fracture does not occur, and crack propagation will involve a high dissipation of energy. However, the highest ductilities are frequently associated with materials which are weak in plastic flow, so that by choosing a material with sufficient ductility to cause blunting of cracks, the stress necessary to cause the cracks to extend may be above the tensile strength of the material. The component would then fail in a normal manner and not because of crack propagation. Where a material of high strength is required, one may also have low ductility and the possibility of crack propagation from inherent flaws needs careful consideration. The quantitative treatment of crack propagation in this manner—the assessment of the property of *fracture toughness*—is known as *fracture mechanics*.

For example, in designing a pressure vessel made from a high strength steel, it is necessary to know what is the critical size of flaw that would propagate at the working stress. Then the inspection techniques used must be capable of detecting any cracks of this size and somewhat smaller so that they can be removed or the component discarded.

13.19. Stress intensity factor

The stress at the end of a sharp crack was given in equation (13.5). This equation is however difficult to apply because of the uncertainty of the value of ρ and for an infinitely sharp crack the stress would be infinite. The stresses near the crack tip in a sheet of material that behaves in a linear elastic manner are known. At a distance x along the line ahead of the crack tip (fig. 13.14) they are given by

$$\sigma_x = \sigma_y = \sigma\left(\sqrt{\left(\frac{c}{2x}\right)} + C\right)$$

where C is a constant. Near the crack tip,

$$\sigma_x \sqrt{(2\pi x)} = \sigma\,(\sqrt{(\pi c)} + \sqrt{(2\pi x C)})$$

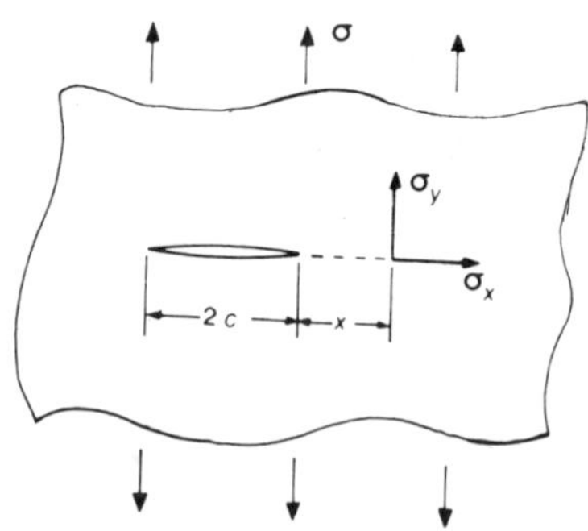

Fig. 13.14. — Stresses ahead of crack tip.

At distances of the order of the atomic spacing, the stress is no longer given by this equation, which relates to an elastic continuum, nor would the stress be realistic if it exceeded either the yield stress or the cohesive stress of the material.

Nevertheless, we see that as x gets smaller, the term $\sqrt{(2\pi x C)}$ will become negligible compared to $\sqrt{(\pi c)}$ so that in the limit

$$\lim_{x \to 0} \sigma_x \sqrt{(2\pi x)} = \sigma \sqrt{(\pi c)}$$

which is known as the *stress intensity factor* and is denoted by K. It characterizes the stress intensity or stress distribution near the crack tip. The dimensions of K are stress $\times$ length$^{\frac{1}{2}}$.

For cracks of any shape in any component we can again define K as

$$\lim_{x \to 0} \sigma_x \sqrt{(2\pi x)}$$

Another parameter introduced into fracture mechanics terminology is G, the *crack extension force* per unit length of crack or the *strain energy release rate.* Using the terminology of Section 13.7

$$G = \frac{dF_{el}}{dc}$$

13.20. Modified Griffith equation

Irwin and Orowan independently modified the Griffith equation to

$$\sigma = \left(\frac{2(\gamma_s+\gamma_p)E}{\pi c}\right)^{\frac{1}{2}} \tag{13.9}$$

where γ_p is the work done in plastic deformation per unit area of crack surface. Although γ_p will greatly exceed γ_s in plastic materials, its range varies greatly from material to material and can increase rapidly with a small increase in temperature at temperatures near the transition temperature. The ratio of γ_p to γ_s is of the following orders in different classes of materials:

metals	10^4–10^6
polymers	10^2–10^3
ceramics	10

It is obvious from the above equation that if the value of $\sigma \sqrt{(\pi c)}$ exceeds a limit typical of the material, the crack becomes unstable. This limiting value is the *critical stress intensity factor* or *fracture toughness* and is denoted by K_c.

At the point of instability, there is also a specific value of G denoted by G_c which is the *toughess* or *critical strain energy release rate.*

$$G_c = \frac{dF_{tip}}{dc}$$

G_c replaces the term $2(\gamma_s + \gamma_p)$ in equation (13.9), i.e., it is a measure of the energy needed to create the two surfaces of unit area of crack. It can be readily seen that

$$K_c^2 = EG_c$$

13.21. Plane stress and plane strain conditions

The Griffith equation derived in Section 13.7 and the preceding discussion in this section relate to a sheet of material so thin that there is negligible stress perpendicular to the sheet—i.e., a *plane stress* condition exists. If the material yields, it can do so by undergoing strain in all three directions and the sheet will become thinner in the plastic zone. Where a crack exists, dimples form on each surface just ahead of the crack tip.

However, with a thick sheet of material, this lateral contraction is impeded (see Section 13.9) and there is a tensile stress in the transverse direction. This is known as a *plane strain* condition. The ratio of stress to strain is altered under plane strain conditions, the effective modulus being $E' = E/(1 - \nu^2)$ and the Griffith equation would become

$$\sigma_c' = \sqrt{\left(\frac{2E'\gamma_s}{\pi c}\right)}$$

In a thin sheet, which is stretched so that a crack in it extends, the fracture surface may start perpendicular to the surface but eventually tends to turn into a plane at 45° to the surface, the material failing in shear (fig. 13.15*a–c*). The force to cause failure will be proportional to the thickness t as also will be the shear strain to failure. Hence the work done for unit length of crack growth will be proportional to t^2 and the work per unit area proportional to t. Hence the apparent value of K_c will increase with plate thickness.

As thicker plates are considered, the plane strain region becomes prominent and gives a fracture surface which is perpendicular to the plate surfaces with 45° shear lips only at its edges (fig. 13.15*d*). Because of the triaxial tension ahead of the crack in the plane strain region, less plastic flow occurs. Less plastic work will therefore be done in creating the new

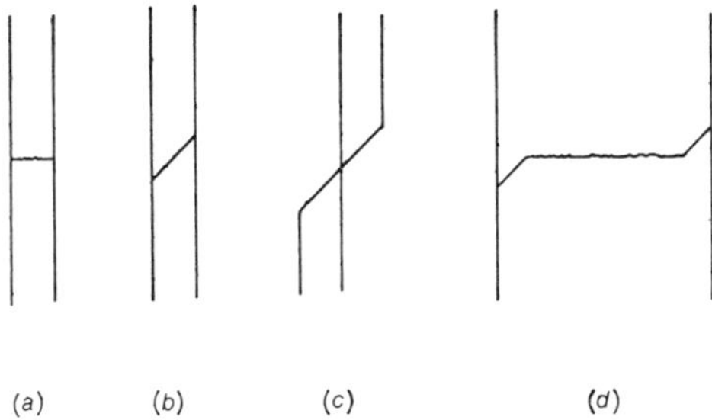

Fig. 13.15.—Cross-sections of cracks in thin and thick plates. (*a*)–(*c*) Crack initially perpendicular to plate surfaces turns on to 45° plane and fails by shear. (*d*) Crack mainly perpendicular to surfaces with narrow shear lips at sides

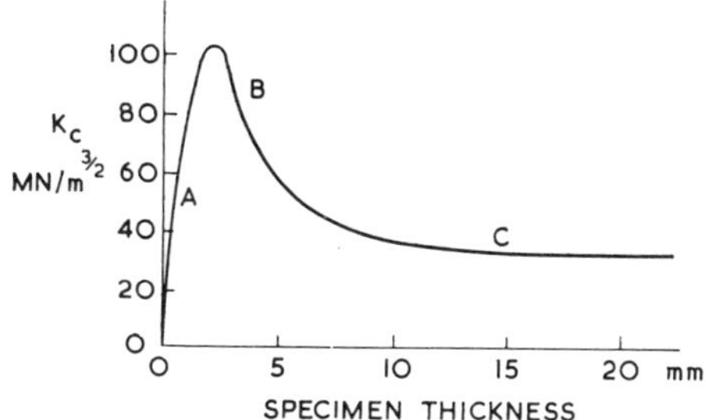

Fig. 13.16.—Variation of K_c with plate thickness for an aluminium alloy

crack surfaces and so the critical stress intensity factor is less. The variation of K_c with plate thickness is shown in fig. 13.16.

The value of K_c for the fully plane strain condition is denoted by K_{Ic}. Whereas K_c is a function of material and plate thickness, K_{Ic} is solely a material property. It is the value commonly used in design because it is the lowest one (except for very thin sheets) and hence is safest to use. Methods for determining its value are given in Section 13.24.

13.22. Failure modes

The surfaces of a crack can be displaced relative to one another in three possible ways as shown in fig. 13.17. In fig. 13.17*a* the crack is opened by a tensile force and in fig. 13.17*b* is sheared by a force acting in the plane perpendicular to the crack edge, i.e., both of these are plane strain conditions. In fig. 13.17*c* the crack is sheared in a direction parallel to the crack edge—a state of anti-plane strain. These are known as Mode I, Mode II, and Mode III deformations, respectively, and for each there is a critical value of the stress intensity factor, namely, K_{IC}, K_{IIC}, and K_{IIIC}, respectively.

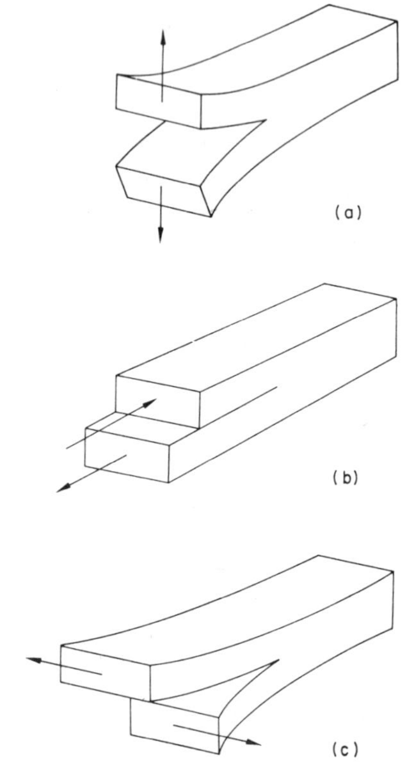

Fig. 13.17.—The three modes of crack extension:
(*a*) Mode I—opening or tensile mode
(*b*) Mode II—forward shear mode or edge sliding
(*c*) Mode III—parallel shear mode or tearing

13.23. Geometrical effects

The stress intensity factors discussed apply to a single straight crack in a very large sheet of material. In a specimen or component of finite size, the stress pattern around the crack can be calculated by stress analysis techniques and the relation of actual failure stress to the appropriate critical stress intensity factor determined. There will be a compliance function correcting for the geometrical shape of the specimen and a flaw shape parameter allowing for the geometry of the flaw so that

$$K = \sigma \sqrt{(\alpha \pi c)}$$

Values of α have been derived for many geometries and are given in various publications. If the flaw shape is unknown, one would have to be assumed, e.g., a surface crack might be semi-elliptical of width $2c$ and depth a, the flaw shape parameter then depending on $a/2c$. Hence, provided that the plastic zone at the crack tip is small relative to the crack length and if the appropriate K_c is known, one can calculate either the safe working stress for a component with flaws (of which the maximum size is known) or the maximum size of flaws permissible for a given working stress.

13.24. Determination of K_{Ic}

The fracture toughness of a material can be measured by a test on a specimen of finite size. One of the standard tests to determine K_{Ic}* utilizes a specimen of the proportions shown in fig. 13.18. After machining, the specimen is loaded, via the holes, in tension in a fatigue machine (see Section 14.7) until a fatigue crack has propagated a definite distance from the bottom of the slot. This crack will be of atomic dimensions at its root. The specimen is then loaded in tension until the crack propagates, this being determined either acoustically or by recording changes in the distance between the notches at the open end of the slot. From the failure load and the thickness of the material between the end of the fatigue crack and the back surface, the value of K_{Ic} is calculated. For high strength materials, the dimensions of the specimen shown in fig. 13.18 are suitable, but with more ductile materials larger specimens have to be used to ensure that a plane strain condition exists at the crack tip. In the case of mild steel, specimens 150 mm or more in thickness have to be used and the results could only be applied to structures of like thickness.

If the plastic zone size is so large that the size of specimen needed to get a K_{Ic} value is impracticably large, a test method taking account of non-linear plastic behaviour of the whole specimen is needed. Such methods exist but are not yet standardised and are beyond the scope of this book.

13.25. Relationship of design to fracture

The relationship between applied stress and flaw size for failure would be as shown in fig. 13.19. If the design of a component is considered for an operating stress equal to 0·5 of the yield stress, the critical flaw depth to cause failure would be c_d. Therefore provided that no defect is present of

* B.S. 5447: 1977. *Methods of test for plane strain fracture toughness* (K_{Ic}) *of metallic materials.*

depth greater than c_d failure should not occur on initial loading. If in a proof test the component were loaded to a stress above the normal operating stress and the test was successful, then there could not be a flaw present of size greater than c_p. During service life crack growth of the order of $(c_d - c_p)$ could be tolerated before failure. Hence it is important to know whether cracks might grow and the rate of growth of flaws under operating conditions. Such subcritical flaw growth can occur due to fatigue and due to stress corrosion which are considered in Sections 14.12 and 18.9, respectively.

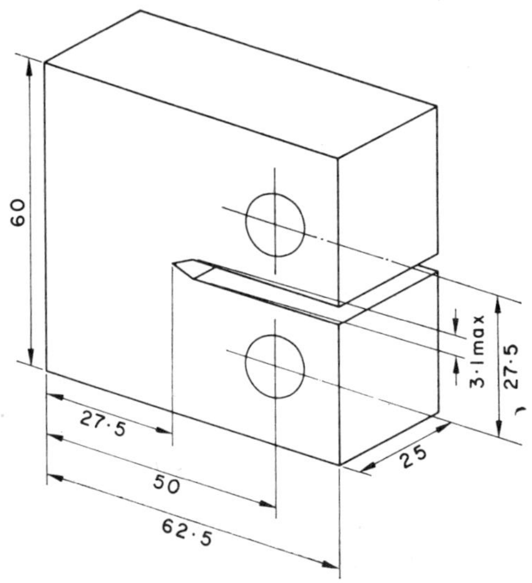

Fig. 13.18.—Standard tensile specimen for determination of K_{Ic}. Dimensions are in millimeters

Example

A thick plate of a high strength steel contains a crack of total length $2c = 10$ mm which runs perpendicular to the applied stress. For this steel, $K_{Ic} = 105$ MN m$^{-3/2}$ and the yield stress = 700 MPa. If the applied stress is increased steadily, will the plate fail by general yielding or by brittle fracture?

Answer

Stress to cause brittle failure

$$= \sigma_c = \frac{K_{Ic}}{\sqrt{(\pi c)}} = \frac{105 \times 10^6}{\sqrt{(\pi \times 5 \times 10^{-3})}} \text{ Pa}$$

$$= 840 \text{ MPa}$$

This is greater than the yield stress, so the plate will fail by yielding.

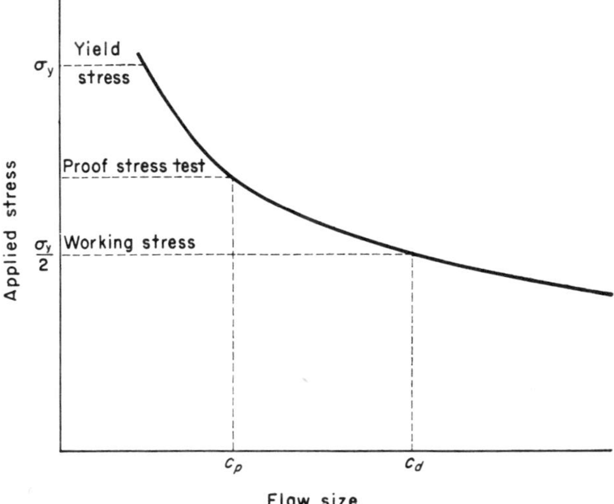

Fig. 13.19.—Applied stress—flaw size relationship

QUESTIONS

1. Define the terms *true stress, nominal stress, true (logarithmic) strain, nominal strain.*

The tensile stress–strain curve of a metal may be represented by $\sigma = A\varepsilon^b$, where σ is the true stress and ε the nominal strain. If a hydrostatic pressure p be superimposed on the material during the test the stress–strain curve is displaced downwards by p, its shape remaining unchanged. Describe the effect of such a pressure upon the value of the strain at which necking begins and hence comment upon the fact that, in wire drawing, it is possible to extend a metal by an amount which greatly exceeds the normal elongation in a tensile test. [MST]

2. Minimum values of yield point, tensile strength and elongation are normally quoted in steel specifications. Discuss the value of these data in selecting steels for given applications, and indicate what additional information is obtained by performing notch impact tests. [MST]

3. Give an account of the phenomenon of *brittle fracture* in metals and explain what is meant by *transition temperature.*

Describe a laboratory test to evaluate the ability of a steel to resist fracture in this manner, and discuss the practical relevance of such tests. [MST]

4. How has the Griffith formula been modified for materials which show macroscopic brittle behaviour but in which there is local plastic deformation in the vicinity of the crack tip?

A sheet of maraging steel has a tensile strength of 1950 MPa. Calculate the percentage reduction in strength due to the presence in the sheet of a crack 4 mm long and perpendicular to the stressing direction. For this steel, $E = 0{\cdot}2 \times 10^6$ MPa, $\gamma_s = 2$J m^{-2}, and γ_p, the work of plastic deformation of each surface at the crack tip, is 2×10^4 J m^{-2}. [E]

5. A cylindrical pressure vessel of diameter 3 m and length 9 m, with closed ends, is to be constructed using butt welded steel plates which are 0·03 m thick and approximately 1 m square. It must be designed to contain a pressure P without failure by yielding or by brittle fracture. Yield occurs when the equivalent tensile stress equals the yield stress, i.e., when

$$(\sigma_1 - \sigma_2)^2 + (\sigma_2 - \sigma_3)^2 + (\sigma_3 - \sigma_1)^2 = 2\sigma_y^2.$$

The butt welds joining the plates are known to contain thumb nail (semi-circular) cracks with a maximum depth of a. For such cracks $K_I = 1{\cdot}128\sigma\sqrt{(\pi a)}$ where σ is the tensile stress across a crack, and brittle failure will occur when

$$K_I \geqslant K_{Ic}.$$

Three steels are available for constructing the pressure vessel. Their yield strengths and fracture toughnesses are:

Steel	σ_y MPa	K_{Ic} MN m$^{-3/2}$
HY 140	965	280
T 1	690	180
HY 180	1240	180

Construct a plot of maximum pressure against crack depth a for each steel, showing the region of P and a which is safe. If non-destructive testing can detect cracks of depth $a_1 \geqslant 20$ mm (allowing welds with larger cracks than this to be repaired) which steel gives the greatest margin of safety? If a more refined technique will detect cracks of depth $a_2 \geqslant 2$ mm, which steel offers the greatest margin of safety? [E]

CHAPTER 14

Fatigue

14.1. Failure due to fatigue

Certain machine components and other structural parts subjected to loads which are repeated a large number of times, may break abruptly without any permanent deformation to herald the fracture, although these components will safely withstand even greater loads if applied once only in a steadily increasing manner, and under such conditions will show considerable plastic deformation.

As an example, a particular steel (containing 0·5% carbon and 5% nickel) is found to have a tensile strength of 835 MPa. Therefore it can be loaded in tension to a stress of 600 MPa with no fear of fracture. If this load is applied repeatedly, it is found that fracture may occur after about 10^4 loadings. Or an even lower load of 450 MPa may fracture the specimen if applied 10^5 times. This phenomenon is known as *fatigue*.

As repeated loading is encountered very frequently in engineering practice, it is necessary for a lot of attention to be paid to fatigue.

As in the case of toughness, the performance of a particular material in a tensile test is not a reliable guide to its behaviour under fatigue loading conditions. This is evident from the following results for two different types of alloy.

Alloy		*Tensile strength* (MPa)	*Stress to cause fracture after* 10^8 *loadings* (MPa)
En 12	(1% Ni steel)	590	290
RR 88	(Al+1% Cu, 2·7% Mg, 0·5% Mn, 5·3% Zn)	590	155

14.2. Characteristic of fatigue failures

The surface of a fatigue fracture often exhibits two zones, one a glossy smooth surface and the other a crystalline or fibrous surface as found in a broken Izod specimen. The crack usually has its origin in a minute fissure which progressively enlarges until insufficient sound metal remains in the section to support the applied load, whereupon sudden fracture occurs.

The glossy smooth surface indicates that there has been rubbing or fretting of the surfaces of the crack against one another during its development. The crystalline or fibrous zone shows the area of the instantaneous final fracture. The crystalline appearance is more commonly found.

If the fracture has been brought about by a number of periods of application of stress separated by periods of rest, the surface of the

Fig. 14.1.—Fatigue fracture in circular shaft showing lines of arrest

fatigue crack is divided by a series of curved lines known as *lines of arrest*. These are thought to be due to plastic deformation due to the stress concentration at the bottom of the notch. Oxidation of the crack may also occur and help to distinguish the lines of arrest. Lines of arrest may be seen in fig. 14.1.

The pattern of the lines of arrest may assist in identifying the point of origin of the fatigue crack.

Fatigue cracks tend to originate from sharp corners, etc., and follow directions perpendicular to the maximum tensile principal stress, typical paths in two components being shown in fig. 14.2.

14.3. Detection of fatigue cracks

If a fatigue crack can be detected before final failure has occurred, it may be possible to replace the damaged part and so avoid the consequence of a fracture.

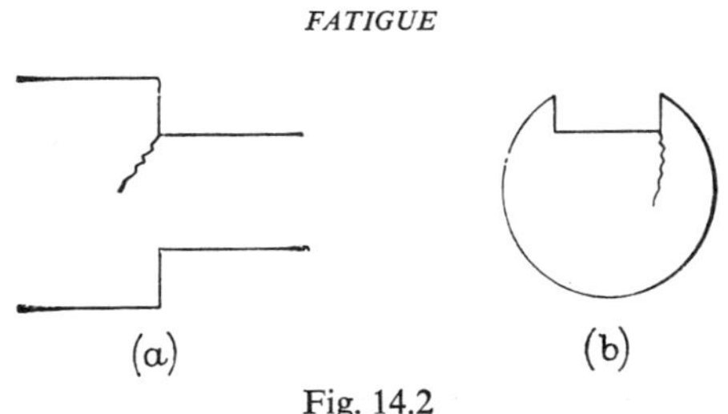

Fig. 14.2

(*a*) Longitudinal section of shaft with sharp reduction in diameter
(*b*) Transverse section of splined shaft
Most probable path of a fatigue crack is shown in each case

The cracks, which are initially very fine, are detectable when about 85 to 90% of the total life of the component has been reached. Hence, once a crack is detected, it is expedient to replace the component at once. If the crack has reached the stage at which it can be seen by the naked eye, the life of the component is almost at an end.

A variety of methods for detecting cracks are available. Surface cracks may be rendered visible by painting the surface with a suitable liquid and then wiping off the excess. Any liquid which has penetrated the crack will subsequently emerge and may be observed in a variety of ways, for example, fluorescent liquids will show up in ultra-violet light. Acoustic and ultrasonic methods may detect both surface and subsurface cracks. An example of the acoustic method is the tapping of railway rolling stock wheels and listening to the " ring ".

14.4. Mechanism of fatigue failure

Under repeated stressing, slip occurs in crystal grains, the slip planes of which are favourably oriented in or near planes of principal shear stress and which are so related to their neighbouring grains that slip is permissible under the applied stress. For example, grains which have a free surface on one side can slip more easily.

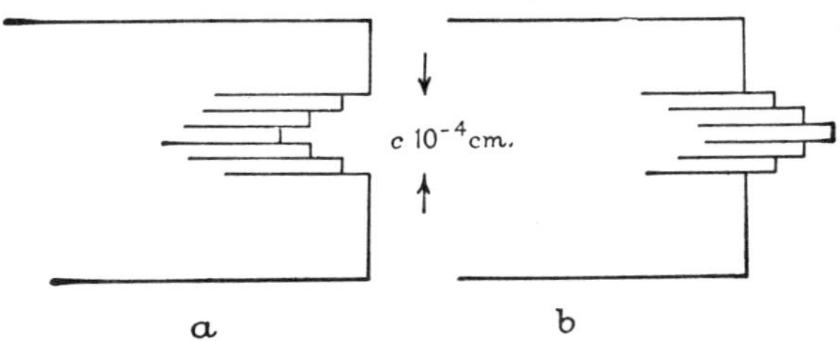

Fig. 14.3.—(*a*) Notch-like contour and (*b*) peak produced at surface by to-and-fro slip movements on adjacent slip planes. [Wood]

Slip may occur one way on one slip plane and the other way on an adjacent slip plane during the reverse half cycle, reverse slip on the first slip plane being inhibited by local work hardening. Such repeated slip on neighbouring slip planes may cause grooves or extruded tongues of metal to form as in fig. 14.3. Extruded tongues have been found by examination under the electron microscope and have an appearance which is shown schematically in fig. 14.4. The grooves cause local stress concentrations and eventually fatigue cracks develop from them.

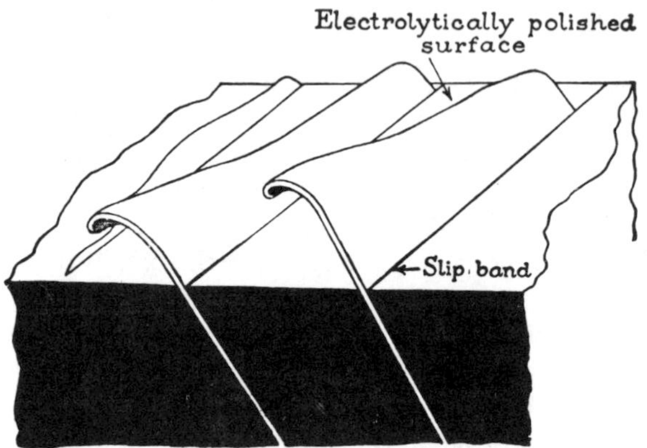

Fig. 14.4.—Diagrammatic representation of slip-band extrusions in aluminium-copper alloy. [Forsyth]

14.5. Fatigue crack growth

In a simple specimen loaded in push-pull, this initial crack growth is in a plane which lies at about 45° to the direction of the applied stress, i.e., it is spreading due to shear stresses. Later the crack turns on to a plane perpendicular to the applied stress and continues to grow until fracture of the remaining material occurs (fig. 14.5*a*). These two stages of crack propagation are designated Stage I and Stage II, respectively. It will be observed that these are, respectively, Mode II and Mode I crack extensions (see Section 13.22).

In the case of torsion applied to a cylindrical tube, there are shear stresses on longitudinal and transverse planes and Stage I cracking occurs on one or other of these planes. The crack extends through the wall thickness by Mode III and along the plane by Mode II. Eventually the manner of cracking may change to Stage II with cracks growing in Mode I on planes at 45° to the longitudinal axis (fig. 14.5*b*).

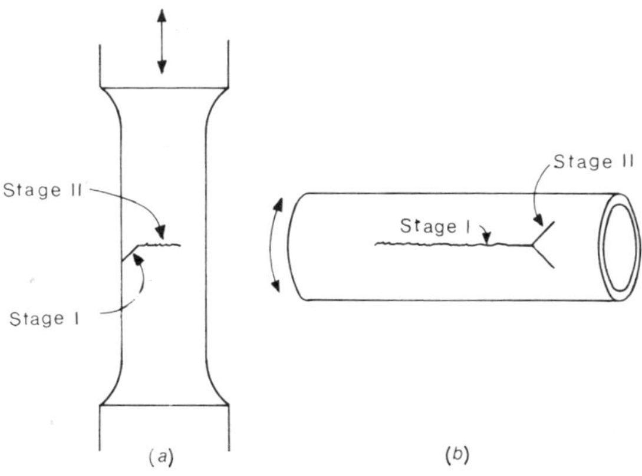

Fig. 14.5.—Directions of fatigue crack growth in (*a*) a specimen subject to axial push-pull, (*b*) a tube subjected to reversed torsion

In a tube subject to a combination of axial loading and torsion the Stage I cracks will again follow the plane of maximum shear stress.

14.6. Fatigue testing‡

As components, in general, are subjected to more than one form of load at the same time, a complete range of fatigue testing must cover a large number of combinations.

The simple type of testing would be the application of alternating tension and compression with or without a simultaneous static direct stress.

The usual form of load cycle is one with a range of stress that remains constant throughout the test, the shape of the load–time curve being either sinusoidal as in fig. 14.6 or triangular. The first curve in fig. 14.6 shows an alternating stress about a mean value of zero, and the other curves show a mean value that is not zero. If the semi-range of the alternating stress is σ_a and the means stress is σ_m, then the stress varies between the limits of $\sigma_{max} = \sigma_m + \sigma_a$ and $\sigma_{min} = \sigma_m - \sigma_a$.

The form of cycle can be defined by

$$R = \frac{\sigma_{min}}{\sigma_{max}}$$

so that $R = -1$ for a reversed cycle and $R = 0$ for a pulsating load.

‡ B.S. 3518. *Methods of Fatigue Testing.*
Part 1: 1962. *General principles.*
Part 2: 1962. *Rotating bending fatigue tests.*
Part 3: 1963. *Direct stress fatigue tests.*
Part 4: 1963. *Torsional stress fatigue tests.*

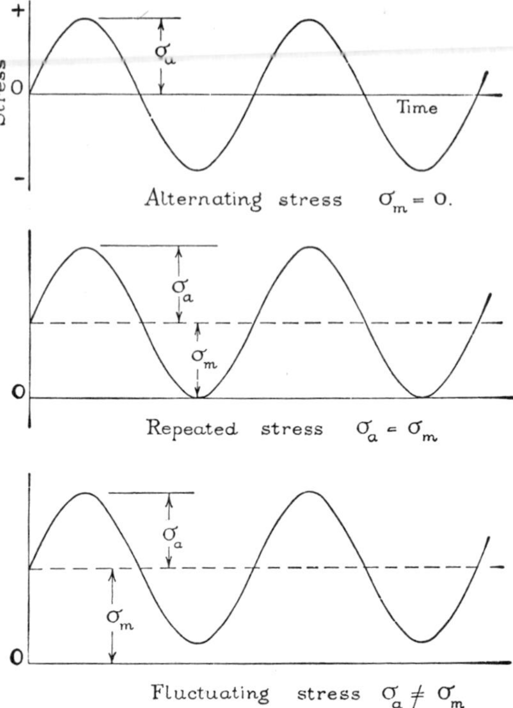

Fig. 14.6.—Forms of stress–time curves in fatigue loading

Similar series of tests can be carried out with torsional loading to give an alternating shear stress.

To simulate more closely actual conditions of service life, it might be necessary to carry out tests combining an alternating and a static stress of different types, e.g., alternating torsion with static tension. Again the two types of stress might be applied as alternating loads of different frequencies.

An example of this is the complicated series of loadings which are carried out on complete aircraft fuselages to simulate flight conditions. When on the ground the weight of the wings causes tension in their upper surfaces. When in flight, the weight of the aircraft is supported by the wings so that there is compression in the upper surface. This represents a static load to which must be added an alternating vertical load to simulate the effect of wind gusts while in flight. If the aircraft has a pressurized cabin, the pressure inside the fuselage is varied at a much lower frequency to represent the ascent into a rarified atmosphere once each flight. In

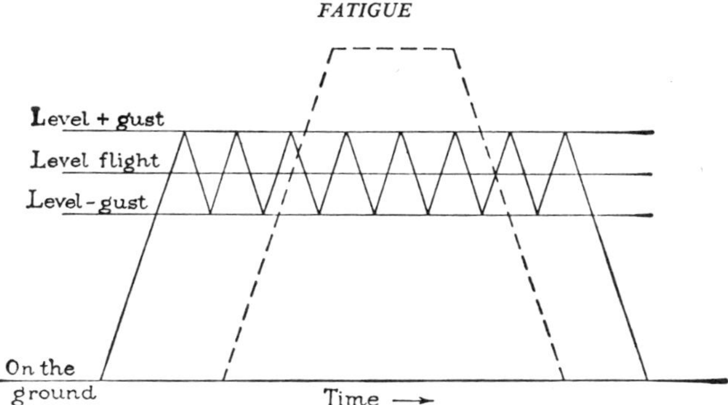

Fig. 14.7.—Typical loading curve for flight simulation in aircraft fuselage
——— wind load, – – – – –cabin pressure

certain regions of the fuselage, the stresses will be due to both these loadings. A typical simple cycle of loading is shown in fig. 14.7.

14.7. Fatigue testing machines

The principles of a few only of the more common types of machine for direct tension-compression testing are described in this section.*

In the rotating bending machines, a specimen of circular cross-section is rotated while a bending moment is applied in a fixed plane through the axis of rotation. The moment may be applied as shown in fig. 14.8 either

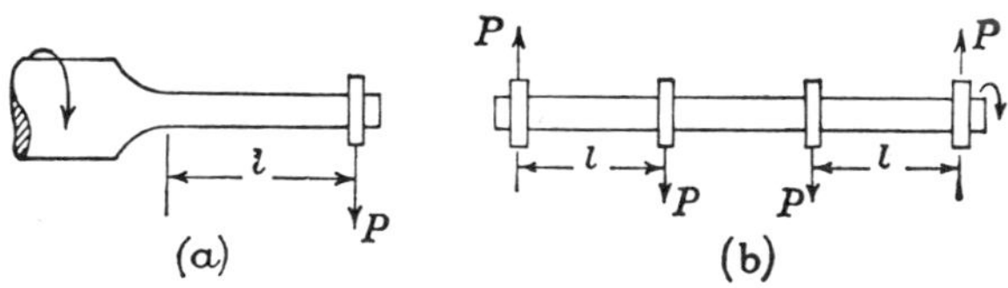

Fig. 14.8.—(*a*) Cantilever and (*b*) four-point loading for rotating bending fatigue tests. In each case the maximum bending moment is Pl

as an end load to a cantilever or by four-point loading. In the latter case there is a uniform bending moment over a length of the specimen. The stress in the outer fibre will alternate between a tensile stress and a compressive stress about a mean value of zero once each revolution and will have a sinusoidal wave-form.

For the application of a static stress simultaneously with the alternating stress, a more complicated type of machine is needed. The static prestress may be applied by a spring between the specimen and the fixed

* A large number of types of testing machines are described in *Manual on Fatigue Testing*, American Society for Testing Materials, Philadelphia, 1949.

anchorage, the alternating load being applied to the end of the specimen adjacent to the spring either by electromagnets or out-of-balance rotating masses (figs. 14.9 and 14.10).

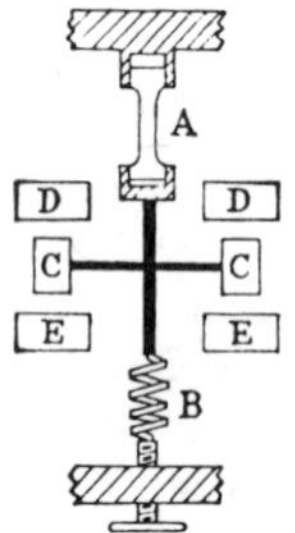

Fig. 14.9.—Haigh type fatigue machine. Specimen A given mean stress by tension spring B and alternating stress by attraction of iron C to alternately excited electromagnets D and E

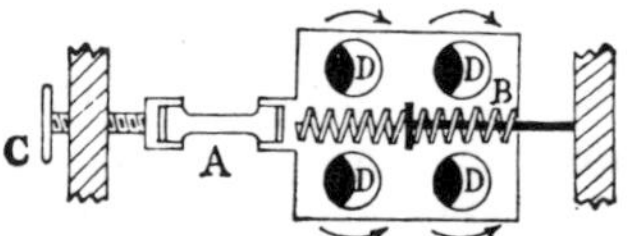

Fig. 14.10.—Schenck type fatigue machine. Specimen A given mean stress by tension spring B using pre-tensioning wheel C. Unbalanced rotating masses D give oscillatory force

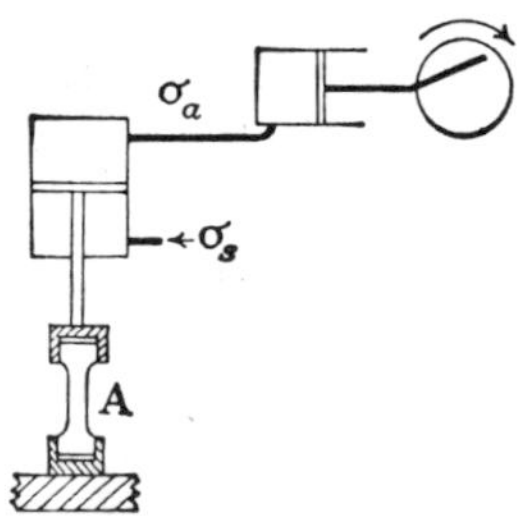

Fig. 14.11.—Losenhausen type pulsating machine. Specimen—A

Earlier types of hydraulic machines (fig. 14.11) have a piston connected to one end of the specimen, an alternating pressure being applied to one side of the piston and a static pressure to the other side. The piston is connected to the specimen. The alternating pressure is derived from another cylinder in which is a driven piston, the stroke being varied to give the required stress amplitude on the specimen.

The most modern machines are hydraulically operated with servo control, usually giving push-pull loading, though there are also special machines combining tension–compression with torsional loading or tension–compression loading on two mutually perpendicular axes. The controlled variable can be selected to be either the load on the specimen or the extension of the specimen or the strain in some part of the specimen. This variable is constrained to follow the control signal which may be derived from a simple function generator or from a more complicated predetermined programme designed to simulate the service life of a component.

14.8. Results of fatigue tests

In completing a test cycle on a material, it is necessary to test different specimens with different values of alternating stress and for each value determine the number of cycles of stress that can be applied before fracture. The number of cycles to fracture is known as the *fatigue life N.*

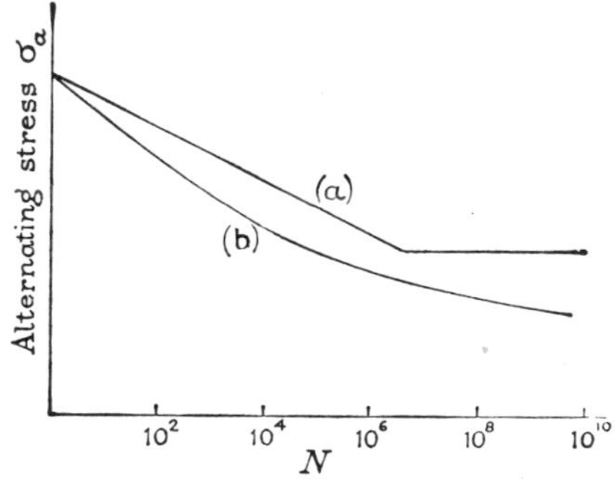

Fig. 14.12.—*S–N* curves (*a*) for a material showing a fatigue limit, (*b*) for a material which does not show a fatigue limit

The alternating stress is the *fatigue strength* for this life. As the stress range is reduced, for some materials a value is reached at which fracture will not occur. This value is known as the *fatigue limit.*

A plot of the semi-range of alternating stress σ_a against N on a logarithmic scale gives curves such as those in fig. 14.12, familiarly known as *S–N* curves. Some materials represented by curve (*a*) have a well-defined fatigue limit. For components made from these materials, which are intended to have only a limited life, it is possible to use a higher permissible design stress than would be the case if the design stress were based on the fatigue limit.

Other materials, aluminium alloys for example, give a curve of the shape shown as (*b*) in fig. 14.12. With such materials it is possible to design for a limited life only.

The scatter of results in fatigue tests is relatively great, the curves shown in fig. 14.12 representing the mean value of the life at each stress range. It is found that the variation in life may be as much as 3:1 for identical polished specimens and as great as 10:1 for complete components. Due consideration must be paid to this statistical variation in accepting a stress value for a limited life.

When the stress range $\Delta\sigma$ is plotted against N on logarithmic scales, the relationship within the elastic region is almost linear for stress ranges less than the fatigue limit. At higher stress ranges which take the material into the plastic region ($N < 10^4$) the relationship becomes curved as shown in fig. 14.13. The linear portion of the curve can be represented by

$$\Delta\sigma N^a = \text{constant}$$

which is known as *Basquin's law*, The exponent a has a value within the range $\frac{1}{8}$ to $\frac{1}{15}$.

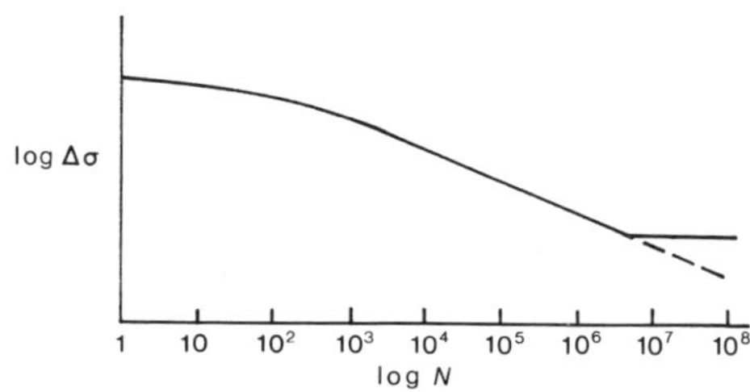

Fig. 14.13.—Log σ–log N curve extended to short lives

For $N < 10^4$, known as *low-cycle fatigue*, the life varies more rapidly with change of stress and it is usual to plot the strain range $\Delta\varepsilon$ against N as in fig. 14.14. This shows that the curve becomes asymptotic to two straight lines. The strain range can be separated into the plastic part $\Delta\varepsilon_p$ and the elastic part $\Delta\varepsilon_e$ such that

$$\Delta\varepsilon = \Delta\varepsilon_e + \Delta\varepsilon_p$$

It is found for most metals that the relationship in the plastic region is

$$\Delta\varepsilon_p N^b = \text{constant}$$

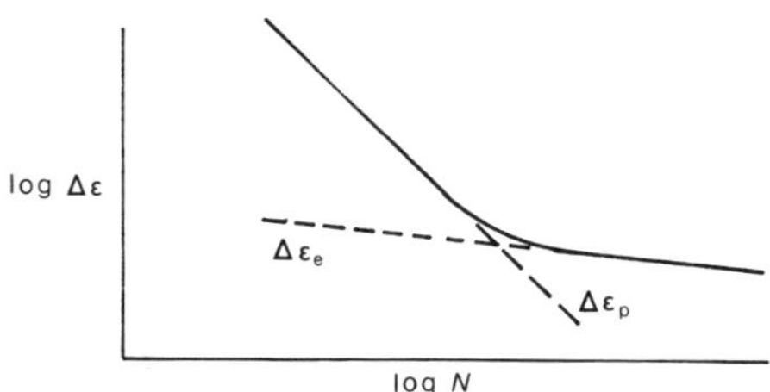

Fig. 14.14.—Log ε–log N curve showing separation into elastic and plastic portions

This is known as the *Manson–Coffin law* and b has a value near 0·5. The complete $\Delta\varepsilon$–N curve is the sum of the two components as shown in fig. 14.14.

14.9. Effect of mean stress

For any value of R or value of the mean stress σ_m, it will be possible to determine a similar S–N curve. As the value of σ_m is raised, the curve will appear lower on the diagram as in fig. 14.15*a*.

The variation of fatigue limit with mean stress is commonly represented on one of three types of diagram:

Rŏs diagram on which $\sigma_{max} = \sigma_m + \sigma_a$ is plotted against $\sigma_{min} = \sigma_m - \sigma_a$.
Haigh diagram on which σ_a is plotted against σ_m.
Smith diagram on which both σ_{max} and σ_{min} are plotted against σ_m.

When σ_m is zero, the value of σ_a is the fatigue limit in the absence of a mean stress, and when σ_m equals the tensile stress, then σ_a is zero. The points A and B corresponding to these two values, respectively, are marked on the three types of diagram in fig. 14.16.

On a basis of certain experimental values for intermediate points, Goodman proposed that the intermediate values should lie on a straight line between A and B. From a study of other results, Gerber proposed that the relationship should be a parabola with its apex at A and passing through B. These curves are shown in the three diagrams in fig. 14.16. It will be noted that the Rŏs diagram is merely the Haigh diagram rotated through 45°. A further method, proposed by Soderberg, is to use a straight-line relationship with the yield or proof stress as the limiting value for the mean stress corresponding to zero alternating stress. Whereas some experimental results for mild steel follow closely the Gerber curve, and some for naval brass follow the Goodman line, in the majority of cases results lie between the two proposed curves. The Soderberg line is usually somewhat conservative and hence safe to use.

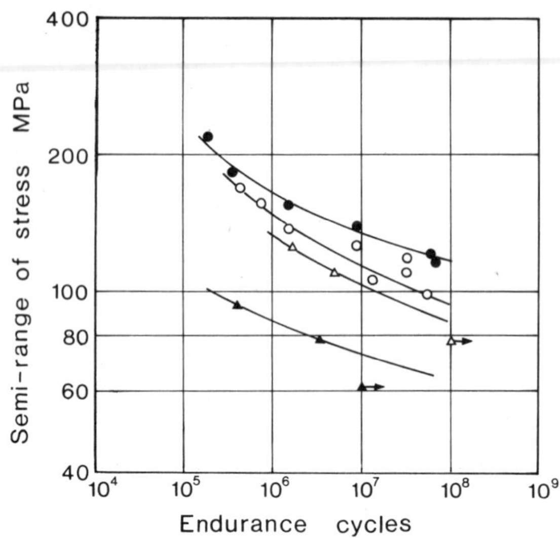

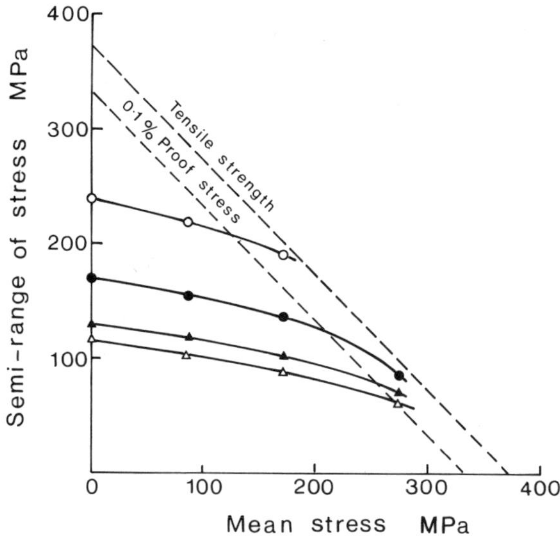

Fig. 14.15.—Results of fatigue tests on an aluminium alloy

(*a*) *S–N* curves for various values of σ_m: ● 0, ○ 90, △ 170, ▲ 270 MPa. Points with arrows signify that tests were discontinued at 10^7 and 10^8 cycles, respectively, failure not having occurred.

(*b*) Haigh diagram showing curves for various lives ○ 10^5, ● 10^6, ▲ 10^7, △ 5×10^7 cycles. [after Woodward, Gunn, and Forrest]

For a material which does not show a fatigue limit, curves of the fatigue strengths for various lives can be shown on any of these diagrams. An example is given in fig. 14.15*b*.

14.10. Variation of fatigue limit

The lack of any close relationship between the fatigue limit and the tensile strength of different materials has already been mentioned (p. 241). An analysis of the results of a large number of tests has produced the frequency distribution curves shown in fig. 14.18. Although the endurance ratio (that is, the ratio of the fatigue limit to the tensile strength) has a mean value of about 0·42, there is a considerable spread of values. The endurance ratio for any material must be determined by experiment. Its value cannot be guessed with any certainty.

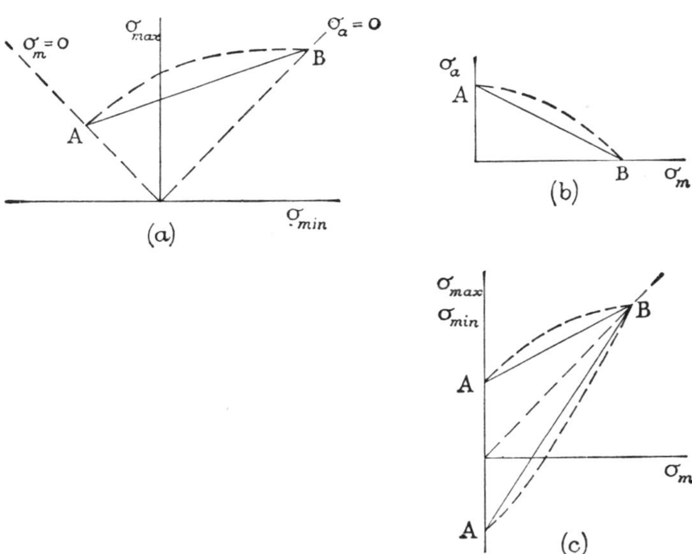

Fig. 14.16.—(*a*) Rŏs, (*b*) Haigh, (*c*) Smith diagrams for plotting fatigue test results, ——— and - - - - are Goodman's and Gerber's proposed curves. All diagrams are to the same scale

Most material variables which affect the tensile strength will also affect the fatigue strength but not necessarily to the same extent. Thus, increasing the grain size causes a greater reduction of fatigue strength than it does of tensile strength.

In low-cycle fatigue, an annealed material strengthens to a limited

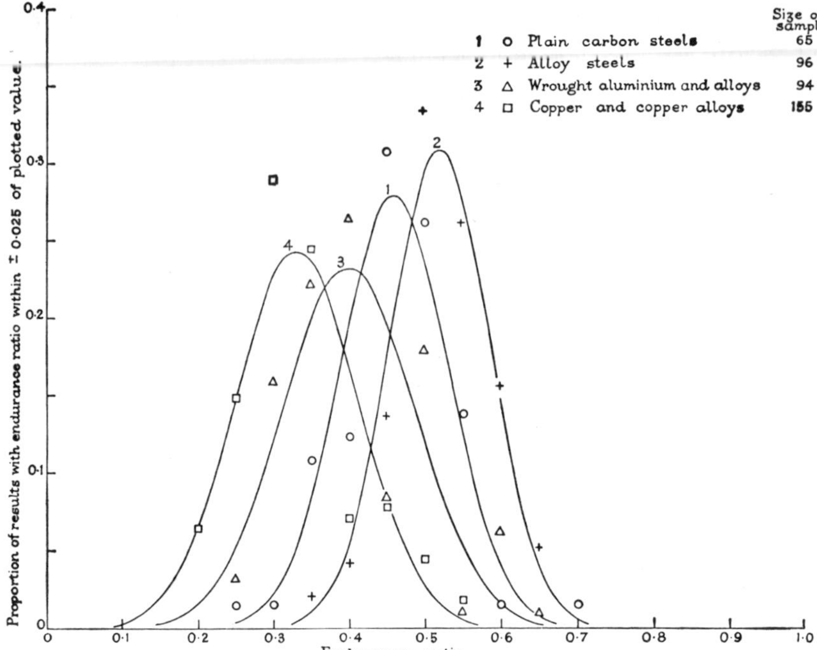

Fig. 14.17.—Frequency distribution curves for fatigue test results [An analysis of collected data from *Fatigue of Metals and Structures* by H. J. Grover, S. A. Gordon and L. R. Jackson (Thames and Hudson, London, 1956). The curves are normal distribution curves with the same mean values and variances as the samples]

extent while a work-hardening material softens. In either case, the material tends to the same saturation level giving a stable cyclic stress–strain loop.

In a two-phase material, the distribution and shape of second-phase particles affect the fatigue behaviour. In a precipitate-hardened alloy, the dispersed second-phase particles may coalesce giving an averaging effect and so reduce the strength. For more massive second-phase regions, rounded particles have a less weakening effect than plate or needle-like particles. For example, compare spheroidal graphite and ordinary cast irons (Section 8.11).

14.11. Cumulative damage effects

Many components will not be subjected to a regular repeated cycle but experience successive loadings of different amplitudes as described in

Section 14.6. It is not always practicable to carry out realistic tests and so a method of predicting the cumulative damage effects of variable pattern loading is needed. The simplest is the *Palmgren–Miner law.* This postulates that if a component undergoes n_1 cycles at a stress range for which the life would be N_1, n_2 cycles at a stress range for which the life would be N_2, etc., then failure would occur when

$$\frac{n_1}{N_1} + \frac{n_2}{N_2} + --- = 1$$

In practice it is found that the sequence of loading also matters. For example, if a large number of low stress range cycles precede a number of high stress range cycles, the summation will be greater than 1, whereas if the high stress range cycles come first, the summation will be less than 1. This is because the high stress range cycles initiate cracks almost immediately while the initiation is delayed in low stress range cycling. Also a single overload in an otherwise constant amplitude series of cycles causes a temporary retardation of crack growth. No satisfactory prediciton method that will allow for all loading patterns has yet been found.

14.12. Crack growth rates

Studies of crack propagation in various materials have shown that under steady loading conditions the rate of growth of a crack is often a unique function of the range of the stress intensity factor ΔK_I (see Section 13.19). A logarithmic plot of crack growth rate, dc/dN, where c

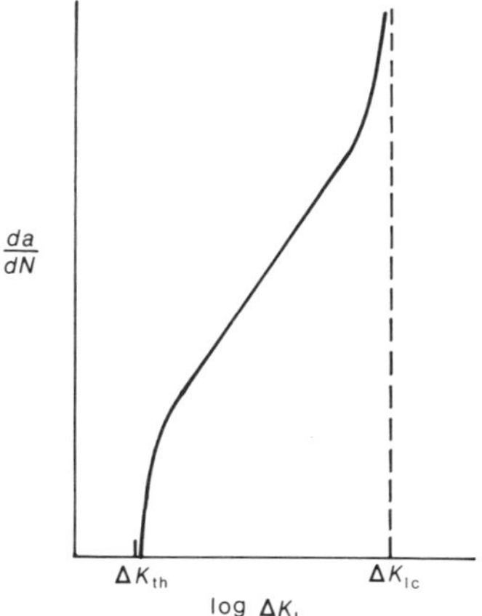

Fig. 14.18.—Crack growth rate as a function of stress intensity factor range

is the crack length at N cycles, against ΔK_I gives a sigmoidal curve (fig. 14.18). At one end there is a threshold stress intensity factor ΔK_{th} below which fatigue crack growth will not occur. At the other, catastrophic failure occurs when ΔK_I approaches K_{Ic}. Over a limited range, there is a parallel portion for which a power law relationship applies

$$\frac{dc}{dN} = C(\Delta K_I)^n$$

The value of n is usually within the range 2 to 10.

If a component containing a crack of known length is subject to a known range of loading, then from the crack growth relationship above, the number of cycles in which the crack may grow to the critical size, at which catastrophic failure will occur, may be calculated.

Example

A component is made of a steel for which $K_{Ic} = 54$ MN m$^{-3/2}$ and cracks grow under fatigue loading according to

$$\frac{dc}{dN} = 4 \times 10^{-13} \, \Delta K_I^4 \text{ m/cycle}$$

It is subject to an alternating stress of range $\Delta\sigma = 180$ MPa. The component is found to contain cracks of length 0·2 mm. Given that $\Delta K_I = \Delta\sigma\sqrt{(\pi c)}$, calculate the number of cycles of loading that would produce failure.

Catastrophic failure will occur when

$$\Delta\sigma\sqrt{(\pi c)} = 54 \text{ MN m}^{-3/2}$$

i.e.,

$$c_f = \frac{l}{\pi}\left(\frac{54}{180}\right)^2 = 0.0286 \text{ m}$$

$$\frac{dc}{dN} = 4 \times 10^{-13} \, \sigma^4\pi^2c^2 = 4{\cdot}14 \times 10^{-3} \, c^2$$

i.e.,

$$\frac{dc}{c^2} = 4{\cdot}14 \times 10^{-3} \, dN$$

Integrating from $c_0 = 0{\cdot}0002$ m to c_f gives

$$\frac{1}{c_0} - \frac{1}{c_f} = 4{\cdot}14 \times 10^{-3} \, N$$

or

$$N = \frac{10^3}{4{\cdot}14}\left[\frac{1}{0{\cdot}0002} - \frac{1}{0{\cdot}0286}\right]$$

$$= 1200000 \text{ cycles}$$

14.13. Effect of notches

If the geometry of a component is such that there is a change of section, a hole, or a notch, which of itself will have a stress concentration factor, then the fatigue limit will be lowered.

The stress concentration factor is defined in Section 13.6. Its theoretical value assuming elastic behaviour is denoted by K_t.

It would appear probable that as plastic deformation is not generally observed at a fatigue fracture, there would be no reduction of stress concentration as far as the varying stress is concerned. Hence if σ_0 were the fatigue limit with no stress concentration present and σ_k the corresponding fatigue strength when a stress raiser is present, we would expect that

$$\frac{\sigma_k}{\sigma_0} = \frac{1}{K_t}$$

In general this is not the case, σ_k/σ_0 being greater than $1/K_t$; that is, the reduction in fatigue strength is less than would be anticipated from theoretical considerations. The ratio σ_0/σ_k is called the *fatigue strength concentration factor* or the *fatigue strength reduction factor*, denoted by K_f. Now K_f may have any value between K_t and 1, the actual value depending upon the material, and the shape and size of the specimen. From this has been developed the *notch sensitivity index* which is defined as

$$Q = \frac{K_f - 1}{K_t - 1}$$

The value of Q will lie between 0 and 1, being 0 for a case which is not notch-sensitive, i.e., $\sigma_k = \sigma_0$, and 1 for a material which is 100% notch sensitive, i.e., $K_f = K_t$.

14.14. Relation of design to fatigue

Crack growth rates for any material can be related to the stress intensity factor range as described in Section 14.12 and are relatively insensitive to changes in the metallurgical structure of the material. Crack initiation, on the other hand, is more sensitive to metallurgical structure and especially sensitive to surface conditions.

Unless $Q = 0$, then it is important to reduce or avoid stress concentrations in the design of any part which will be subject to repeated loading. Blended curves should be substituted for sharp corners. Threaded bolts or rods are most likely to suffer fatigue damage from the roots of threads, so particular care must be taken to use overall sizes and thread shapes which will reduce the stresses to safe limits.

Any process that introduces residual compressive stresses into the surface layers delays the initiation of fatigue cracking and is beneficial. These include plastic deformation by such processes as shot or hammer peening or cold rolling and surface hardening by flame hardening, carburizing or nitriding (see Section 17.13). These last two processes would involve structural volume changes due to the diffusion into the surface layers of carbon or nitrogen and so induce high compressive stresses.

A polished specimen has a higher fatigue limit than a rough turned one.

14.15. Corrosion fatigue

When a fatigue test is carried out in the presence of a corrosive medium, there is a reduction in the fatigue strength for any particular life, and the fatigue limit may disappear. Even fresh water is sufficiently corrosive to have an effect. Also when two mating surfaces are subject to small relative motion, fretting corrosion occurs which may become a source of fatigue cracks. These phenomena will be dealt with at greater length in Chapter 18.

The usual protective coatings and corrosion inhibitors have a beneficial effect.

QUESTIONS

1. Discuss briefly how a connecting rod should be designed and manufactured so that it would have a high resistance to failure by fatigue. What tests should be performed to assist in the selection of a suitable material?

In order to assess the resistance to fatigue of an aluminium alloy, groups of 20 specimens were tested for up to 10^6 cycles at six different stress levels. The results were as follows:

Stress MPa	At 10^6 cycles	
	Broken	Unbroken
±165	0	20
±172	3	17
±179	8	12
±186	14	6
±193	18	2
±200	20	0

Determine the median fatigue strength and the fatigue strength for 97·7% survival of the specimens at 10^6 cycles. It may be assumed that 97·7% survival corresponds to two standard deviations. Arithmetic probability paper may be used if desired.

[MST]

2. It is required that a component, made from an aluminium alloy, should not fail by fatigue in less than 10^7 cycles whilst it is subjected to a tensile mean stress of 200 MPa.

Laboratory tests on the material under alternating fatigue conditions show that (*a*) the cyclic yield stress of the material is 400 MPa, and (*b*) when the range of cyclic stress $\Delta\sigma$, is 340 MPa the number of cycles to failure, N_f, is 10^8 and when $\Delta\sigma$ is 520 MPa, N_f is 10^5.

Assuming that the fatigue behaviour of the material, under alternating stress conditions, is given by

$$(\Delta\sigma)^b.N_f = C,$$

where b and C are material constants, calculate the permissible amplitude of cyclic tensile stress to which the component may be subjected.

Discuss the effect of surface finish on fatigue life and briefly describe surface treatments that enhance fatigue resistance. [E]

3. Briefly describe the various regimes that may occur in the fatigue lifetime of a component.

A fatigue crack may be stated to have been initiated in a particular material when a surface crack of length 10^{-6} m is created. The percentage of cyclic lifetime required for this stage may be calculated from the equation

$$1000N_i = 1{\cdot}421\ N_f^{1{\cdot}421}$$

where N_i is the number of cycles to initiate a crack and N_f is the total number of cycles to failure. Determine N_f for two specimens of this material tested at stress ranges $\Delta\sigma_1$ and $\Delta\sigma_2$ for which the N_i/N_f ratios are 0·01 and 0·99, respectively. If the crack at failure is 1 mm deep determine the *mean* crack propagation rate at $\Delta\sigma_1$ and the *mean* crack nucleation rate at $\Delta\sigma_2$.

Compare the differences in cyclic lifetime that may be expected for two speimens of the same material tested as follows:

Specimen 1

50% of the number of cycles to failure at $\Delta\sigma_1$ are initially applied at $\Delta\sigma_1$, followed by cycles at $\Delta\sigma_2$ until failure.

Specimen 2

50% of the number of cycles to failure at $\Delta\sigma_2$ are initially applied at $\Delta\sigma_2$, followed by cycles at $\Delta\sigma_1$ until failure.

Give reasons for your answers and state any assumptions used. [E]

4. Connecting rods for a marine engine are to be made of a cast iron for which $K_{Ic} = 25$ MN m$^{-3/2}$. Non-destructive testing will detect cracks or flaws of length $2a$ greater than 2 mm, and rods with flaws larger than this are rejected. Independent tests on the material show that cracks grow during fatigue such that

$$\frac{da}{dN} = 2 \times 10^{-15}\ (\Delta K_I)^3\ \text{m cycle}^{-1}$$

(K_I in MN m$^{3/2}$). The minimum cross-section of the rod is 0·01 m^2, its section is circular, and the maximum tensile force it experiences in service is 1 MN. You may assume that $K_I = \sigma\sqrt{(\pi a)}$, where σ is the tensile stress. The engine runs at 1000 rev min^{-1} and has a design life of 20000 h. Will the connecting rods survive? [E]

CHAPTER 15

Creep

15.1. Creep phenomena

In the preceding chapters it has been assumed that there exists for any material a stress–strain relationship which is independent of time. In a tensile test, on applying a load in the plastic region, there is an immediate extension, but this is followed in time by further extension. In mild steel, for example, at room temperature after applying an increment of load, the strain varies rapidly at first and then at a decreasing rate. In lead, the strain can continue for a long time at a steady rate even under very low stresses. This is seen in the slow movement of lead on sloping roofs and the gradual sagging of unsupported lead pipes. The strain–time variation at constant load is known as *creep*. At high temperatures, steel and other materials show this steady creep, which becomes of importance in components which are exposed to high temperature during service.

15.2. Creep testing machines

Creep properties of materials are determined by keeping specimens under constant stress for long periods during which strain measurements are made by a suitable extensometer.* The majority of the tests are conducted at high temperatures, necessitating a furnace surrounding the specimen and associated equipment to maintain the temperature constant within very close limits. Also to ensure constant loading, it is usual to house the apparatus in a temperature-controlled room.

The usual type of machine is a lever machine loaded by weights. One type of extensometer employed has two arms, one connected to each end of the specimen, which are parallel and long enough to extend beyond the furnace. The relative movement is determined, for example by the rotation of a small hardened steel rhomb carried between them, the rotation being measured by a telescope and scale.

15.3. Characteristics of creep curves

The extension with time during creep is greater at increased stress or

* BS. 3500: 1969. *Methods for Creep and Rupture Testing Metals*

temperature. A set of short-time creep curves for some lead specimens tested at room temperature is shown in fig. 15.1.

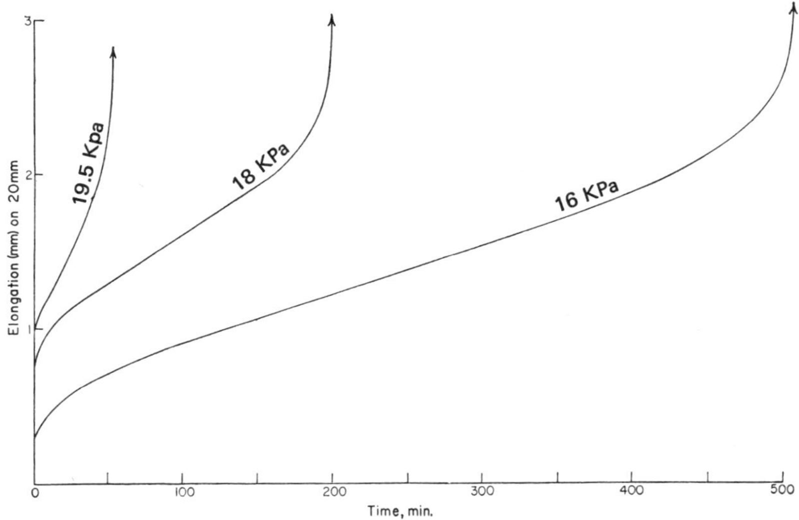

Fig. 15.1.—Creep curves for lead tested at room temperature

All creep curves are of the type shown idealized in fig. 15.2. There are four parts to the curve:

(i) An initial instantaneous extension.
(ii) A stage of creep at a decreasing rate.
(iii) A stage of creep at an approximately constant rate.
(iv) A stage of creep at an increasing rate ending in fracture.

Stages (ii), (iii), and (iv) are usually referred to as *primary*, *secondary*, and *tertiary* creep, respectively.

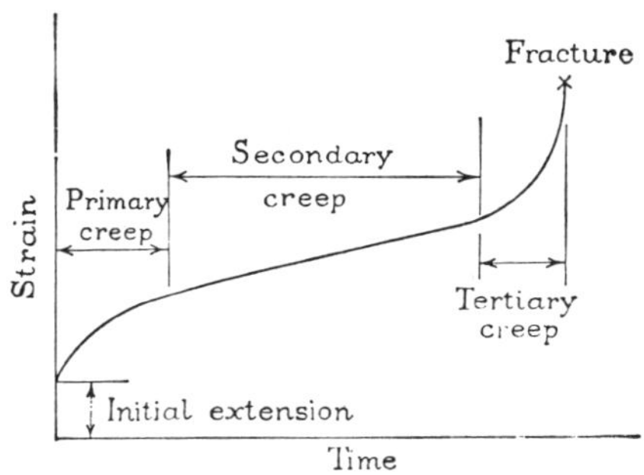

Fig. 15.2.—Idealized creep curve for test at constant load

The types of flow in stages (ii) and (iii) have been called *transient* and *quasi-viscous* flow, respectively.

If tests are conducted at constant stress, that is, the load is decreased as the cross-section decreases, the tertiary stage of increase of creep rate may not occur. In tests at constant load, the decrease of cross-section which accompanies the extension causes the stress to increase, so that in the tertiary stage the strain rate increases rapidly up to the point where final failure occurs by fracture or rupture (see Section 15.8).

15.4. Presentation of creep results

The variables involved are the strain, temperature and time. In two dimensions data are presented as:

(i) variation with temperature of the stress to produce a stated amount of strain or rupture in various times, and

(ii) variation with time of the stress to produce various amounts of strain or rupture at a stated temperature.

Examples for Nimonic 90 are shown in figs. 15.3 and 15.4.

15.5. Creep and design

As already referred to in Section 15.1, creep is an important factor to be considered in the design of structures and components that are subject to stress while at elevated temperatures, e.g., nuclear reactors, steam-turbine power plant, and gas turbines. Many of these have to be operational for

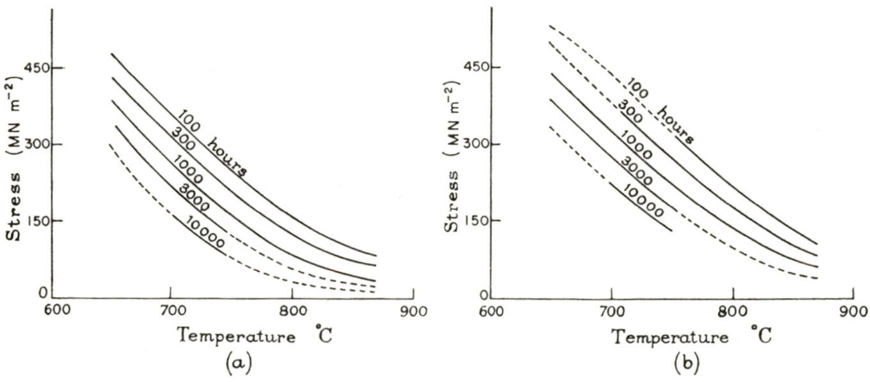

Fig. 15.3.—Typical creep properties of Nimonic 90. Derived isochronous curves. (*a*) Stress to produce 0·2% creep strain, and (*b*) stress to produce rupture in certain times at various temperatures

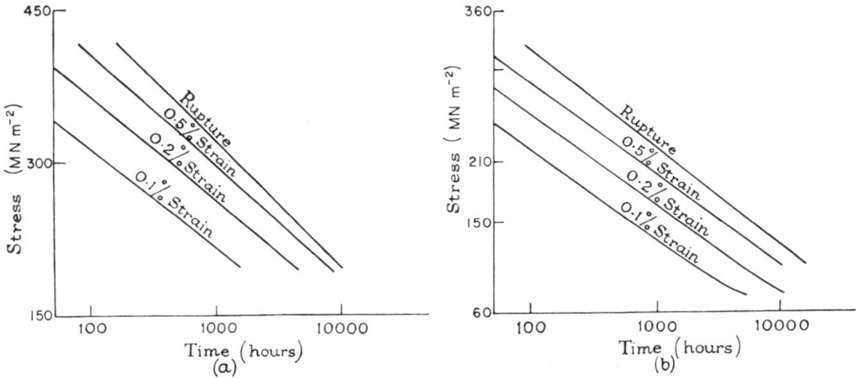

Fig. 15.4.—Typical creep properties of Nimonic 90. Stress-log. time curves. Stress to produce certain strains or rupture in various times at (*a*) 700 °C, (*b*) 750 °C [Figs. 15.3 and 15.4 redrawn from *The Nimonic Alloys: Design Data* (Henry Wiggin and Co. Ltd., Birmingham)]

periods of many years. To get the necessary information for satisfactory design, creep tests that would reproduce operating conditions of load and temperature are necessarily long term, so it would be desirable to predict the creep behaviour from short-term tests. Usually, this is done by the graphical extrapolation of steady-state creep rate data, plotted against stress on double logarithmic scales as in fig. 15.5. Such extrapolation may be an unreliable procedure. If, however, the mechanisms of creep are known, then it is possible to decide when such extrapolations would be reasonable.

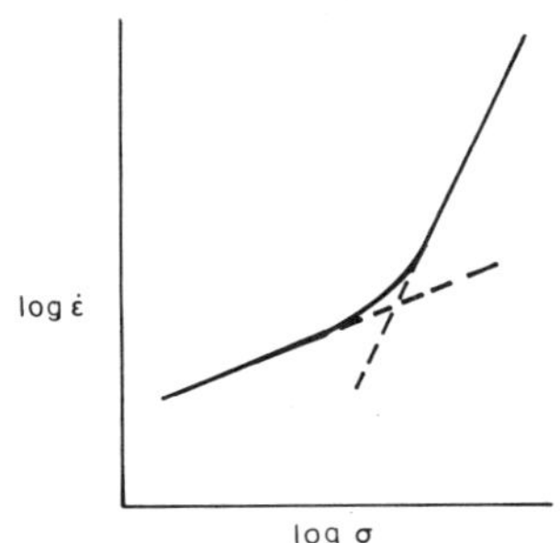

Fig. 15.5.—Steady-state creep rate as a function of stress at constant temperature

The dependence of steady-state creep rate, $\dot{\varepsilon}$, upon stress, σ, and temperature, T, can be determined by plotting log $\dot{\varepsilon}$ against log σ for constant T and log $\dot{\varepsilon}$ against $1/T$ for constant σ. Typical results would take the

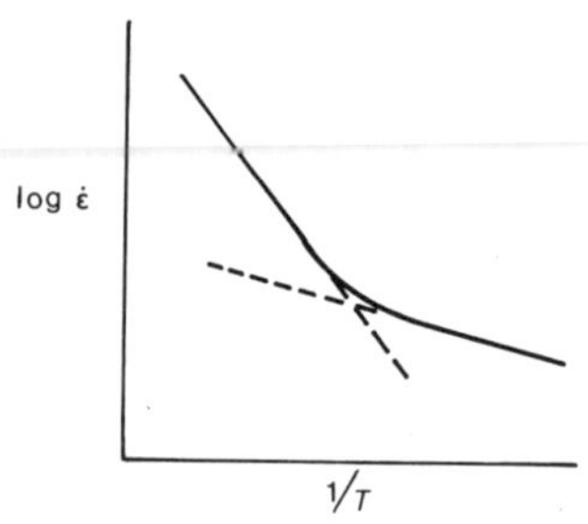

Fig. 15.6.—Steady-state creep rate as a function of temperature at constant stress

forms shown in figs. 15.5 and 15.6. These show that

$$\dot{\varepsilon} = a\sigma^n e^{-q/kT}$$

applies but that a, n, and q take different values for different regimes of stress and temperature. Presumably different mechanisms dominate the creep behaviour in each regime.

15.6. Mechanisms of creep

In discussing the various mechanisms, it is convenient to refer to the temperature as a function of the melting point temperature T_m on the absolute scale of temperature.

At temperatures below $\sim 0{\cdot}3\ T_m$ creep is governed by the ability of dislocations to bypass obstacles by cross slip or even to overcome the resistance of the lattice to dislocation movement.

At higher temperatures (above $0{\cdot}3\ T_m$), diffusion plays an important role. Diffusion within the lattice may be vacancy diffusion (see Section 9.11) or diffusion of interstitial atoms while grain boundaries act as high conductivity paths. When the material is stressed, atoms diffuse both through the lattice and along grain boundaries from the boundary regions which are less stressed to those which are more highly stressed in tension, as shown diagrammatically in fig. 15.7.

This process causes a continuous elongation of the grains in the direction of the maximum tensile stress. Diffusion flow results in creep behaviour for which $n \approx 1$. Obviously for a smaller grain size, the length of the diffusion path is smaller and the creep rate will be higher.

Another mechanism is the movement of dislocations to form dislocation cells, the size of which is smaller for higher stress. These cells form during primary creep. They may act somewhat as the grains do in diffusional flow but with the cores of the dislocations acting as high con-

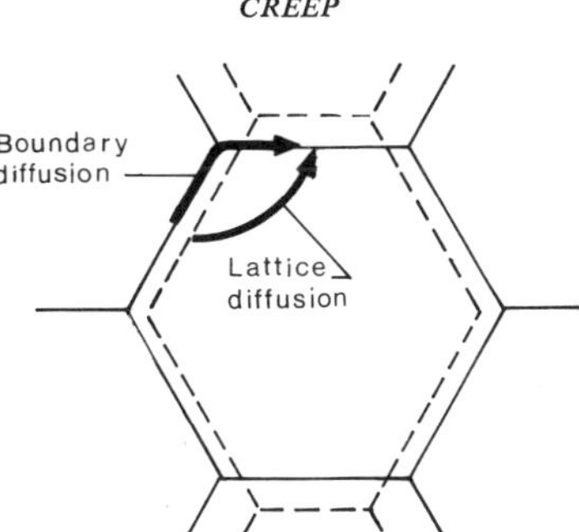

Fig. 15.7.—Diffusion paths in a strained crystal at elevated temperature

ductivity paths. Because the cell size is dependent upon creep stress, the macroscopic creep behaviour shows a higher value of n (usually 4–6) and the mechanism is known as *power-law creep.*

15.7. Deformation mechanism maps

While the form of the $\dot{\varepsilon}$, σ, T relationship for each mechanism can be predicted by theory, the exact values of the numerical constants must be determined by experiment. It is then possible to predict which mechanism will dominate at any value of σ and T. Ashby* has developed *deformation mechanism maps* which show as functions of stress and temperature the ranges over which each mechanism is dominant (fig. 15.8). Both stress and temperature are shown in non-dimensional form, the shear stress as a function of the elastic shear modulus, G, and temperature as a homologous temperature, i.e., T/T_m.

A further refinement is to add lines along each of which creep rate is constant as in fig. 15.9. The effect of a change of grain size by a factor of about 17 is clearly seen from the two diagrams.

If a deformation mechanism map has been prepared for a material under active consideration, then for a known stress and temperature condition it can be seen from the map what creep mechanism dominates and whether extrapolation to that condition from the results of short term tests is or is not likely to be reliable.

Such maps cover a wide range of deformation rates from the high rates ($\dot{\varepsilon} \sim 10\text{–}10^{-1}$ s^{-1}) used in metal-working processes such as rolling, forging, and extrusion to the low rates which must not be exceeded in structures that have to sustain loads for many years (a strain of 1 % would be reached in 30 years at a steady-state strain rate of 10^{-11} s^{-1}).

* M.F. Ashby, A first report on deformation-mechanism maps *Acta Met.*, Vol. 20, 1972, pp. 887–897.

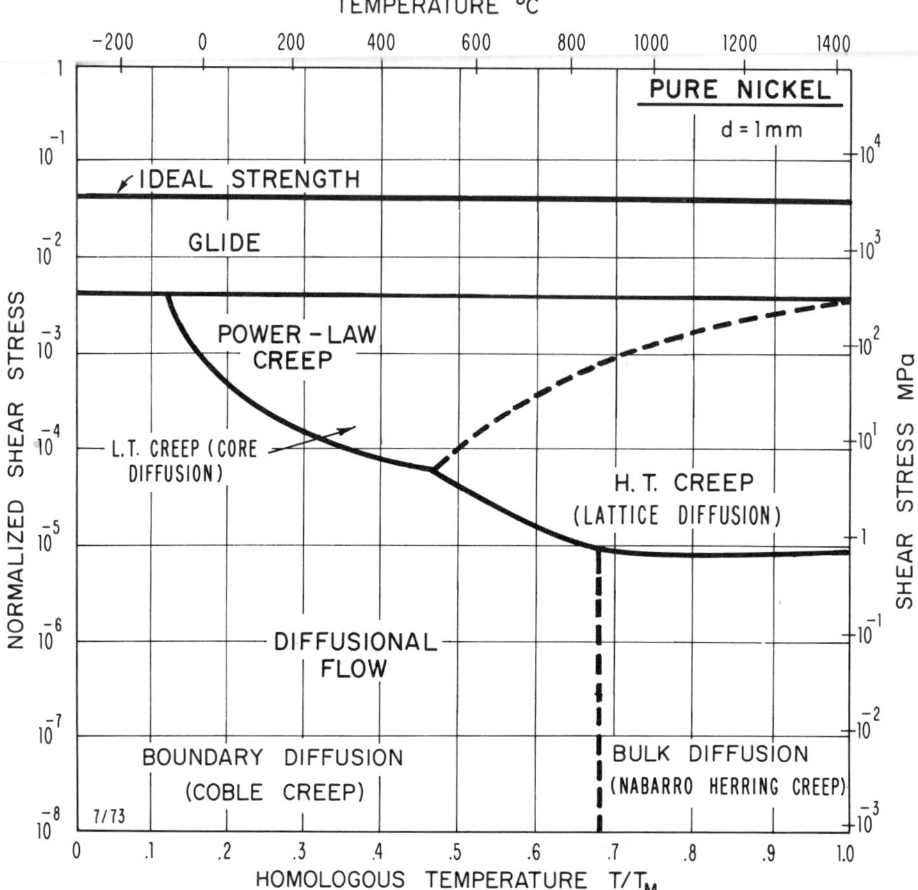

Fig. 15.8.—Deformation mechanism map for pure nickel showing dominant creep processes. [after Ashby]

15.8. Creep fracture mechanisms

When a material deforms rapidly by power-law creep, so that the time to fracture is short, the final fracture mechanism, *transgranular creep fracture*, resembles that for ductile fracture observed at lower temperatures (Section 13.3) where plasticity is almost rate independent. Holes nucleate at inclusions within the grains and grow during creep until they coalesce to give a fracture path.

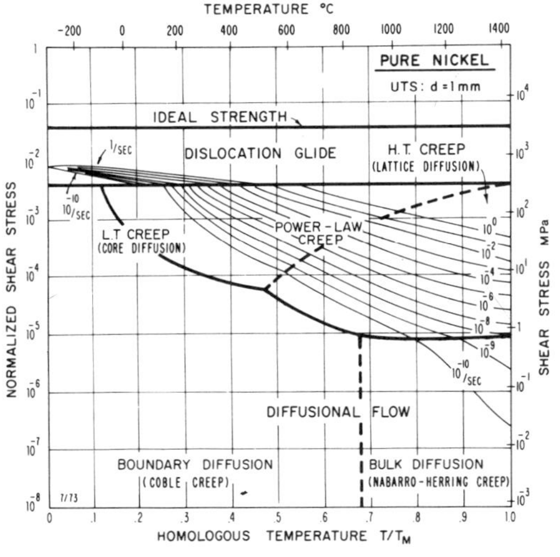

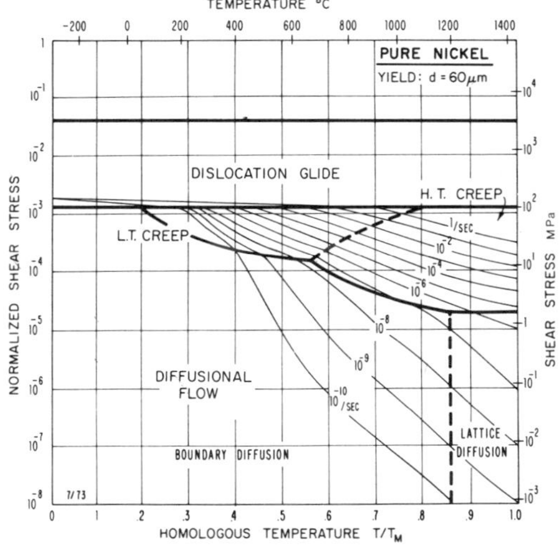

Fig. 15.9.—Deformation mechanism maps for two grains sizes of pure nickel showing contours of constant creep rates. [Ashby]

At high temperatures but at lower stresses so that the time to fracture is greater, a transition from transgranular to *intergranular, creep-controlled, fracture* is observed. Grain boundaries slide which stimulates the nucleation of cavities on the grain boundaries. These cavities grow, eventually linking to form fracture paths.

If no other fracture mechanism intervenes, a material loaded in tension will eventually become mechanically unstable and neck (see Section 12.4) or shear until the cross-sectional area has reduced to zero. This failure mechanism is known as *rupture*.

These mechanisms together with low temperature cleavage which is found in most crystalline solids, except face-centred cubic metals and their alloys (see Section 13.8) are classified diagrammatically in fig. 15.10.

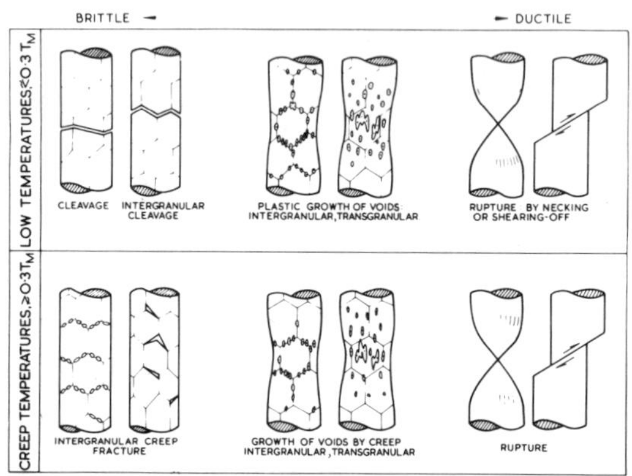

Fig. 15.10.—Simple classification of fracture mechanisms. The upper row refers to low temperature ($<0{\cdot}3\ T_m$) where plastic flow does not depend strongly on temperature or time; the lower row refers to the temperature range ($>0{\cdot}3\ T_m$) in which materials creep. [Ashby]

15.9. Fracture mechanism maps

Ashby* has analysed the various fracture modes both from actual fractographic studies of failed specimens and by modelling each process and presented his results as *fracture mechanism maps*. These show on axes of normalized stress (σ/E) and homologous temperature (T/T_m) boundaries between regions in each of which one mechanism dominates (fig. 15.11). The field boundaries have finite width within which *mixed modes* of failure occur. Lines along each of which time to fracture t_f is constant are also shown.

15.10. Development of creep-resisting alloys

It has been shown that the rate of creep increases rapidly with homologous temperature, so that for use at a given temperature it will be advan-

* M. F. Ashby Fracture mechanism maps, *Acta Met.*, to be published.

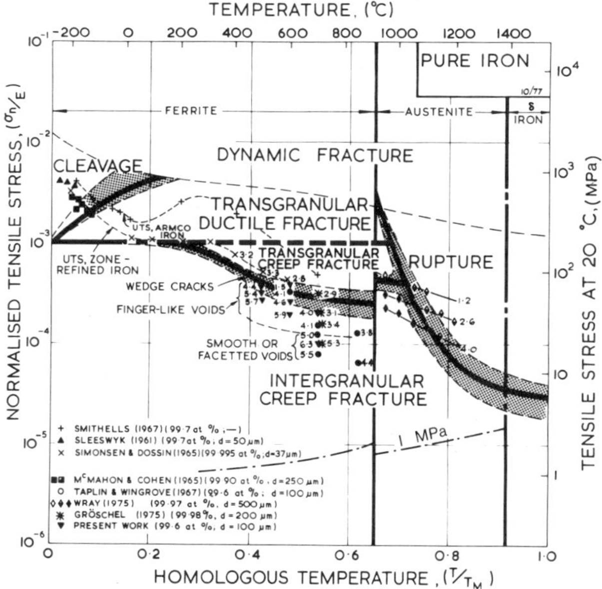

Fig. 15.11.—Fracture mechanism map for nominally pure iron. It shows five mechanism fields: cleavage, ductile fracture, transgranular creep fracture, intergranular creep fracture and rupture. The numbers against data points are $\log_{10}. t_f$ [Ashby]

tageous to choose a material with high T_m.

Power-law creep is the result of movement of dislocations. Hence any form of obstacle used for blocking dislocations will reduce creep. Many of the obstacles used to give room-temperature strength do not remain at high temperatures. For example, in heat-treated plain carbon steels, martensite breaks down on tempering and the carbide, cementite, becomes less dispersed, producing softening; at higher temperatures eventually all the carbide dissolves in the austenite, so that these have poor creep resistance. In alloy steels containing strong carbide-forming elements such as chromium, tungsten, vanadium, and molybdenum (see Chapter 17) the carbides remain in a finely dispersed condition to much higher temperatures, and creep-resistance properties are satisfactory up to about 550°C in low-alloy steels and to higher temperatures in some high-alloy steels. For use at higher temperatures, the type of steel used must also be resistant to oxidation.

A series of creep-resistant alloys have been developed for use in gas turbines. Most of these are nickel-based and are collectively referred to as the *super-alloys*, the best known being the Nimonic series. These are based on an 80/20 nickel-chromium alloy which has excellent resistance to oxidation at high temperatures. Addition of elements which by suitable heat treatment will form dispersed precipitates stable at high temperatures give the desired creep-resistant properties. One of the more important is a titanium-aluminium phase which is obtained by a prolonged solution treatment at about 1100 °C followed by air cooling and precipitation treatment at 700 °C.

To reduce diffusional flow, the grain size should be as large as possible. This requirement is in conflict with the need to refine grain size to improve other mechanical properties such as achieving high yield strength and lowering the ductile–brittle transition temperature.

15.11. Creep–fatigue interactions

When a component is subjected in service to elevated temperatures and also to fluctuating loads, then both fatigue and creep will contribute to the ultimate failure.

It was seen in Section 14.11 how life under variable amplitude fatigue loading is estimated by a cumulative damage law, of which the Palmgren–Miner law is the simplest but, in practice, gives answers which are often only correct within an order of magnitude. A similar life-fraction cumulative damage equation may be applied where both fatigue and creep occur together. The simplest form of equation is

$$\sum\left(\frac{n_i}{N_{fi}}\right) + \sum\left(\frac{t_j}{T_j}\right) = 1$$

where n_i = number of cycles of loading at load condition i
N_{fi} = cycles to failure at load condition i
t_j = time duration at stress and temperature condition j
T_j = time to failure at stress and temperature condition j

As with the Palmgren–Miner law, this may only give an order-of-magnitude estimate.

QUESTIONS

1. Sketch typical creep curves for a series of tests under the same load over a range of temperatures and explain what is meant by *minimum creep rate* and *creep limit*.

A series of creep tests at constant load on an aluminium alloy gave the following data for minimum creep rate (strain h^{-1}):

Temperature (°C)	140	180	220	260	300
Minimum creep rate h^{-1}	0·04	0·33	1·27	3·26	6·57

2. A radial blade of an axial-flow gas turbine is of uniform cross-section and made from a material of density p. Show that, when the turbine is running at an angular

speed ω, the stress due to centrifugal force at a radius r is given by

$$\tfrac{1}{2}\rho\omega^2(r_0{}^2 - r^2),$$

where r_0 is the tip radius of the blade.

A particular blade of tip and root radii, 150 mm and 100 mm, respectively, is made of a material of specific gravity 8·27 which has the creep characteristics at 815 °C shown in fig. 15.12. Estimate, graphically, or otherwise, the elongation of the blade due to creep if the turbine were to run at 12500 rev min^{-1} for 1000 h at a temperature of 815 °C.

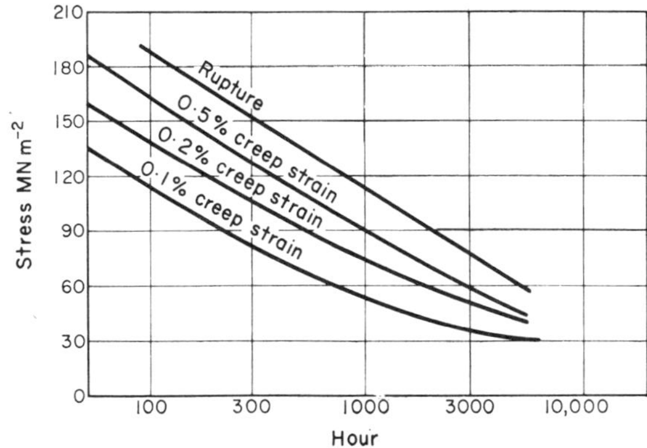

Fig. 15.12

3. Describe the typical creep behaviour of a metal and the manner in which it varies with temperature and with stress.

The secondary stage creep behaviour of a 12% Cr steel at 500 °C is shown approximately in Fig. 15.13. Investigate whether there is a simple power-law relationship between the creep rate and the stress. [E]

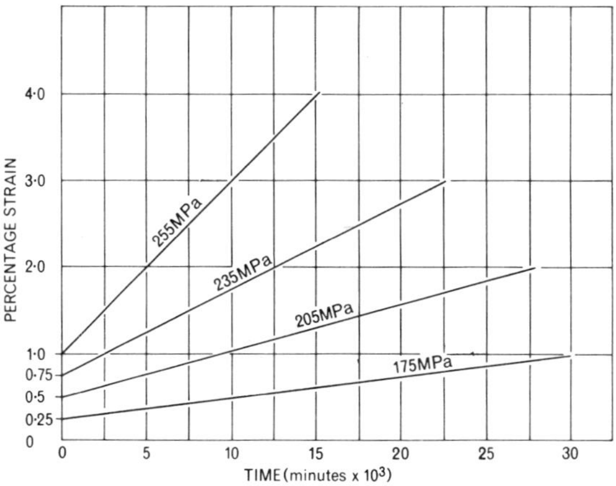

Fig. 15.13.

4. What is *secondary stage creep* in metals and how is it affected by temperature?

The data below refer to the dependence of secondary stage creep rate $\dot{\varepsilon}$ on tensile stress σ for a steel specimen tested in tension at 780 °C. Verify that the relationship $\dot{\varepsilon} = A\sigma^n$ is consistent with the data and find the values of the constants A and n.

Stress σ (MPa)	110	150	200	250
Creep rate $\dot{\varepsilon}$ (strain h^{-1})	$1{\cdot}00\times10^{-5}$	$1{\cdot}35\times10^{-4}$	$1{\cdot}50\times10^{-3}$	$1{\cdot}00\times10^{-2}$

A tie-bar manufactured from this steel is loaded in pure tension and maintained at a working temperature of 780 °C. The initial stress in the bar at the working temperature is 50 MPa. Assuming that its ends remain rigidly clamped,what is the relaxed stress in the bar after 30 000 h? The value of E at 780 °C is 160 GPa. [E]

5. What are the physical mechanisms responsible for creep and hence for stress relaxation in a strain metal?

A metal creeps according to $\dot{\varepsilon} = A\sigma^n$, where σ and ε are conventional stress and strain and A and n are constants. Show that, if a material of modulus E be stressed in tension to a fixed strain ε_0 under stress σ_0, the stress after a time t is given by

$$\frac{1}{\sigma^{n-1}} = \frac{1}{{\sigma_0}^{n-1}} + AE(n-1)t.$$

In laboratory tests at 550 °C a material gave $\dot{\varepsilon} = 4{\cdot}0\times10^{-8}\ \mathrm{h}^{-1}$ at 30 MPa and the value of n was 3·8. The elastic modulus for this material is 200 GPa and is assumed to be unaffected by temperature. Compute the stress, after one year, in a bolt of this material clamping two stiff plates at 550 °C and initially tightened to a stress of 70 MPa. [E]

CHAPTER 16

Heat Treatment of Steel

16.1. Introduction

Heat treatment is the term that describes in a general manner an operation or series of operations which involve the heating and cooling of a metal or alloy in the solid state, carried out for the purpose of obtaining certain desired properties.

The properties of steels can be varied over a very wide range by heat treatment, which is one of the reasons for the great usefulness of steels. The heat treatment of steels is generally taken to embrace the martensitic change which occurs on quenching from the austenitic state and the softening which results from the subsequent tempering. These changes have already been mentioned in Section 11.8. The actual heat treatment employed depends upon the values of mechanical properties desired. Thus if high hardness is needed, the treatment may consist solely of quenching, but if high toughness is needed, then a high-temperature tempering must also be given. Generally, a full heat treatment is employed to give an all-round improvement in properties.

The behaviour of pure iron-carbon alloys will be discussed in this chapter, the effect of alloy additions being considered in Chapter 17. The transformations that occur on heating and cooling are first considered separately.

16.2. Transformations on heating

A hypoeutectoid steel which has been slowly cooled under almost equilibrium conditions consists at room temperature of grains of pro-eutectoid ferrite and grains of pearlite as in fig. 8.11*a–b*. If the steel is heated to any temperature below the A_{c1}-temperature and then cooled, no alteration of structure will occur, but if it is heated to just above this temperature, the pearlite will transform to austenite.

The mechanism of the change is one of nucleation and growth, the nucleation occurring at points on the ferrite–cementite interfaces in the eutectoid. Once formed, these nuclei of austenite grow by absorbing both ferrite and cementite until each pearlite grain has completely changed to austenite. As the change is proceeding, the carbon diffuses from the original cementite regions until it is homogeneously distributed within each austenite grain. Because the diffusion of carbon is not instantaneous, but occurs at a finite rate, time plays a part and the transformation

$$\text{ferrite} + \text{FeC}_3 \rightleftharpoons \text{austenite}$$

is slow compared with the allotropic change from α- to γ-iron when carbon is absent.

As the temperature is slowly raised above the A_{c1}-temperature, the proeutectoid ferrite becomes absorbed by the growth of the austenite grains, the transformation being complete at the A_{c3}-temperature.

In hypereutectoid steels, a similar process occurs, the pearlite being transformed to austenite at the $A_{c1\,3}$-temperature and the proeutectoid cementite dissolving in the austenite between the $A_{c1\,3}$ and the A_{cm}-temperatures.

Each pearlite grain may give rise to several austenite nuclei, so that the austenite grain size just above the A_{c3}-temperature is smaller than the previous ferrite–pearlite grain size. On heating the austenite to higher temperatures, the grain size increases by a process of absorption of some grains by others. The character of the grain-size/temperature relationship is discussed in Section 16.10.

16.3. Transformations on cooling

As a hypoeutectoid steel is slowly cooled from above the A_{c3}-temperature, proeutectoid ferrite will begin to separate from the austenite at the A_{r3}-temperature. Since the α-iron dissolves only a very small amount of carbon, most of the carbon diffuses into the remaining austenite, thereby lowering the temperature for further change, until the eutectoid composition is reached. The remaining austenite then transforms to pearlite at the eutectoid temperature. The proeutectoid ferrite separates at nuclei on the austenite grain boundary.

When the eutectoid temperature is reached, nuclei of cementite form at the austenite grain boundary. These contain a higher proportion of carbon than the eutectoid austenite, so that carbon diffuses from the adjacent austenite, which immediately transforms to ferrite. Alternate layers of cementite and ferrite thus form and grow edgewise into the austenite grain. The method of growth is shown diagrammatically in fig. 16.1.

A hypereutectoid steel cooled from the austenite region will first precipitate cementite which migrates to the grain boundaries and forms a network structure which is visible at room temperature (fig. 8.11*d*). As this proceeds, the carbon content of the remaining austenite decreases until it is of eutectoid composition, when it transforms at constant temperature to pearlite.

The pearlite that forms from the eutectoid austenite does so by separation of ferrite and cementite, which involves diffusion of carbon, and therefore its formation takes a finite time. If the steel is cooled more rapidly, a lower temperature is reached before the change is complete. At lower temperatures, diffusion is less rapid, so that the spacing of the pearlite lamellae is smaller. The finer distribution results in a modification of the physical properties.

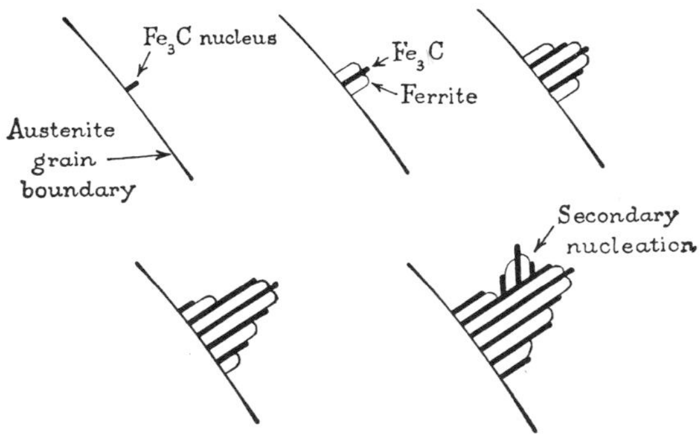

Fig. 16.1.—Diagrammatic representation of pearlite formation

16.4. Quenching temperature and rate

When austenite is cooled extremely rapidly, the change to ferrite and cementite is suppressed and martensite is formed. As stated on p. 181 this is extremely hard and brittle and may be softened by tempering. If the steel is quenched from a temperature at which it is not 100% austenite, then the ferrite or cementite present at that temperature will be unchanged on quenching, and during any subsequent tempering. Hence, to get the best effects from heat treatment the quenching should be carried out from above the A_{c3}-temperature for hypoeutectoid steels and above the A_{cm}-temperature for hypereutectoid steels. The cementite present in hypereutectoid steels between the A_{c13} and A_{cm}-temperature is, however, hard and can contribute significantly to the final hardness in the quenched and in the tempered states. The danger of cracking on quenching high-carbon steels is great and is reduced by quenching from lower temperatures. It is usual to quench hypereutectoid steels from just above the A_{c13}-temperature. In practice, to ensure that the temperature throughout the test piece has reached the required value and that the desired changes on heating have been completed, it is usual to quench from a temperature about 50 K above the A_{c1} or A_{c13}-temperature as shown in fig. 16.2.

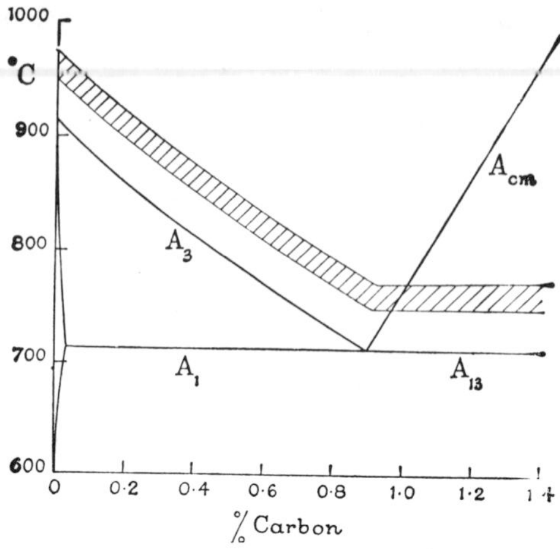

Fig. 16.2.—Shaded band is approximate temperature range used for austenizing carbon steels prior to quenching

In order to transform the austenite present at the quenching temperature to 100% martensite, it is necessary to cool the steel faster than some minimum rate—the *critical cooling rate.* The critical cooling rate depends upon the carbon content, varying in the manner shown in fig. 16.3.

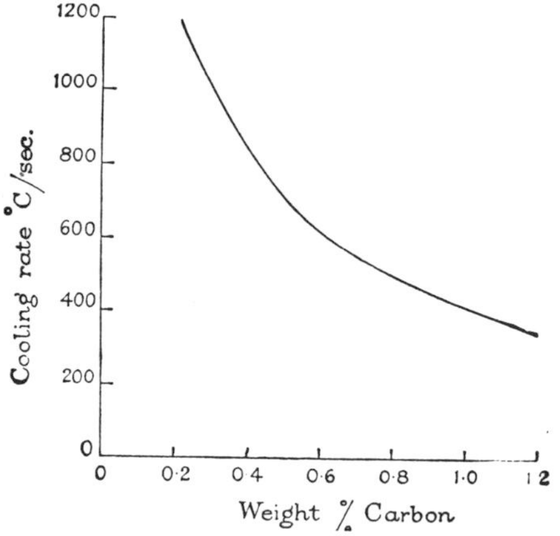

Fig. 16.3.—Variation of critical cooling rate with carbon content of carbon steels

If the rate of cooling is less than the critical cooling rate, the structure may contain some martensite together with either ferrite or pearlite, or bainite (see p. 278), or a mixture of these.

16.5. Isothermal transformation diagrams

In cases where the austenite is not transformed completely to martensite, the other changes involve nucleation and grain growth processes which occur at varying temperatures. As both nucleation and grain growth rates are temperature-dependent, a fundamental investigation should study the rates at which the changes would occur at a constant temperature.

Isothermal investigations involve the study of the times required for initiation and completion of changes at various sub-critical temperatures, and the nature of the transformation products.

The method originally used by Davenport and Bain* was to study the progress of transformation microscopically. Small pieces of the steel under investigation (about the size of a halfpenny is reasonable) are used so that their temperature can be changed rapidly. These are heated to a temperature above the upper critical temperature for long enough to be austenized, and then in turn cooled instantaneously by quenching in a suitable liquid bath held at the temperature under investigation. Each piece is held at this temperature for a different length of time and then quenched in cold water. The liquid bath may be a molten metal such as lead, or a neutral salt of the type used for tempering baths. This procedure is carried out over a range of sub-critical temperatures.

At the sub-critical temperature, some or all of the austenite may transform to the transformation products appropriate to that temperature. Upon quenching in water, any unchanged austenite will form martensite, while the transformation products do not change. The final structure of the specimen can be found by suitable etching and examination under a microscope.

It is found that definite times are required for the initiation and completion of the transformation and that these times vary with the temperature. The progress of the transformation can also be studied and times for (say) 10%, 50% and 90% transformation found.

Any other property that changes with proportion of austenite can also be used as a guide to the progress of transformation, although the nature of the transformation products cannot be identified other than by microscopic examination. In an investigation utilizing the method of Davenport and Bain, it is convenient to follow the progress of the transformation by hardness tests. A hardness test is quickly carried out on each specimen as it is quenched. The results obtained can be used to indicate whether transformation has started, is continuing, or has completed at the trans-

* E. S. Davenport and E. C. Bain, *Trans. Amer. Inst. Min. Met. Engrs.*, Vol. 90, 1930, pp. 117–154.

formation time investigated, and guidance can be immediately obtained as to the next time of quench required without awaiting the laborious preparation for microscopic examination, which can follow later. A typical set of results is given in fig. 16.4.

Dilatation (see p. 107) is another property which is commonly used for transformation studies,* particularly for transformations taking longer times, such as one hour or more.

The complete diagram for a eutectoid steel is shown in fig. 16.5. As the transformation temperature is lowered from the A_{13}-temperature to about 550 °C, the size of stable nuclei decreases and so nucleation and completion times decrease and the pearlite lamellae become finer. As the temperature decreases further, the time for nucleation increases again due to the decreasing rate of diffusion of carbon in iron.

At temperatures considerably below A_{e1}, any austenite still present transforms to an acicular (needle-like) structure known as *bainite*. The temperature below which this happens is known as the bainite shelf or B_s temperature. Bainite is composed of the two equilibrium phases, ferrite and cementite, but differes from pearlite in appearance and in the method of growth. The ferrite nucleates first and forms elongated grains known as laths. The cementite is then deposited between these laths at higher temperatures and within them at lower temperatures. The microstructure appears somewhat different at the upper and lower ends of the temperature range, the two types of structure being known as *upper bainite* and *lower banite*, respectively.

The transformation to bainite gives ferrite with a high dislocation density, the density being greater the lower the transformation temperature. Also the lower the transformation temperature, the more finely is the cementite distributed. These factors both contribute to the strength which is greater than that of the ferrite–pearlite structures.

Figure 16.6 shows the development of a lower bainite structure during isothermal transformation.

The point of minimum nucleation time is known as the *nose* or *knee*.

Below about 250 °C, martensite is formed. This is not a nucleation process and takes place almost instantaneously, but the amount formed depends upon the temperature. The upper and lower limits of the martensite transformation range are called the M_S and M_F-temperatures. The experimental procedure for determining these temperatures, introduced by Greninger and Troiano† is as follows. The specimen is quenched to the temperature being investigated then reheated to some elevated

* N. P. Allen, L. B. Pfeil, W. T. Griffiths, *Second Report of the Alloy Steels Research Committee*, Iron and Steel Institute, Special Report No. 24, 1939, Section XIII, pp. 369–390.

† A. B. Greninger and A. R. Troiano, *Trans. Amer. Soc. Metals*, Vol. 28, 1940, pp. 537–562.

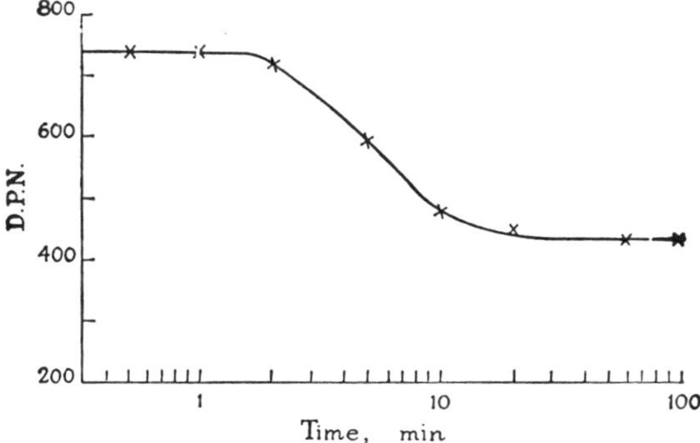

Fig. 16.4.—Typical hardness-time curves for isothermal transformation of En steel at 340 °C. Microstructures (fig. 16.6) show more bainite and less martensite for longer transformation times

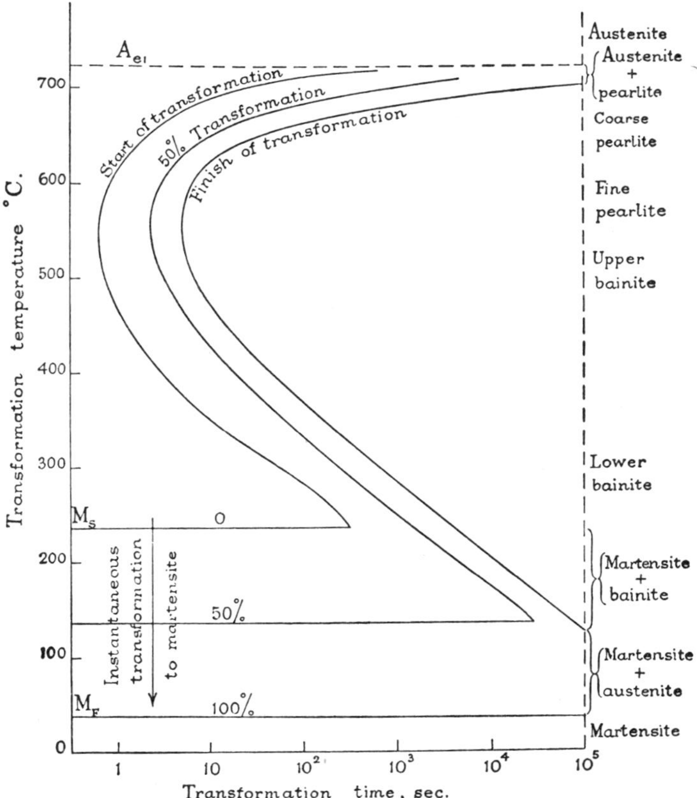

Fig. 16.5.—Isothermal transformation diagram for a eutectoid steel. Structures present after 10^5 seconds are given on the right-hand side

Fig 16.6—Four stages in the formation of lower bainite in a $2\frac{1}{2}$% Ni-Cr-Mo high-carbon steel (En 26), austenized at 835 °C, and isothermally transformed for the times indicated at 340 °C (×850). [International Nickel]

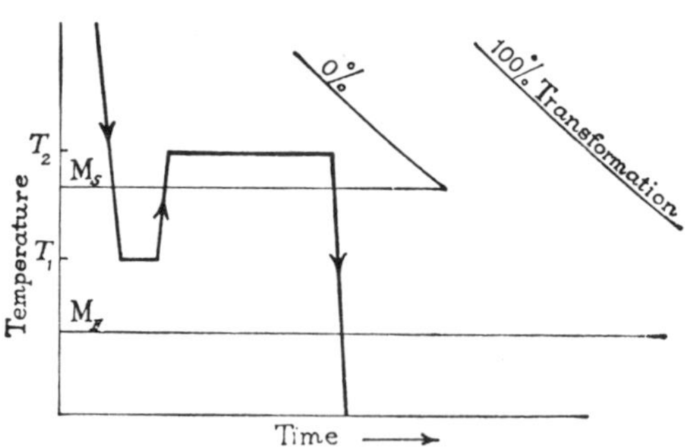

Fig. 16.7.—Temperature-time cycle for determination of martensite transformation temperatures. Martensite formed at temperature T_1 is tempered at temperature T_2 and can be distinguished from martensite formed in final quench

temperature that is below the A_{13}-temperature and held for a time to temper the martensite already formed. It is then quenched in cold water, when the unchanged austenite transforms to martensite. When etched and examined under the microscope, the tempered and untempered martensite are readily distinguished. Care must be taken that the tempering temperature and time are chosen so that no other transformation products are formed. The temperature-time cycle is shown in fig. 16.7.

The complete isothermal transformation diagrams are also known as Time–Temperature–Transformation diagrams, TTT or triple-T diagrams and, because of their shape, S–curves.

16.6. Isothermal transformation diagrams for hypoeutectoid and hypereutectoid steels

The isothermal transformation diagram for a typical hypoeutectoid steel is shown in fig. 16.8. Between the A_3 and A_1-temperatures, ferrite

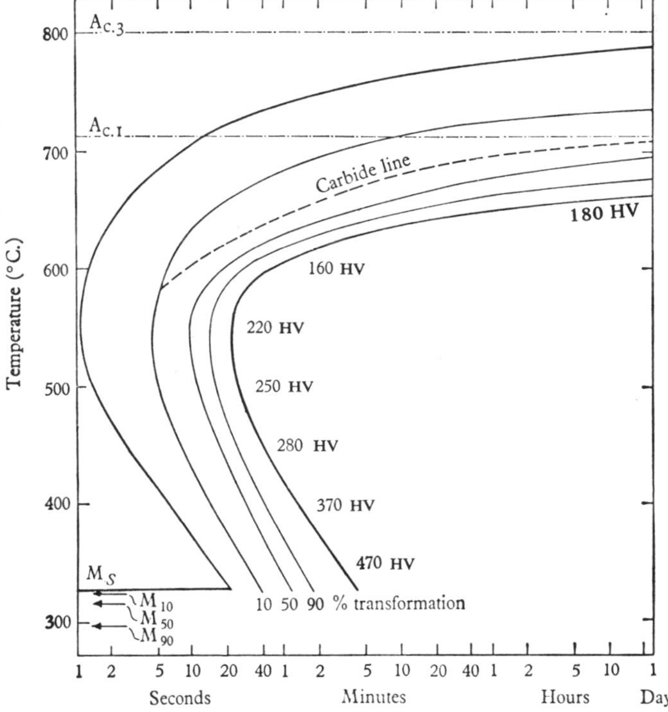

Fig. 16.8.—Isothermal transformation diagram for an En 12 steel (0·34% C, 1·06% Mn, 0·75% Ni). The hardness values are those of the fully transformed steel. [After International Nickel]

nucleates and grows, and austenite transforms to the extent indicated by the equilibrium diagram. Below the A_1-temperature, ferrite nucleates first, followed by pearlite. The time at which pearlite starts to form at any particular temperature is shown by the *carbide* line. The start of the pearlite formation does not mean the end of proeutectoid formation, but is a close approximation to it. Thus at lower temperatures, the amount of proeutectoid ferrite decreases, and the ratio of ferrite to cementite in the pearlite alters, so that the microstructure suggests that the steel has a higher carbon content. The bainite formation is not preceded by ferrite formation.

The diagram for a hypereutectoid steel would be similar, only cementite being nucleated at first and pearlite later.

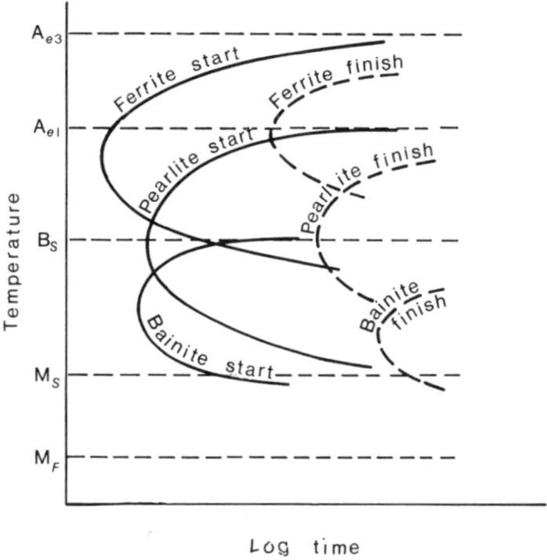

Fig. 16.9.—Nucleation and transformation times for ferrite, pearlite and bainite structures

16.7. Separation of pearlite and bainite knees

Each type of transformation (except to martensite) has its own knee or temperature of maximum rate of nucleation as shown in fig. 16.9. At any sub-critical temperature, the mechanism which has the shortest nucleation and growth time will give the product. In alloy steels, the knees become much more separated than in plain carbon steels, giving curves of the type shown in fig. 16.10.

16.8. Continuous cooling transformation

In a similar manner it is possible to plot on a temperature–time graph points marking the beginning and end of transformation in the case of continuous cooling of steels. The results so obtained cannot be derived quantitatively from an isothermal transformation diagram, but a qualitative correlation is possible and helpful. A set of typical continuous cooling transformation curves is shown in fig. 16.11. A slow cool in the furnace, such as curve (*a*), would yield coarse pearlite, while slightly faster cooling in air such as curve (*b*) would give fine pearlite. Such treatments

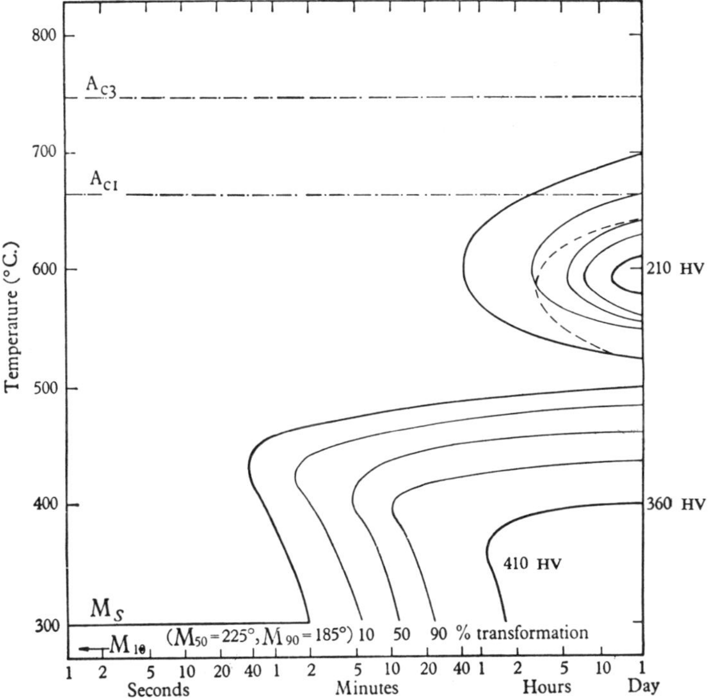

Fig. 16.10—Isothermal transformation diagram for En 23 steel (0·33% C, 0·57% Mn, 3·26% Ni, 0·85% Cr). [After International Nickel]

are known as *annealing** and *normalizing*, respectively. At faster rates of cooling, the transformation occurs at lower temperatures, that is, the A_r-temperatures fall as the cooling rate is increased.

Curve (*c*) would correspond to a cooling rate such as that obtained on

* This is not the same as *process annealing*, which is heating to a temperature just below A_{c1}. No phase change occurs, but internal stresses due to mechanical working are relieved, producing some softening.

quenching a large piece of steel in oil. Some of the austenite would transform to fine pearlite, but when the temperature had fallen below that of the pearlite knee, pearlite formation would cease. There is insufficient time for bainite to form, and the remaining austenite will transform to

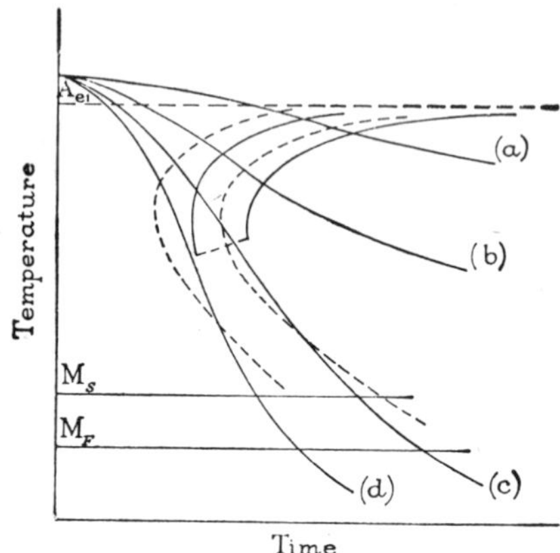

Fig. 16.11.—Continuous cooling transformation curves and cooling curves. Dashed lines show relative position of isothermal transformation curves

martensite between the M_S and M_F-temperatures. This is called a *split transformation* and occurs at cooling rates just slower than the critical.

Curve (*d*) corresponds to the critical cooling rate and is the slowest rate at which only martensite is formed.

The continuous cooling transformation curves lie at somewhat longer times and lower temperatures than the isothermal transformation curves for the same steel.

16.9. Retained austenite

The variation of the M_S and M_F-temperatures with carbon content is shown in fig. 16.12. Above 0·7% carbon the M_F-temperature is below 0 °C, so that quenching higher-carbon steels into ice-cold water will not produce 100% martensite. The *retained austenite* is much softer than martensite and is an undesirable factor in the usual heat treatment cycle.

It has been found that if cooling is interrupted between the M_S and M_F-temperatures, not only does the transformation to martensite cease, but the remaining austenite acquires an increased stability, so that on subsequent cooling, the transformation does not start until the temperature has been lowered by a distinct amount. The extent of this *stabilization*

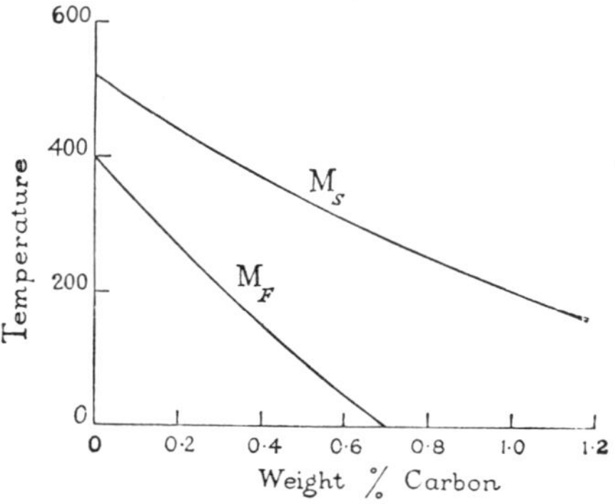

Fig. 16.12.—Variation of M_S and M_F-temperatures with carbon content

From Donald S. Clark and Wilbur R. Varney, *Physical Metallurgy for Engineers*. Copyright 1952, D. Van Nostrand Co., Inc., Princeton, New Jersey.

of austenite depends upon the temperature and on the duration of the interruption.

A large-sized specimen, even with the most severe quenching, may not cool faster than the critical cooling rate, so that it may not be possible to get 100% martensite. This *mass effect* and related phenomena are dealt with further in Sections 16.15–16.20

16.10. Grain size of steel and its effects

The rate of transformation of a steel at any given temperature is also dependent upon the grain size of the austenite. The transformation to ferrite and cementite structures involves the growth of the new structures from the austenite grain boundary, and the rate of growth normal to the grain boundary is dependent upon the diffusion rate of carbon. At any constant temperature the diffusion rate is constant, and so it is to be expected that the transformation time is longer for coarse-grained austenite. It is found that this is so and also that the time for transformation to start is longer. Isothermal transformation diagrams for two grain sizes of the same steel are compared in fig. 16.13. The coarser-grained steel has the lower critical cooling rate.

The grain size of the ferrite–pearlite structure at room temperature is dependent upon the austenite grain size, a coarser-grained austenite giving a coarser room-temperature structure. The relationship is, however, only qualitative, and for hypoeutectoid steels, the austenite grain size cannot be determined directly from the ferrite grain size.

As stated on p. 274 each pearlite grain gives several austenite grains when heated through the critical temperature range, so that the grains are refined. Grain refinement of a steel is normally done by heating and cooling through the critical range a sufficient number of times. There is a limit to the smallness of grain that can be attained by this means, the A.S.T.M. value being about 5 to 8.

On heating further into the austenite region, the equilibrium grain size increases by the process of large grains absorbing the smaller ones. There are two distinct patterns to the grain-size–temperature relationship

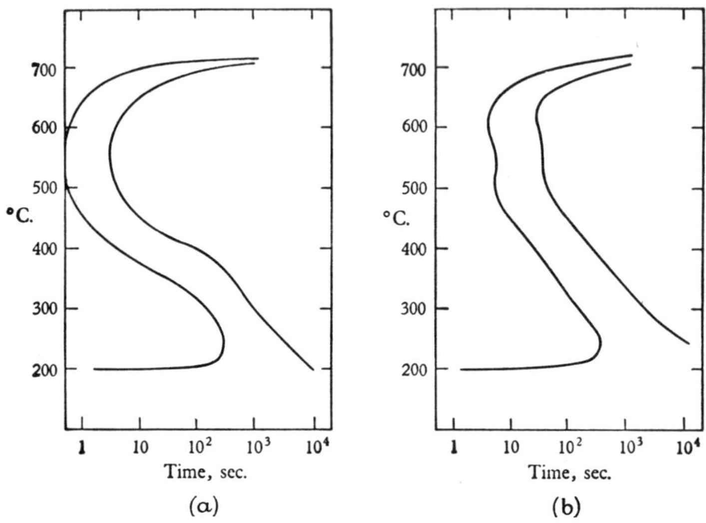

Fig. 16.13.—Effect of grain size on isothermal transformation diagram of a eutectoid steel

(*a*) Grain size 11 (fine), (*b*) grain size 2–3 (coarse)

From Donald S. Clark and Wilbur R. Varney, *Physical Metallurgy for Engineers*. Copyright 1951, D. Van Nostrand Co., Inc., Princeton, New Jersey.

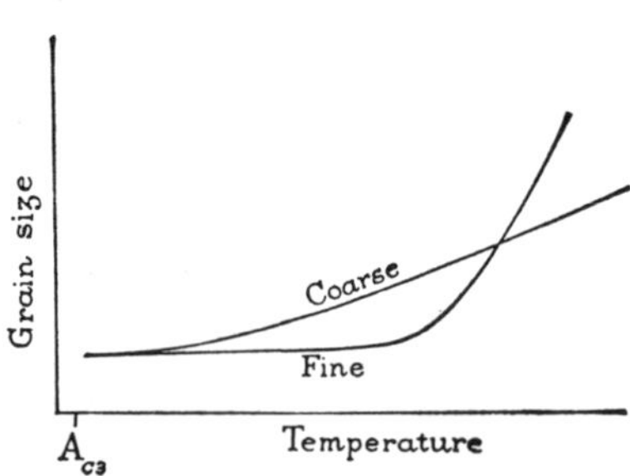

Fig. 16.14.—Grain growth of inherently fine and coarse-grained steels

which are shown in fig. 16.14. An *inherently coarse-grained* steel shows grain growth in a steady manner from the A_{c3}-temperature, upwards, whereas an *inherently fine-grained* steel remains at constant grain size for 100 to 200 K above the upper critical temperature, after which the rate of growth is high.

The difference in the grain-growth tendency of different steels has been found to depend upon small alloy additions, particularly those used in killing (see p. 230). Aluminium produces fine-grained steels and silicon gives coarse-grained steels.

As the austenizing of steels is usually carried out at about 50 to 100 K above the upper critical temperature, the inherent grain size of the steel is important because of the relationship of various properties to grain size.

Although coarse-grained austenite gives a lower critical cooling rate, small austenitic grain size produces steels which are tougher and more favourable for heat treatment, being non-warping and having less possibility of cracking and less retained austenite on quenching. Hence, for heat treatment and general purposes, fine-grained steels are favoured, while lower critical cooling rates, when required, are obtained by other means (see 303).

16.11. Determination of austenite grain size*

As the size of austenite grains cannot be observed directly, indirect methods have to be used, that is to say, by observing the transformation products. The method used depends upon the carbon content.

A *hypoeutectoid steel* is either cooled slowly for sufficient time for ferrite separation to start and then quenched, or else cooled at a steady intermediate rate which gives a split transformation. The ferrite formed will have migrated to the austenite grain boundaries before the remainder changes to martensite, and so outlines the original austenite grains.

During the slow cooling of *near-eutectoid* and all *hypereutectoid steels*, proeutectoid ferrite or cementite forms at the grain boundaries before the remainder changes to pearlite.

When *eutectoid steels* are quenched at a rate slightly less than the critical cooling rate, transformation to pearlite begins at the grain boundaries before the remainder changes to martensite. An alternative method for eutectoid steels is to quench to 100% martensite, temper at about 300 °C, and quench in water, after which etching with a special solution will reveal the former austenite grain boundaries.

* B.S. 4490: 1969. *Methods for the Determination of the Austenitic Grain Size of Steel.*

16.12. Tempering

In the tempering cycle, the structure and mechanical properties are determined by the tempering temperature and the time at that temperature. The changes proceed rapidly at first and then slow down, the rate of change being very slow after about 30 minutes. At the tempering temperature the carbon atoms are able to diffuse and give a precipitate of cementite, leaving areas of ductile ferrite. The size of the cementite particles increases, and their number decreases as the tempering temperature is raised, so that there are fewer obstacles to dislocation movement. The structure is called tempered martensite (see fig. 16.15*b*), but may also be known as *troosite* or *sorbite*, or (for tempering temperatures just below the A_1-temperature) *spherodite*.

Recent research work using electron-microscope techniques* has shown that the initial precipitation of carbide is a hexagonal close-packed ε-carbide, which then forms thin platelets of orthorhombic cementite. The structure of the ε-carbide has not been definitely established, but it has a composition between Fe_3C and Fe_2C. From the platelets of cementite a microstructure of spheroidal cementite develops which then coarsens by diffusion of carbon from the smaller spheroids to the larger spheroids. Thus large particles grow at the expense of the small. The process gets slower and slower, but changes are still occurring after 100 hours at the tempering temperature. However, raising the temperature by about 50 K produces the same mechanical properties in 1 hour as would 100 hours at the lower temperature. In practice, tempering times of $\frac{1}{2}$ to 2 hours are used.

The structure of lower bainite is somewhat similar to that of a lightly tempered martensite and subsequent tempering will proceed in a manner very similar to that described for martensite. In upper bainite, the carbides in the lath boundaries will tend to spherodize on heating above 450 °C, but there is little gain in the mechanical properties.

16.13. Mechanical property changes with tempering

Tempering increases the ductility and toughness above that of the as-quenched structure and decreases the hardness. As the tempering temperature is raised, all the mechanical properties alter. It is usual to show these changes in a *tempering diagram* or *property chart*, a typical example of which is shown in fig. 16.16.

The general features are that increasing the tempering temperature increases ductility and toughness, while decreasing strength and hardness.

* E. D. Hyam and J. Nutting, *J. Iron and Steel Inst.*, Vol. 184, 1956, p. 148.

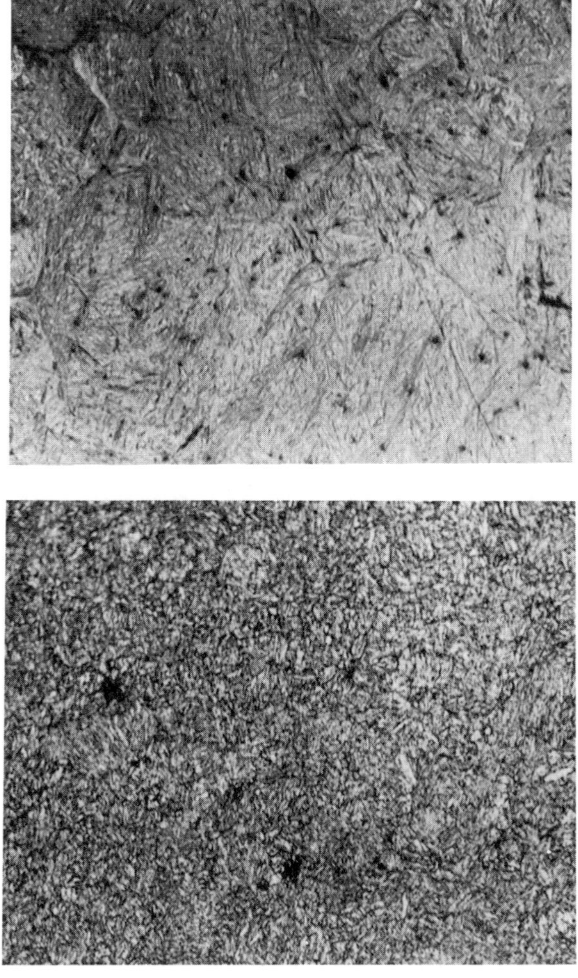

Fig. 16.15.—Microstructures of (*a*) martensite (×200), (*b*) 0·53% C Steel, water quenched and tempered at 350 °C (×200)

The tempering temperature is chosen to give a desired combination of properties.

In certain cases, tempering can produce brittleness as shown by an Izod test. Some steels show an increase of Izod energy with tempering temperature up to 200 °C and then a drop in the range up to 450 °C, after which it rises in the normal way. This is known as *brittle temper*, and for successful heat treatment the brittle tempering range must be avoided. The cause is not fully understood. It may be due to the conversion of

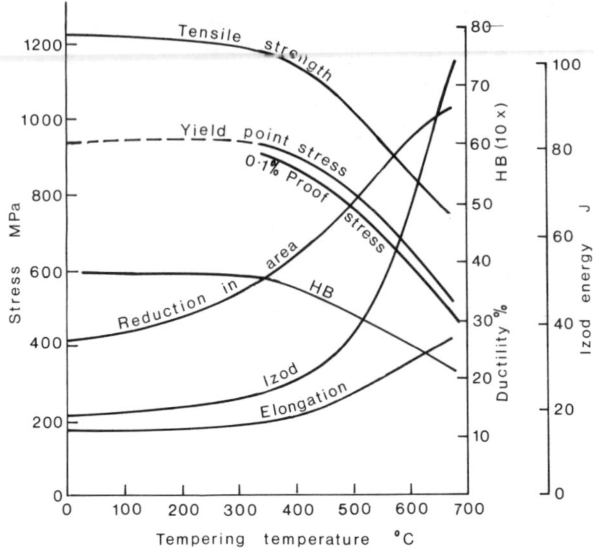

Fig. 16.16.—Tempering diagram or property chart for water-quenched 30 mm diameter bars of En 12 steel. [After International Nickel]

retained austenite to martensite or bainite. It is more pronounced in alloy steels.

Temper brittleness is a trouble found with many alloy steels. After tempering at 550 to 600 °C, a slow cool leads to a low Izod value, but a rapid quench from the tempering temperature gives a satisfactory value. It may be due to the larger grain size produced by a slow cool. In steels where this is likely to occur, it can be prevented by incorporating $\frac{1}{2}\%$ of molybdenum in the composition.

At about 300 °C mild steel shows a loss of toughness and ductility, and so is unsuitable for working at this temperature. This *blue brittleness* is due to a combination of straining and strain ageing. At that temperature, the diffusion rate of the impurities responsible for locking the dislocations (see Section 11.4) matches the velocity of the dislocations themselves so that moving dislocations may become locked and a higher stress is necessary for further movement.

16.14. Special heat treatments

Two special heat treatments have been devised for cases involving large sections, where a water quench would be likely to produce cracking and a slower quench would not produce sufficient martensite. On quenching a large specimen to room temperature, the outside cools rapidly and transforms to martensite. The centre cools more slowly and transforms later, accompanied by dilatation, which can crack the brittle outer skin. The treatments which avoid this cracking are based on the isothermal transformation diagram.

In *martempering* (fig. 16.17*a*), the steel is cooled rapidly to a temperature just above M_S and held until the temperature becomes uniform through the specimen. As the time for initiation of bainite formation is so long at this temperature, there will be no transformation, and hence no stresses set up which could cause cracking. The steel is then allowed to cool slowly to room temperature during which process martensite forms uniformly through the specimen. The steel can then be tempered if desired.

In *austempering* (fig. 16.17*b*), the steel is also quenched to a temperature just above M_S and is then held at that temperature until transformation to lower bainite is complete. The properties of lower bainite are almost as good as, and in some cases superior to those of tempered martensite.

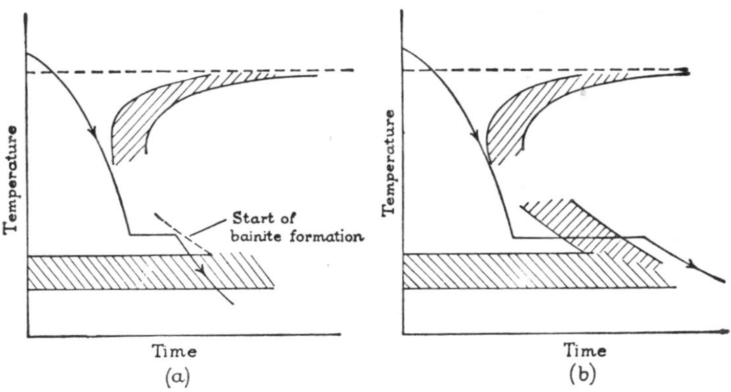

Fig. 16.17.—Time-temperature curves for (*a*) martempering, (*b*) austempering

16.15. Hardenability

The ability to form martensite in a steel is dependent upon the critical cooling rate as well as on the size of the specimen and the manner in which it is quenched. If a series of bars of the same steel but of different diameters are quenched and sectioned, the variation in hardness across the section changes with size. A typical set of results is shown in fig. 16.18*a*.

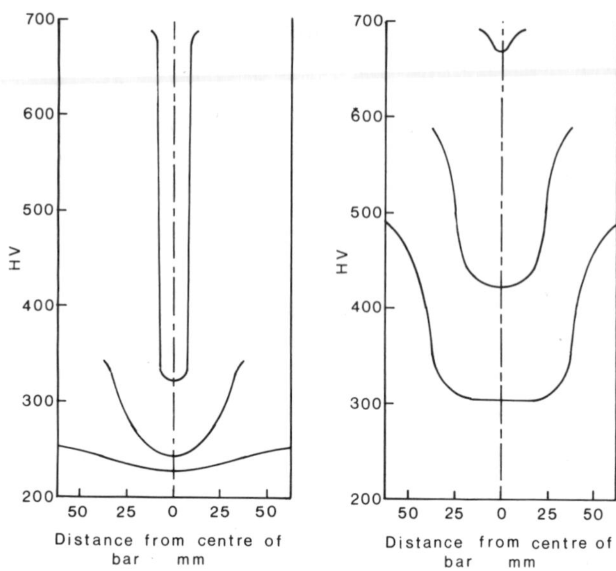

Fig. 16.18.—Variation of hardness with depth in water-quenched cylindrical bars of (*a*) a plain carbon steel, (*b*) a 1% Cr-V alloy steel

For the smallest bar, the centre is considerably softer than the outside. As the bar size increases, the centre gets softer, and eventually the outside also. This is because the rate of cooling is lower towards the centre of each bar and also decreases as the size of bar is increased, so that a smaller proportion of martensite is formed. A similar series of results for a steel of lower critical cooling rate is shown in fig. 16.18*b*. The hardness is more uniform across the smallest bar, and there is less decrease of hardness as the size of bar is increased.

The effect of size of specimen on the as-quenched hardness of a steel is known as the *mass effect*. The ability of a steel to form martensite when cooled at various rates is measured by its *hardenability*. A material with a higher hardenability will form martensite in larger sections on quenching. It is not related to the actual hardness of the martensite formed, which is almost solely dependent upon the carbon content of the steel. Hardenability is in effect the reciprocal of the critical cooling rate.

The properties of a tempered steel as shown in a property chart usually refer to the effect of tempering a structure that is 100% martensite. If the cooling rate during the quench is less than the critical cooling rate, so that the structure is only partly martensite—a *slack-quenched* structure—only the martensite portion will be affected on tempering. The resulting combination of properties is not so good as that obtained by tempering a structure of 100% martensite. Hence the potential mechanical properties of a steel are not fully exploited if the size of a component and the manner of quenching are such that the as-quenched structure is not 100% martensite. The steel to be chosen for a particular application should therefore be one with a sufficient hardenability.

16.16. Measurement of hardenability

Two types of test for the measurement of hardenability will be described here. They are both appropriate for use with steels of medium hardenability. The first is the cylinder series test which gives a single value for hardenability, that being the diameter of the bar which gives a certain percentage of martensite at the centre when quenched in a certain manner, and the second is the Jominy end quench test which gives a curve.

16.17. Cylinder series test

In the cylinder series test devised by Grossman and others,* a series of round bars of different diameters are austenized and quenched in oil or water. The length of each bar must be sufficiently large for the cooling effect at the mid-length to be unaffected by the ends. After quenching, each bar is cut in half and a hardness survey made along a diameter, or the microstructre examined after etching. Figure 16.18 shows typical hardness traverses for two steels. As it is not possible to identify with any precision the depth to which 100% martensite is formed, the hardened zone is specified as that region in which the martensite forms more than a certain percentage of the microstructure. The value usually specified is 50%. From a graph in which the diameter at which 50% martensite occurs is plotted against bar diameter, we can estimate the diameter of bar that would show 50% martensite at the centre. This is called the *critical diameter* for that quenching medium. Because the rate of cooling is less for an oil quench than for a water quench, the critical diameter of any steel will be less for oil quenching than for water quenching.

The quenching condition can be denoted quantitatively by the severity of quench H which is given by

$$H = \frac{\text{heat transfer coefficient between steel and quenching fluid}}{\text{thermal conductivity of steel}}$$

The most severe quench would be one with an H-value of infinity, corresponding to the surface layer of the bar immediately reaching the

*M. A. Grossmann, M. Asimow, and S. F. Urban, "Hardenability, Its Relation to Quenching and some Quantitative Data," *Hardenability Symposium*, Amer. Soc. for Metals, Cleveland, Ohio, 1938, pp. 124-196.

temperature of the quenching medium. The critical diameter for such an idealized and unrealizable condition is called the *ideal critical diameter*. By considering the thermal flow problem involved in the cooling of a quenched bar, the sizes of bar that would give the same rate of fall of temperature at the centre under quenches of different H, the relation between critical diameter, ideal critical diameter, and severity of quench can be determined. Some of the results obtained by Grossmann and others are shown in fig. 16.19. The values of severity of quench for different quenching conditions referred to a still water quench as unity are given in Table 16.1.

TABLE 16.1

Agitation of quenching medium	Movement of piece	Severity of quench			
		Air	Oil	Water	Brine
None	None	0·02	0·3	1·0	2·2
None	Moderate	—	0·4–0·6	1·5– 3·0	—
None	Violent	—	0·6–0·8	3·0– 6·0	7·5
Violent or spray		—	1·0–1·7	6·0–12·0	—

The relationship between ideal critical diameters for different specified percentages of martensite is shown in fig. 16.20.

16.18. Jominy end-quench test*

This is a simpler test requiring one specimen only. A specimen of standard dimensions, as shown in fig. 16.21, is austenized and then transferred very quickly from the furnace to an apparatus in which it is supported by one end and the other end is subjected to a jet of water which can be supplied via a quick-acting valve immediately the specimen is in place. The specimen is cooled, the rate of cooling varying along the length of the bar. When the specimen is cold, a flat, 0·38 mm deep, is ground along the length of the specimen. Hardness measurements are made along this flat and presented in a graph plotted against distance from the quenched end. Some typical curves are shown in fig. 16.22.

* B.S. 4437: 1969. *Method for the End Quench Hardenability Test for Steel (Jominy Test).*

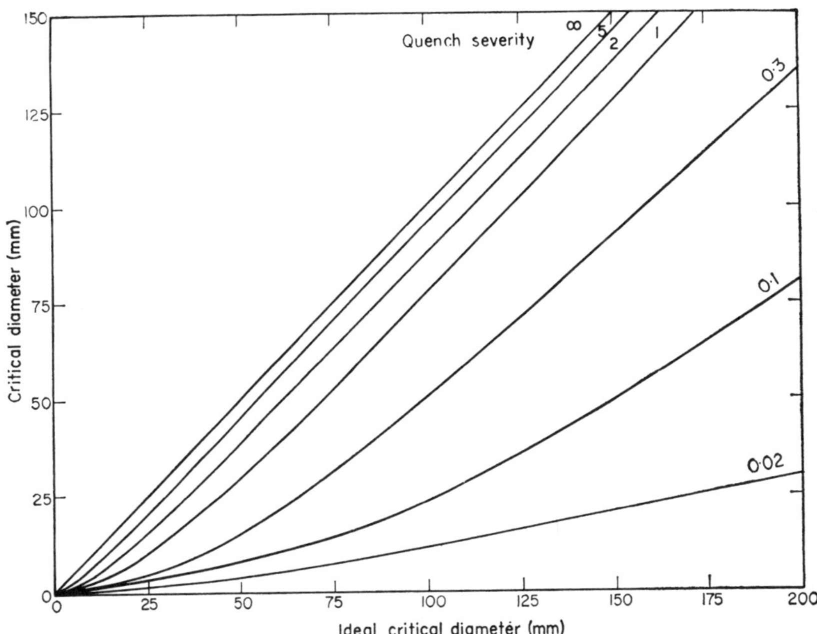

Fig. 16.19.—Relation between critical diameter, ideal critical diameter, and severity of quench

From Donald S. Clark and Wilbur R. Varney, *Physical Metallurgy for Engineers*. Copyright 1952. D. Van Nostrand Co., Inc., Princeton, New Jersey.

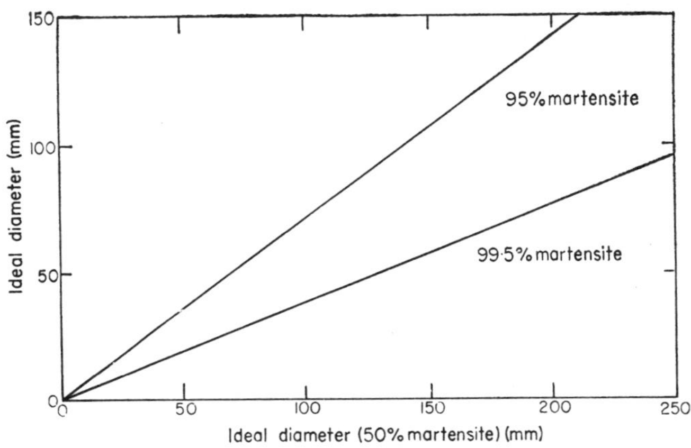

Fig. 16.20.—Relationship between ideal critical diameters for different martensite microstructures. [After *Metals Handbook*]

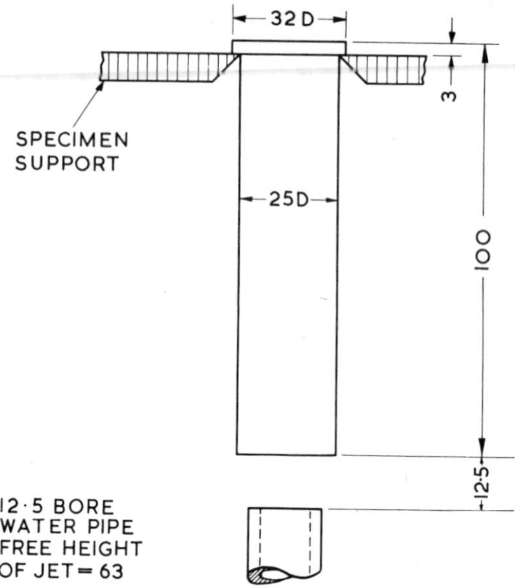

Fig. 16.21.—Jominy end-quench specimen mounted in quenching rig. Dimensions in mm

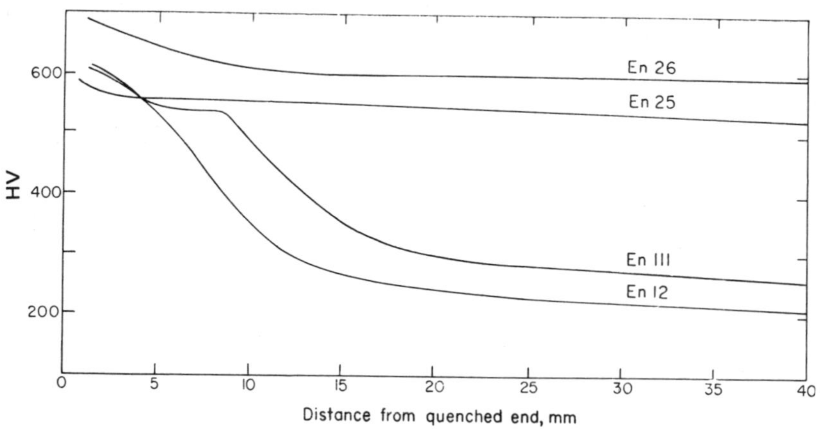

Fig.16.22.—Some typical Jominy end-quench curves. [After International Nickel]

16.19. Relation of ideal critical diameter to end-quench curve

The rate of cooling at any point along the length of a Jominy bar can be determined either by thermal flow calculations or by measurement.

Also the rate of cooling during quenching at various points on sections of round bars can be determined. It may be assumed that points that have the same rate of cooling will develop the same hardness, and hence the hardness variation across the section of a round bar can be predicted from the end-quench curve. Because the cooling-curve shapes differ in the two cases, the relationship depends on the cooling-rate criterion used. For equal average rates of cooling over the temperature range 700 to 500 °C, the curves in fig. 16.23 apply.

If the position on the Jominy bar at which the structure is 50% martensite can be identified, the size of round bar that would give this structure at its centre can be deduced from fig. 16.23. This value is the critical diameter for oil quenching. The 50% martensite point can be taken as the point of inflexion of the end-quench curve or may be found by use of fig. 16.24, which shows the relationship between hardness value and percentage of martensite. The hardness at the water-quenced end would be taken as the hardness of 100% martensite.

Other factors make the relationship not entirely quantitative, as for example the inhomogeneities that exist in all steels, particularly between the surface and centre of round bars.

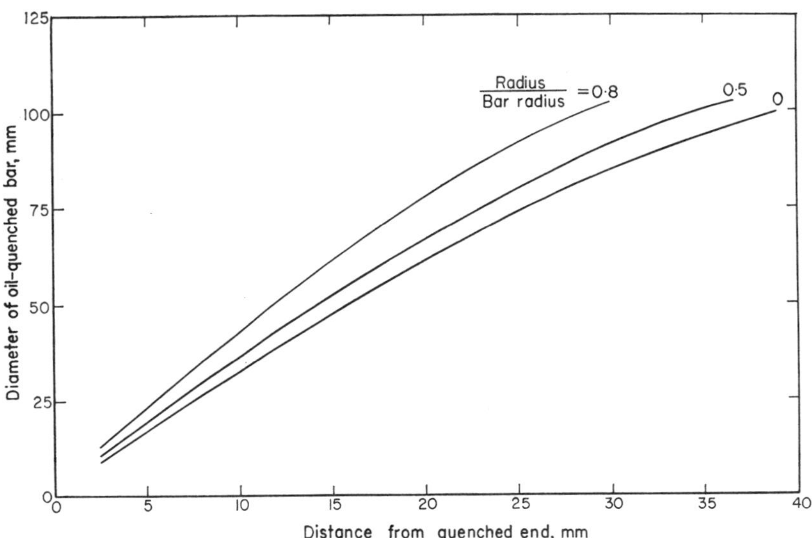

Fig. 16.23.—Positions on oil-quenched round bars and on Jominy specimen having the same cooling rates. [After International Nickel]

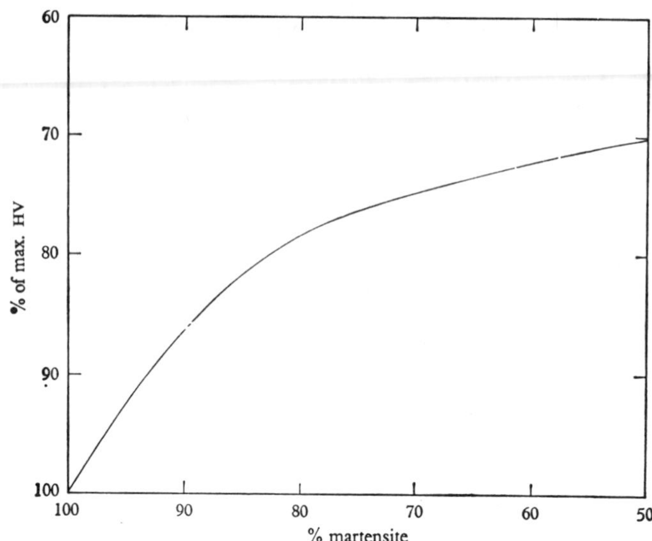

Fig. 16.24.—Relationship between hardness and martensite content of microstructure

16.20. Ruling section

It is possible to assess hardenability in terms of mechanical properties, the critical diameter then being the largest diameter of bar which can be hardened and tempered to develop a selected combination of mechanical properties in specimens machined from the axis of the bar. The critical diameter is referred to as the maximum *ruling section.* Thus, in an actual machine part, the portion that is most difficult to harden will be that which will cool most slowly on quenching, i.e., the part with the largest dimension that limits cooling. That diameter of a round bar that would have the same cooling rate at its centre is known as the *equivalent ruling section*, and under similar quenching and tempering treatment should produce the same mechanical properties as the worst in the actual machine part. Both hardenability assessment and control in terms of ruling section are commonly used in Great Britain, as for example in the British Standard Specification* that relates to wrought steels. For each steel, several maximum ruling sections and the associated mechanical properties are quoted. A knowledge of the percentage of martensite obtained on quenching is not involved in this method.

The relationship between cooling rates of plates and rectangular

* B.S. 970: *Specification for Wrought Steels.*
Part I: 1972. *Carbon and carbon manganese steels.*
Part II: 1970. *Direct hardening alloy steels.*

sections and those of round bars can be assessed from sets of conversion tables and curves given in B.S. 5046.†

QUESTIONS

1. How would you determine the isothermal transformation diagram of a 0·4% carbon steel? Sketch the diagram, label the areas and explain its shape. [P]

2. What is an *isothermal transformation diagram*, and how is it used to determine the appropriate heat treatment for a carbon steel in order to obtain the desired properties in the heat-treated condition?

Discuss the influence on the diagram and on the corresponding heat-treatment procedure of (*a*) grain size, and (*b*) carbon and alloy content. [MST]

3. What is meant by (*a*) austenitic grain size, and (*b*) ferritic grain size? Why is the austenitic grain size important, and how may it be evaluated for a 0·2% carbon steel?

Distinguish between an inherently fine-grained steel and an inherently coarse-grained steel. Show how the former may develop a coarser grain size than the latter. What method can be used for refining the grain size of metals which are not polymorphic? [MST]

4. Describe the structures produced in a small sample of En 12 steel at the end of each stage of the following treatment, giving quantitative results where possible:

(*a*) Heat to 830 °C for 0·5 h.
(*b*) Quench to 600 °C and isothermally transform for 5 s.
(*c*) Continue isothermal transformation for 15 s (20 s total time).
(*d*) Quench to room temperature.
(*e*) Heat to 500 °C for 15 min and cool to room temperature. [MST]

5. Describe the effects of different heat treatments upon the mechanical properties of a plain carbon steel containing about 0·5% carbon.

What are the relative merits of the structures obtained by tempering slack-quenched and fully-quenched samples of a steel to give the same tensile strength? [S]

6. Discuss the influence of carbon content, heat treatment, and the temperature of test upon the results of Izod tests on plain carbon steels. Why has the effect of the last variable received much notice in recent years? [MST]

7. Define the terms *critical diameter* and *ideal diameter* as used in relation to the hardenability of steels.

A steel bar, of 150 mm diameter, quenched in still water, was found to give 50% martensite and a hardness of 360 HV at the centre. What diameter bar of the same steel would be expected to give 95% martensite at its centre on air cooling? What would be the probable hardness at the centre of this bar? [MST]

8. In certain steels martensite can be produced by moderate rates of cooling. Under what circumstances may this be advantageous?

Describe an experiment whereby the effect of different cooling rates can be investigated by a test on a single specimen, and discuss the application of the results of such a test. [MST]

9. Sketch and explain the difference between the isothermal transformation diagram and the continuous cooling transformation diagram for a mild steel. [P]

† B.S. 5046: 1974 *Method for the estimation of equivalent diameters in the heat treatment of steel.*

CHAPTER 17

Alloy Steels

17.1. Purposes of alloying

Carbon steels, as made commercially, always contain certain amounts of other elements which are present in the ore and complete removal of which would be extremely difficult and expensive. Sulphur and phosphorus, when more than 0·05% of either is present, tend to make steel brittle, so that during steel making the amounts present are reduced at least to this value. Silicon has little effect on strength and ductility if less than 0·2% is present. As the content is raised to 0·4%, the strength is increased without impairing the ductility, but above 0·4% it causes a decrease in ductility. Silicon is frequently used as a deoxidizer in steel making, and the fraction that does not form silicon dioxide, which passes into the slag, will remain in solution in the steel. The quantity of silicon used has to be controlled so that, except in special cases, there is not more than a total of 0·4% left in the steel finally.

Manganese has a strengthening effect, but in quantities greater than 1·5% reduces the ductility. As mentioned on p. 231 the manganese/carbon ratio has an important effect on the notch ductile–brittle transition temperature. Because of its beneficial effect, excess manganese is usually added to the molten metal to bring the content up to the desired value, as well as being used as a deoxidizer.

By intentionally adding further elements to iron-carbon alloys, the structure can be modified to improve the properties in a number of ways. The same effect may be obtained in different ways by using different alloying elements or by different combinations of elements. Steels to which these intentional additions have been made (including those steels with manganese in excess of 1% or silicon in excess of 0·3%) are known as *alloy steels*.

One particular effect of alloying is that it enables martensite to be produced with low rates of cooling and permits larger sections to be hardened than is possible with plain carbon steel.

17.2. Effect of alloying elements on iron-carbon equilibrium diagram

The body-centred cubic metals when alloyed with steel (chromium, tungsten, vanadium, and molybdenum are amongst the more important) tend to form carbides which may rob the iron of some of its carbon, so that the proportion of Fe_3C is changed.

The face-centred cubic metals (nickel, aluminium, zirconium, and copper) do not form carbides. As would be expected, the b.c.c. elements are more soluble in α-iron and the f.c.c. elements more soluble in γ-iron (except aluminium). In all these cases, the element forms a substitutional solid solution.

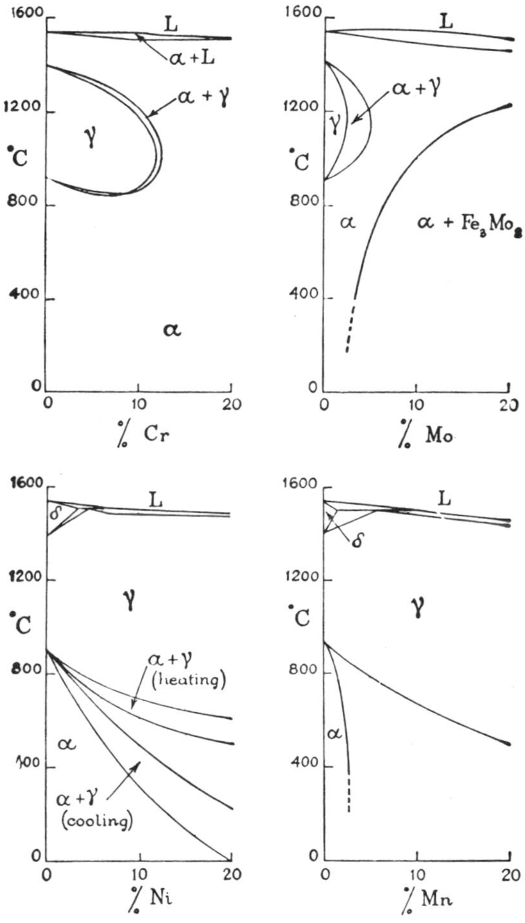

Fig. 17.1.—Iron-rich end of the equilibrium diagrams of iron with chromium, molybdenum, nickel, and manganese

Manganese, which is allotropic and has three complex structures, forms carbides readily and is more soluble in γ-iron than in α-iron.

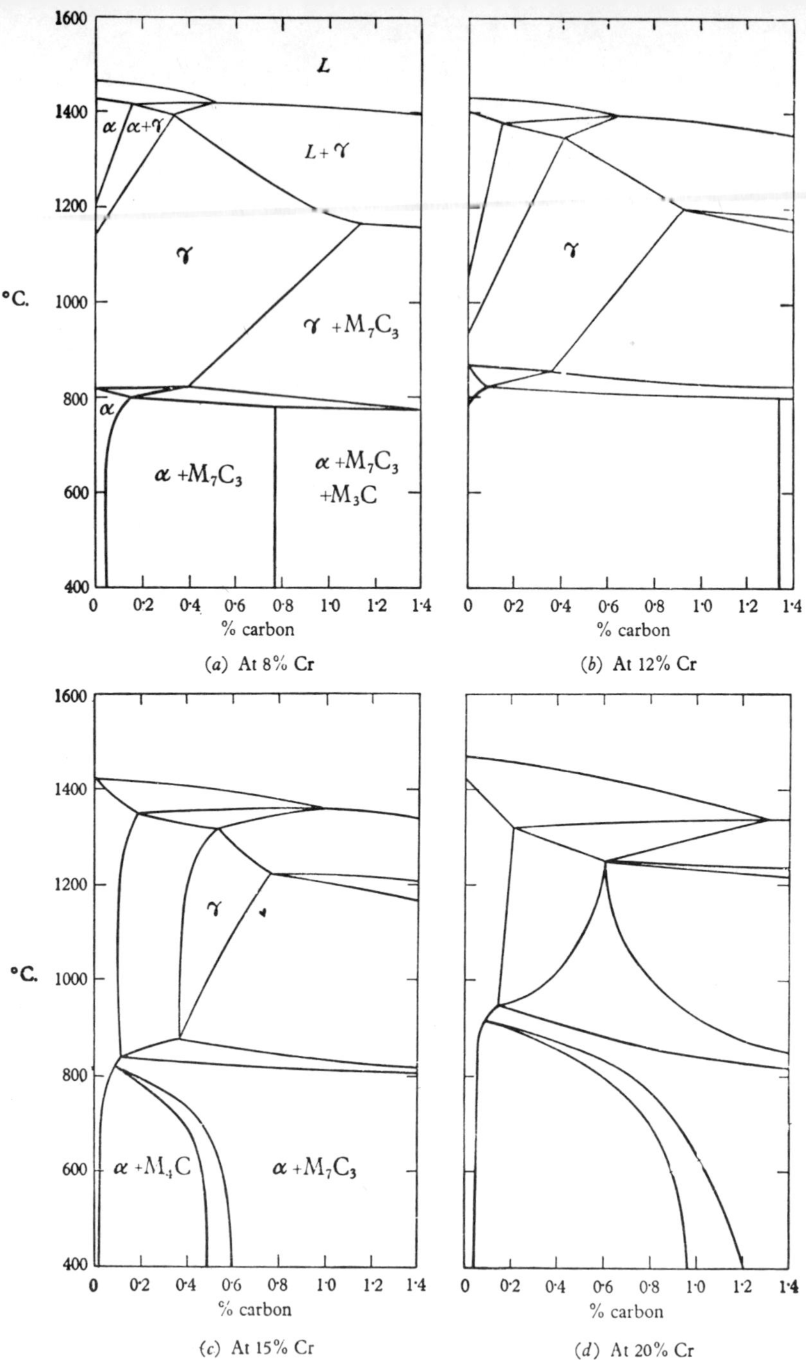

(*a*) At 8% Cr

(*b*) At 12% Cr

(*c*) At 15% Cr

(*d*) At 20% Cr

Fig. 17.2.—Sections of the iron-chromium-carbon equilibrium diagram at various chromium contents

An element that is more soluble in γ-iron will tend to stabilize the γ-phase by lowering the γ–α change temperature, and raising the γ–δ change temperature. An element that is more soluble in α-iron will have the reverse effect. With the exception of chromium, this is found to be so, as may be seen from the binary equilibrium diagrams in fig. 17.1. Chromium lowers the γ–α change temperature slightly when present in amounts up to 8%, but above that amount the temperature is raised sharply. It will be noticed that for more than 13% chromium or 5% molybdenum the γ-phase does not exist. The α–γ changes in iron-nickel and iron-manganese alloys are extremely sluggish, so that for all practical heating and cooling rates the change temperatures are different.

When carbon is present, the transformation temperatures are similarly affected, but the eutectoid has a lower carbon content. The size of the austenite region is increased by γ-stabilizing elements and decreased by α-stabilizing elements. The latter effect may be seen from fig. 17.2, which shows vertical sections of the Fe–C–Cr ternary equilibrium diagram. Although the γ-phase does not exist in pure iron-chromium alloys for more than 13% chromium, in the presence of carbon a γ-phase can exist up to 20% chromium.

17.3. Effect of alloying elements on isothermal transformation diagram*

The general effect of alloying elements is to reduce the diffusion rate of carbon and so slow down the transformation of austenite to ferrite and pearlite structures, so that the isothermal transformation curves are moved to longer times as illustrated in fig. 17.3. There is also a tendency to separate the pearlite and bainite knees as mentioned on p. 282. Because of the slowing down of transformations, the critical cooling rate is reduced and hardenability increased. This effect is only fully realized when the alloying element is dissolved in the austenite. When alloys containing carbide-forming elements are austenized prior to quenching, the treatment must be sufficiently prolonged for the carbides to have dissolved in the austenite. This may involve a higher soaking temperature and a soaking time of two to three hours in place of the half to one hour sufficient for plain carbon steels.

Any added element lowers M_S-temperature.

17.4. Effect of alloying elements on tempering

Because the presence of alloy elements tends to reduce the diffusion rate of carbon, processes are also slowed down in tempering. Higher

* The isothermal transformation diagrams of various alloy steels can be found in *Atlas of Isothermal Transformation Diagrams of B.S. En steels*, 2nd Ed. Special Report No. 56, Iron and Steel Inst., 1956.

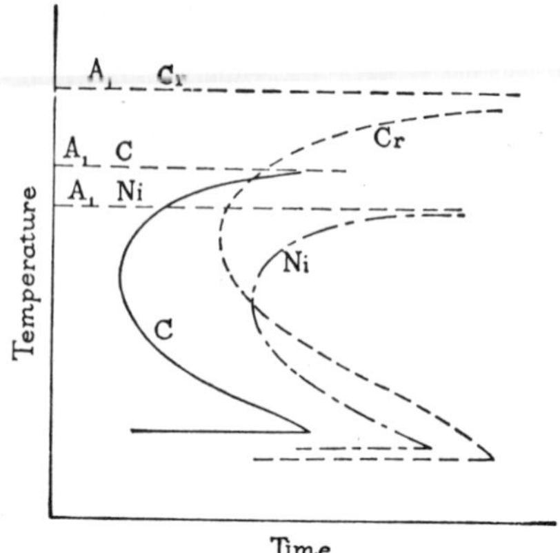

Fig. 17.3.—Shift of isothermal transformation curve due to addition of nickel and of chromium to a plain carbon steel

tempering temperatures and longer times may therefore be necessary when heat-treating alloy steels.

17.5. Effect of alloying elements on mechanical properties

In the normalized condition the alloy elements dissolved in the ferrite cause strengthening, interstitial solute atoms having a much greater effect upon the yield strength than have substitutional solute atoms. Any carbides present also increase strength and hardness. In a quenched condition, the hardness is solely dependent upon the carbon content if 100% martensite is present. After tempering, elements dissolved in the ferrite will again have a strengthening effect.

17.6. Classification of alloy steels

Alloy steels can be classified according to the elements they contain (e.g., nickel steels, chromium-vanadium steels, etc.) or according to their uses (e.g., high-tensile structural steels, stainless steels, etc.) or according to the structure produced by the heat treatment usually employed (e.g., pearlitic, martensitic, austenitic). The most reasonable classification, which is adopted in the following summary, is that based primarily on use.

17.7. High-tensile structural steels

High-tensile structural steels are steels of low carbon content which are used in the normalized or as-rolled condition. They are thus in a pearlitic state and the improved properties are due to solute hardening and to carbides dispersed in the ferrite. Ferrite can dissolve about 0·03% carbon at the eutectoid temperature but only 0·0001% at 20 °C. On slow cooling the extra carbon is precipitated as Fe_3C. If this were precipitated in a dispersed condition in the ferrite grains it would have a strengthening effect. However, it adds to the cementite lamellae of the pearlite. By quenching from the eutectoid temperature and ageing, precipitation hardening does occur. Ageing at 20 °C for 100 hours will give an increase of 60–75 MPa in the yield stress, but such a process, as with any heat treatment, is expensive.

By adding an element which forms a carbide with a smaller solubility in ferrite than cementite at temperatures below the eutectoid, this carbide can be precipitated on slow cooling from the rolling or normalizing temperature. The increase of strength is then obtained without any special heat treatment.

Niobium, titanium, and vanadium are all suitable elements—niobium in quantities between 0·005 and 0·03% being the best because NbC also restricts austenite grain growth during normalizing thus ensuring a small ferrite grain size, even in thicker sections. By careful control of the rolling practice to make the final austenitic grain size as small as possible, the strength can be maximised. Merely reducing the final rolling temperature of a 0·03% Nb, 0·1% V steel from 780 °C to 680 °C will increase the yield stress by 20%.

Normalizing at higher temperatures up to 1250 °C increases σ_i (see Section 11.3), but at the expense of having a coarser grain size. Hence the treatment which gives a greater strength also gives a higher ductile-brittle transition temperature.

Tensile strengths of the order of 600 MPa are common in this group,* compared with 450 MPa for mild steel.

The general object of using these stronger steels is to give a reduction of weight, which can be an important factor in structures such as ships and bridges, but care must be taken that the notch ductility is sufficient at the operating temperatures. The saving of weight has to be balanced against the increased cost of using more expensive materials.

17.8. Alloy steels used in a fully heat-treated condition

This group, which accounts for the greatest part of the tonnage of alloy steels, owes its importance to the high hardenability and to better all-round properties in the fully heat-treated condition. Because of the former reason, large sections can be fully quenched and slower cooling

* Steels of this type are covered by B.S. 4360: 1972. *Weldable Structural Steels.*

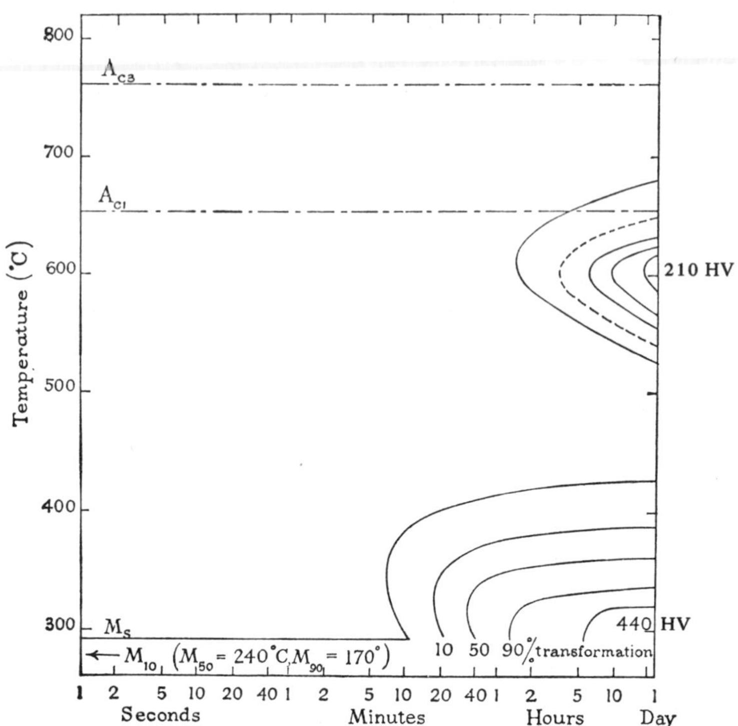

Fig. 17.4.—Isothermal transformation diagram for an En 30 steel (0·35% C, 4·23% Ni, 1·43% Cr). [After International Nickel]

rates can be used. The latter is important where there is danger of cracking in a water quench. For *air-hardening steels*, the critical cooling rate is so low that a quenching bath is unnecessary for smaller sections. Thus En 30, a steel containing 3·9–4·3% nickel and 1·1–1·4% chromium, whose isothermal transformation diagram is shown in fig. 17.4, can be air-hardened in sections up to about 65 mm diameter. This treatment may give a little lower bainite with the martensite, but not sufficient to impair the final properties. Fig 17.5 shows a property chart for this steel. En 30 is intended for use as-cooled or tempered at a temperature not exceeding 250 °C. In this condition it will show a tensile strength of 1500 MPa.

There are manufactured a very large number of steels that fall into this group of heat-treated steels. Selection of the correct steel, after the required hardenability and mechanical strength are met, will then depend upon secondary requirements, such as suitability for welding, when a low carbon content is desirable, or low-temperature notch toughness when a nickel steel is needed, or wear resistance, which can be given by the carbides in chromium steel, etc. Such selection is a matter of consulting the manufacturers' lists of available steels and a matter of experience and is beyond the scope of this text.

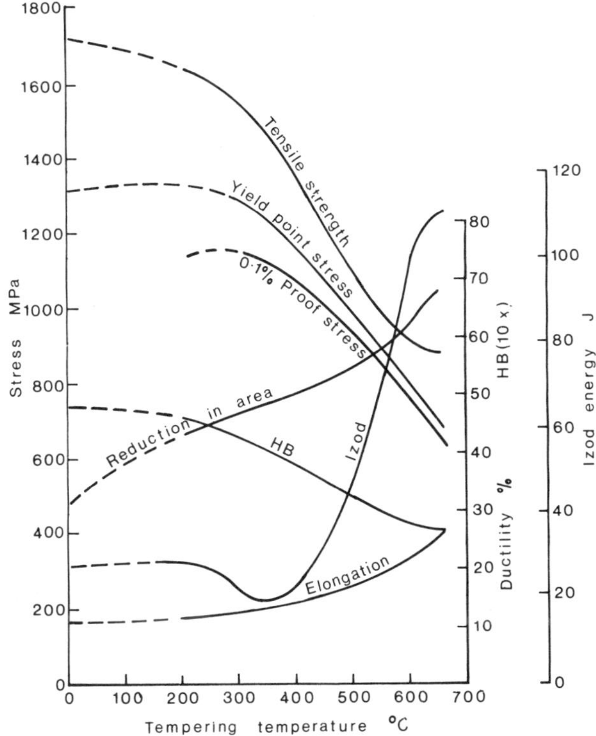

Fig. 17.5.—Tempering diagram for air-hardened, 30 mm diameter bars of En 30 steel. [After International Nickel]

17.9. Boron steels

Although not a separate class, these steels are of special note. Boron is used as an alloying element in low-carbon steels in quantities of about 0·002–0·004%, producing higher strengths in air-cooled parts and a higher hardenability.

The presence of these small quantities of boron inhibits ferrite formation but has negligible effect on the pearlite and bainite transformations. Hence it tends to separate the ferrite and bainite transformations and produce structures with a high proportion of bainite. Molybdenum also delays ferrite formation and lowers the B_S and M_S temperatures so that the presence of both elements affects the isothermal transformation diagram in the manner shown in fig. 17.6.* Air cooling thus gives a bainite structure with its superior strength. The effect of variation of boron content on mechanical strength is shown in fig. 17.7. It is thought

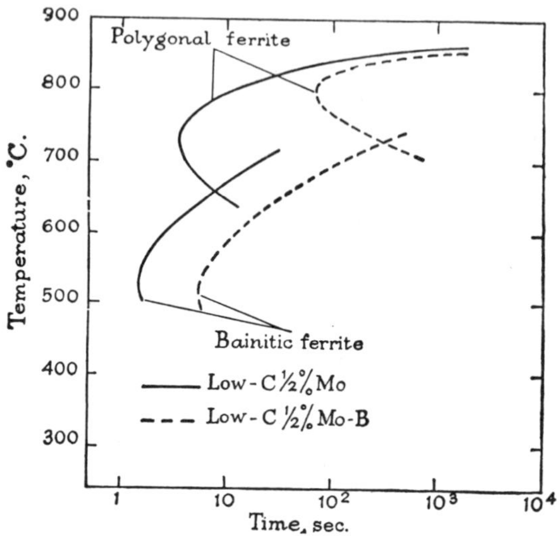

Fig. 17.6.—Isothermal transformation diagrams for low-C ½% Mo steels with and without boron (showing start of transformation only). [Irvine, Pickering, Heselwood, and Atkins]

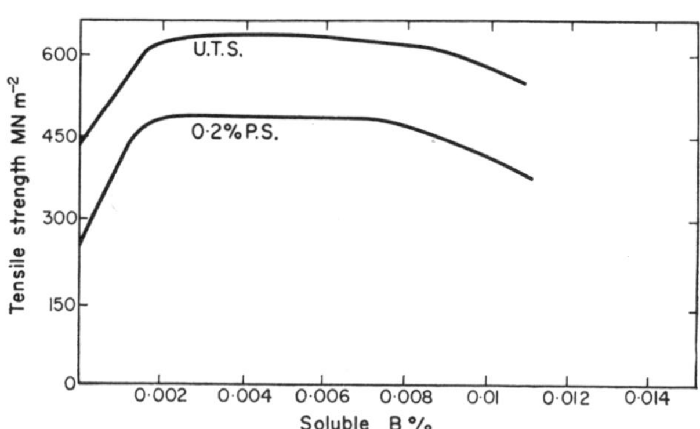

Fig. 17.7.—Effect of soluble boron content on steels containing 0·13–0·15% C and 0·40–0·60% Mo. [Irvine, Pickering, Heselwood, and Atkins]

* K. J. Irvine, J. B. Pickering, W. C. Heselwood, M. Atkins, " The Physical Metallurgy of Low-carbon Low-alloy Steels containing Boron," *J. Iron Steel Inst.*, Vol. 186, 1957, pp. 54–67.

that the boron, which forms an interstitial solid solution, segregates to the grain boundaries where, by occupying the most favourable interstices, it retards the interstitial diffusion of carbon, thereby retarding the formation of proeutectoid ferrite during cooling.

The carbon content in these bainitic boron steels can be considerably less than that of a mild steel and yet give greater mechanical strength. This makes such steels eminently suitable for welding.

At higher boron contents, Fe_2B is precipitated at the austenite grain boundaries, and for more than 0·007% boron forms continuous films which have a bad effect upon hot workability.

17.10. Ausformed steels

Ausforming is a combination of mechanical and thermal treatments which gives very high strength to steels. The quench is interrupted and the steel is subjected to extensive plastic deformation (up to 80% reduction in cross-sectional area has been used) while it is still austenitic. This plastic deformation must be carried out without any transformation products, pearlite or bainite, being formed. Hence it can only be applied to alloy steels which have a large bay between the pearlite and bainite curves as in fig. 17.4.

The dislocations present at the time of the final quench act as barriers to the growth of the martensite units. These small units, the very high dislocation density ($>10^{18}$ lines m^{-2}), and possibly some precipitation hardening all contribute to the final strength. The deformation should be carried out at the lowest temperature possible so as to minimize recovery and recrystallization. At such low temperatures the flow stress of the material is considerably higher than at normal hot working temperatures so that greater forces are needed.

Strengths of 3000 MPa have been realized. The process is particularly suitable for components which can be shaped by the deformation process, e.g., punches, before the final hardening. These steels have excellent wear properties.

17.11. Maraging steels

The interstitial solute elements, carbon and nitrogen, strengthen a steel, but also affect the notch ductility adversely. High strength steels with appreciable notch ductility can be produced by dispersed precipitates of compounds which do not contain either of these elements.

Alloys of iron with very low carbon content and high nickel content (18–22%) will change from austenite to ferrite on cooling and titanium or molybdenum, if present, will form intermediate compounds Ni_3Ti and Ni_3Mo. On quenching the alloy a martensite is formed, which is of low strength due to the low carbon content and is machinable. Subsequent precipitation hardening at 450–500 °C, the *maraging* process, causes the compounds to separate as particles of the order of 50 nm in size.

Although these steels will be more expensive than conventional heat-treating steels, because of the high nickel content, their use can lead to economies for various reasons apart from the possible superiority of mechanical properties. There is so little change of volume or shape during the hardening process, far less than in a conventional quenching and tempering process, that parts can be machined to the finished dimensions before the final heat treatment while still in a soft state. High temperature heating and quenching are not needed and because the hardening temperature is low, a controlled atmosphere furnace to prevent undue surface oxidation is not necessary.

Three basic types of 18% nickel maraging steels are now marketed having compositions and properties as listed in Table 17.1. Steels of higher strengths are under development.

TABLE 17.1.—COMPOSITIONS AND PROPERTIES OF TYPICAL MARAGING STEELS

Nominal 0·2% *proof stress*	MPa	1400	1750	2000
Composition				
Nickel	%	18	18	18
Cobalt	%	8·5	8	9
Molybdenum	%	3	5	5
Titanium	%	0·2	0·4	0·6
Properties in annealed state				
0·2% proof stress	MPa		650–800	
Tensile strength	MPa		1000–1080	
Hardness	HV		280–320	
Properties of fully heat-treated steels				
0·2% proof stress	MPa	1310–1450	1600–1810	1770–2080
Tensile strength	MPa	1370–1510	1650–1880	1810–2130
Elongation on $5{\cdot}65\sqrt{A}$	%	9–13	8–10	6–9
Reduction of area	%	35–70	35–60	30–50
Charpy V-notch at 20 °C	J	34–68	20–40	13–27
Hardness	HV	430–480	500–560	540–620

17.12. Corrosion and heat-resistant steels

Stainless steels have corrosion-resistant properties due to a high proportion of chromium, which forms an oxide film that is impervious to oxygen and highly resistant to the passage of ions. Thus once the film is formed, the metal ions and the oxygen of the atmosphere cannot continue to meet, and further oxidation is prevented. The corrosion resistance is improved as the chromium content increases, provided that the chromium is in solid solution in the iron and not combined as carbides. The corrosion resistance is enhanced by the presence of a certain amount of nickel. These steels may be sub-divided according to the structure obtainable at room temperature. If chromium is the only alloying element and more than 13% is present, in the absence of carbon the alloy will be α-phase at all temperatures. With carbon present, more chromium is necessary to give an α-structure at all temperatures. Such steels are known as *ferritic stainless steels*.

When alloys contain at least 24% chromium and nickel combined, and not less than 8% of either element, the γ-phase is retained on cooling

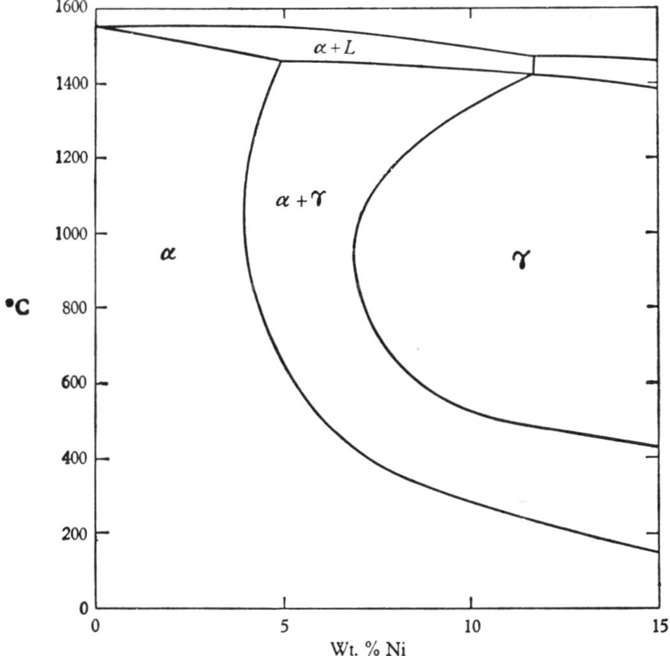

Fig. 17.8.—Section at 18% chromium of the iron-chromium-nickel diagram

Metals Reference Book, second edition edited by Smithells, Butterworth.

at normal rates. This is not immediately obvious from the ternary equilibrium diagram section in fig. 17.8, but the separation of the α-phase occurs only on very slow cooling. These are *austenitic stainless steels,* the most well known being the 18/8 stainless steel which contains about 18% chromium and 8% nickel.

Neither of the above groups is heat-treatable. Any steel which contains chromium and nickel in such proportions that it has a γ-phase at high temperatures and an α-phase on cooling at normal rates can be quenched to give a martensite structure. Such heat-treatable steels are known as *martensitic stainless steels*, even when not in a heat-treated condition. Whereas the mechanical properties of a martensitic stainless steel can be changed by heat treatment, austenitic and ferritic stainless steels can be strengthened only by cold working. Various precipitation-hardening stainless steels have also been developed.

Heating a stainless steel in the region of 500 to 700 °C can lead to the precipitation of carbides at the grain boundaries. The chromium is removed from the adjacent solid solution, resulting in a lessening of the corrosion resistance. This can be particularly troublesome in the heat-affected zone (see p. 334) of welds in these materials, when the subsequent corrosion is known as *weld decay*. The defect can be cured by reheating the steel after welding to a sufficiently high temperature to redissolve the carbides—about 900–1000 °C dependent upon the steel—and quenching. If such a treatment is impracticable, as for example due to size, it is necessary to use a *stabilized stainless steel*, that is, one which has niobium or titanium included in its composition. These elements have stronger carbide-forming tendencies than chromium, so that the chromium is not removed from the solid solution.

Steels which are to be used at high temperatures require resistance to oxidation, and high strength at the working temperature. Chromium is again the principal alloy element used, with several other elements which help to give the desired resistance to creep (see Section 15.10).

17.13. Case-hardening steels

Some equipment parts are required to have hard wear-resisting surfaces, and also toughness. The two are not possible in the same heat-treated steel, as if it is treated to give the greatest surface hardness, it will be martensitic and hence brittle, while if treated to develop maximum toughness, it will not be hard enough. The desired combination can be obtained by *case hardening*.

The surface layers of a low-carbon steel are *carburized*, that is, the carbon content is increased by heating in a suitable atmosphere, which causes carbon to diffuse into the surface layers. Afterwards, the part has to be heat-treated to give a martensitic structure in the high-carbon surface layers, and a tough ferritic structure in the core.

In nitriding, the nitrogen content of the surface layers is similarly increased, giving iron nitrides, which are extremely hard and require no further heat treatment.

Certain combinations of alloy elements are found to produce steels which are more suitable for carburizing and for nitriding.

17.14. Tool steels

Metal-cutting tools require high hardness and resistance to wear. For this purpose, steels containing carbides are necessary, but there must be a matrix tough enough to hold the hard carbide particles. During use, the tool tip may become hot due to friction, and steels are needed which do not soften appreciably with a rise in temperature.

For many purposes, near-eutectoid and hypereutectoid plain carbon steels suffice, but these are not suitable under more severe service conditions. In particular they must not be allowed to run hot.

Low-alloy tool steels have the advantage that they can be hardened by oil quenching or air-hardening and would be less likely to crack than would plain carbon steels when used for intricate tools.

For high-speed steels, which can be used at temperatures up to 500 °C, tungsten, chromium, and vanadium, with sometimes cobalt, are used. These are carbide-forming elements and give mainly Fe_4W_2C, with the other elements sometimes replacing the iron or tungsten. A typical composition is 18% tungsten, 4% chromium and 1% vanadium (an 18/4/1 steel).

17.15 Steels for cryogenic applications

The notch brittleness of steels was discussed in Section 13.13. The ductile-brittle transition temperature for plain carbon steels is usually in the range of − 50 °C to + 30 °C. Such steels are obviously useless for structural uses at cryogenic temperatures. The addition of nickel is very effective for lowering this transition temperature. A 9% Ni steel is notch ductile down to 77 K and is used for the fabrication of containers for storing liquid natural gas (B.P. 111 K). The growth of superconducting technology means that materials for service at liquid hydrogen temperature (20 K) or liquid helium temperature (4·2 K) will be required in the future. Recent work* has shown that steel containing 13·3% Ni will remain notch ductile down to 4·2 K.

* K. Ishikawa and K. Tsuya, " Fracture and Toughness of bcc Iron Alloys at Cryogenic Temperature," *Proc. 4th International Conference on Fracture*, Waterloo, Canada, 1977.

17.16. Special-purpose steels

Special-purpose steels are mainly steels employed for magnetic purposes. Alloys for use as transformer cores, etc., should have a high permeability with little hysteresis loss, while those to be used as permanent magnets should have a high remanence and coercive force.

QUESTIONS

1. Describe some ways in which alloying elements vary the properties of steels.
For what puposes are (*a*) manganese, (*b*) nickel, (*c*) chromium, and (*d*) molybdenum used in low-alloy steels? [S]

2. For what purposes may chromium be used as an alloy element in steels? [MST]

3. Discuss the primary function of the alloying elements in alloy steels of the machine steel group.
Determine the hardness at the centre and at 0·8 × radius of a 32-mm diameter bar of En 12 steel after oil quenching. Is oil quenching sufficiently rapid if a minimum of 50% martensite is specified? [MST]

4. Distinguish between *hardness* and *hardenability*. What factors affect these properties in steels?
Figure 17.9 shows the curves obtained in a Jominy end-quench test for two low-alloy steels which are of identical composition except that one contains 0·003% boron.
Determine the size of bar of each steel that will give a structure of 80% martensite at the centre after a still-water quench.
Explain each step of your method. [MST]

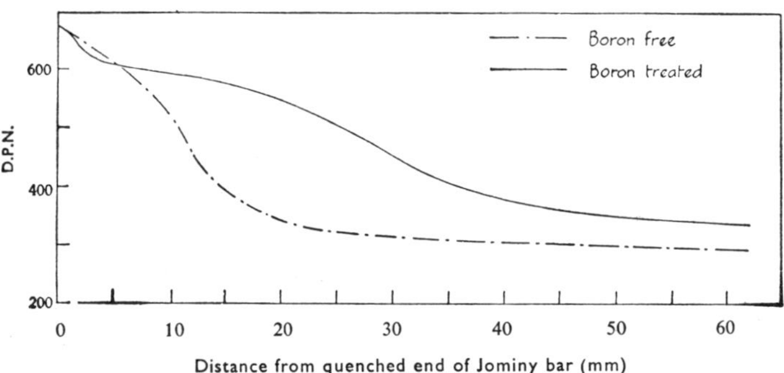

Fig. 17.9

5. Describe briefly the various types of stainless steels available and state, with reasons, those which can be strengthened by heat treatment and those which are magnetic.
What is *weld decay* and how may it be avoided? [MST]

CHAPTER 18

Corrosion

18.1. Introduction

Corrosion is a chemical or electrochemical reaction between a metal and its environment, which involves removal of the metal or its conversion to an oxide or other compound. It may also be defined as the passage of the metal atoms from the metallic to the ionic state. In some cases the compound will form a protective layer which reduces and prevents further corrosion, but in others this is not so and further corrosion is not inhibited.

The effects of corrosion are of serious importance in the engineering world. The cost of replacement of damaged material and of protective measures, such as painting and plating, is substantial and has been estimated at not less than £1 300 000 000 annually* in Great Britain alone.

Many of the basic principles of corrosion are now understood, and although corrosion theory cannot be applied rigorously to all practical problems, the basic principles can frequently give a guide to corrosion control.

18.2. Oxidation and reduction

As many corrosion processes are closely linked with oxidation, either directly or indirectly, it is as well to consider more closely the meanings of these chemical terms.

When charcoal burns in air, it forms carbon monoxide and carbon dioxide.

$$2C + O_2 \rightarrow 2CO$$
$$C + O_2 \rightarrow CO_2$$

Also when hydrogen burns in air, it forms water, and magnesium strip will burn rapidly when ignited to form magnesium oxide.

$$2H_2 + O_2 \rightarrow 2H_2O$$
$$2Mg + O_2 \rightarrow 2MgO$$

These and any other reactions in which the element combines with oxygen are termed *oxidation*.

* Department of Trade and Industry, *Report* of the Committee on Corrosion and Protection, H.M.S.O., 1971.

When these elements react with certain other non-metallic elements, the processes closely resemble that of burning in oxygen, and are also known as oxidation. Examples are the burning of carbon in fluorine.

$$C + 2F_2 \rightarrow CF_4$$

and the combination of iron with sulphur when heated

$$Fe + S \rightarrow FeS$$

In converting a metal to its oxide, the atom of the metal is changed to a positive ion, the electron passing into the oxygen atom. Thus oxidation may also be considered as a removal of electrons.

The reverse process to that of oxidation is known as *reduction.* Metallic ores, usually the oxides of metals, are reduced to the metals by removal of the oxygen or some other combined element by providing another element which has a higher affinity for the oxygen or other element. Thus iron ore is reduced to iron in the blast furnace by heating with carbon. The carbon forms carbon monoxide or dioxide which passes off as a gas, leaving behind the uncombined metal.

Reduction is the addition of an electron to the atom. This occurs directly in the electrolytic extraction of aluminium from its oxide by passage of a current through a solution of aluminium oxide in molten cryolite (Na_3AlF_6)

$$Al^{+++} + 3e^- \rightarrow Al$$

Oxidation or reduction of a substance could be carried out without simultaneous reduction or oxidation of another substance only if a large store of or reservoir for electrons is at hand. Accordingly, oxidation of one substance is always accompanied by reduction of another.

An element or compound which takes up the electrons removed from the oxidized material, thereby itself being reduced, is called an *oxidizing agent*, and one which liberates electrons is a *reducing agent*.

Some elements exhibit more than one valency, for example, iron is divalent in ferrous compounds such as ferrous chloride ($FeCl_2$), and trivalent in ferric compounds such as ferric chloride ($FeCl_3$). Tin is divalent in stannous compounds ($SnCl_2$) and tetravalent in stannic compounds ($SnCl_4$). A change of the -ous to the -ic compound is in effect

$$Fe^{++} \rightarrow Fe^{+++} + e^-$$

or

$$Sn^{++} \rightarrow Sn^{++++} + 2e^-$$

and so is an oxidation process. Conversely the -ic compounds can be reduced to -ous compounds.

18.3. Direct chemical corrosion by dry gases

Direct chemical corrosion is the reaction of the metal with dry oxygen or other gas. In the case of direct oxidation, the metal reacts to form an oxide, there being at the same time a decrease of the free energy, which acts as a driving force. Thus

$$M + \tfrac{1}{2}O_2 \rightarrow MO - \Delta G$$

where M is the metal atom and ΔG is the change in the free energy (see Appendix II). In the case of the noble metals such as gold and platinum, oxidation would involve an increase of free energy. Hence the metal is the stable form, which is why these metals occur in nature in an uncombined state.

The main feature of reactions of this type is not the driving force, but the rate at which it proceeds. Whereas aluminium has a large driving force, a very thin layer of the oxide is protective, and once a film of about 10^{-5} m thickness has formed, further corrosion in air at room temperature ceases.

18.4. Growth of oxide layer

The oxide can continue to form only where oxygen and metal atoms meet. Once a layer has started, it grows by a diffusion mechanism; the metal and oxygen form ions at the respective surfaces and the ions then diffuse through the film as shown diagrammatically in fig. 18.1. This

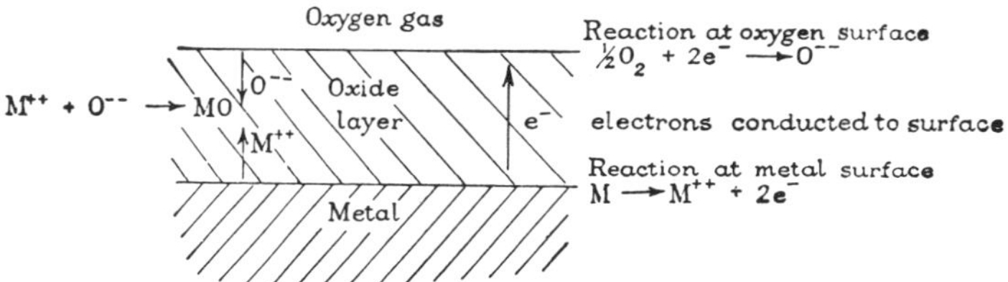

Fig. 18.1.—Movements and reactions during growth of oxide layer on a metal surface

mechanism requires the presence of lattice defects, such as vacancies, in the film, and such vacancies usually arise during film growth. The rate of oxidation is then controlled by the sum of the diffusion rates of the metal and oxygen ions, and these depend upon the size of the hole in the lattice of the ions of opposite sign. With a monovalent metal, e.g., sodium, there are two sodium ions to every oxygen ion, so that the hole in the oxygen lattice is much larger than that in the metal lattice and the metal atoms will diffuse easily, while the oxygen ions will diffuse very little. For a metal of high valency such as titanium (TiO_2) or tungsten (WO_3), the oxygen ions will diffuse relatively easily. Metals ions are smaller than oxygen ions to such an extent that, in oxides of trivalent metals, the two diffusion rates are approximately equal and the sum of the rates is less than for other metals. This helps to explain the high oxidation resistance of aluminium and chromium.

An increase of temperature will cause an increase of diffusion rates in accordance with Arrhenius' rate law (see p. 136). A thicker film must therefore develop to give protection. This may be seen in the films that form on iron when it is heated in the temperature range 200–300 °C. As the temperature is raised, the thin film of ferric oxide (Fe_2O_3) gives various colours due to optical interference effects, and the top temperature reached may be judged from the final *temper colour*.

On many metals the film breaks down before protection is complete. This may be due to evaporation, as occurs with the oxides of arsenic at 200 °C, and molybdenum and tungsten at high temperatures. Although the surfaces of these metals will continue to appear bright, corrosion continues at a constant rate.

The more usual cause of film failure is by cracking. The *Pilling-Bedworth rule* states that if the volume of the oxide is greater than the metal from which it was formed then the oxide gives protection, but that if the volume is less, then the oxide film will crack. Although this is true in many cases, it does not apply universally. If the film is sufficiently plastic, any stresses set up due to volume difference can be relieved by plastic flow. If the film is more brittle and cracks, protection will be lost because the oxygen will then have direct access to the metal surface. Film growth in these cases will continue at an approximately uniform rate. Since the plasticity of a substance varies with temperature, the liability to film breakdown is temperature-dependent. In uranium, for example, the oxide film does not crack at high temperatures, and the protection is better than at low temperatures.

18.5. The ionization of water

The other form of corrosion—electrochemical corrosion—occurs only in the presence of water or aqueous solutions, and is dependent upon the concentration of ions present in the solution.

Pure water ionizes into positive hydrogen ions and negative hydroxyl ions.

$$H_2O \rightleftharpoons H^+ + OH^-$$

It is found that the degree of ionization is such that pure water contains

10^{-7} kilomole of each sort of ion per cubic metre. By application of the laws of chemical equilibrium it can be shown that the product of the concentrations of the two sorts of ions will be constant for all aqueous solutions at the same temperature. Thus in any solution, the *ion product*

$$[H^+] \times [OH^-]$$

where $[H^+]$ and $[OH^-]$ are the concentrations of the two sorts of ion measured in kmol of ions m^{-3}, will have the same value as for pure water, i.e., 10^{-14}.

When acid is added to water, the concentration of H^+ ions will increase and so that of hydroxyl ions will decrease to keep the ion product constant. If hydrogen ions are removed from an aqueous solution in a corrosion process then more water will ionize and the $[OH^-]$ will increase so that the solution becomes more alkaline.

The acidity or alkalinity of a solution can thus be expressed solely in terms of its $[H^+]$ value and is usually given as the *pH value*, which is x where

$$[H^+] = 10^{-x} \text{ kmol of ions m}^{-3}$$

A neutral solution has a pH value of 7. A smaller value denotes acidity and a larger value denotes alkalinity.

18.6. Electrochemical corrosion

Corrosion at room temperature occurs mostly in the presence of water or aqueous solutions of ionic compounds (acids, bases, and salts). The reactions in these cases are electrochemical, that is, electric currents flow through those parts of the metal not actually being corroded.

The passage of the metal into the ionic state is an anodic reaction which is represented by

$$M \rightarrow M^{++} + 2e^-$$

The electrons left behind in the metal must be satisfied if the reaction is to proceed, and they pass through the metal to be neutralized elsewhere in a *cathodic reaction*, an example of which is the deposition of metal in an electroplating bath.

For each coulomb of electricity that passes from anodic to cathodic regions, one electrochemical equivalent of the metal at the anodic region should corrode.

When two dissimilar metals are placed in an aqueous solution so that there is no metallic contact between them, as in fig. 18.2, it is found that there will be an electromotive force between them, which would cause a current to flow if they were connected by an electrical conducting path.

Under standard conditions, the metals can be arranged in a series such that any metal will be anodic to any metal lower in the series, and cathodic to any metal higher in the series. The commoner metals arranged in the electromotive series are given in Table 18.1. Thus for the two metals shown in fig. 18.2, zinc would be anodic to copper and if they were connected externally, as in fig. 18.3, a current would flow causing the zinc to corrode.

The potentials of the metals under standard conditions are also given in Table 18.1, these being the potentials at which the element is in equilibrium with its ions in a "normal" solution, i.e., one which contains one kilomole of the ions per cubic metre. The decrease of free energy* per mole ($-\Delta G$) of a reaction and the potential E are related by

$$-\Delta G = nEF$$

where n is the valency and F is the Faraday, i.e., the quantity of electricity that will liberate or deposit one mole of a substance during electrolysis.

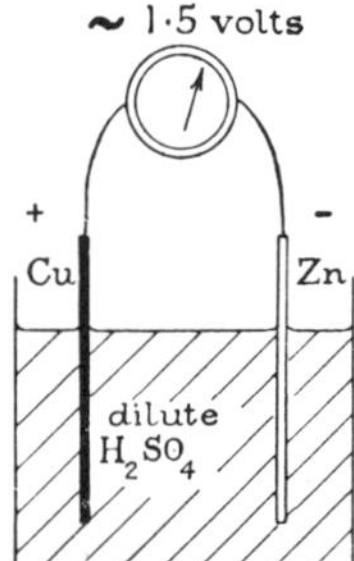

Fig. 18.2.—Simple voltaic cell

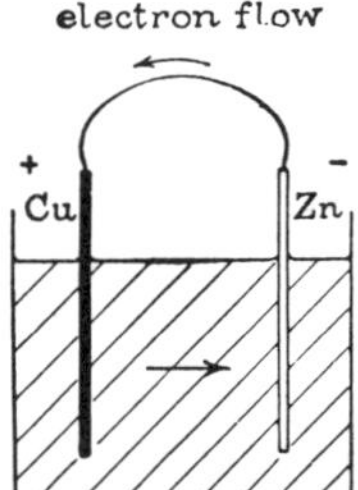

Fig. 18.3.—Current flow from cell of fig. 18.2

If the electrode is raised above its " normal electrode potential ", the metal will corrode, while if lowered below that value the reaction will be reversed.

The potential of the hydrogen electrode, i.e., hydrogen gas in equilibrium with a " normal " solution of hydrogen ions, is taken as the reference voltage. The potential of an electrode is modified if the ion concentration in the solution is not normal, being given quantitatively by

$$E = E_0 + \frac{RT}{nF} \log_e C$$

* See Appendix II.

where E is the potential, E_0 is the standard potential, and C is the concentration in kmol of ions m^{-3}. This will give a potential change of $0{\cdot}059/n$ volts for every change in concentration by a factor of ten.

When a current is flowing, there is also a change of potential due to *polarization*. For example, at a cathode where hydrogen is liberated by the reaction of hydrogen ions in solution with electrons

$$2H^+ + 2e^- \rightarrow H_2$$

the greater the current density, the greater is the potential necessary to continue to produce evolution of hydrogen.

Under any conditions other than the standard conditions described above, the actual potential difference that would be set up between two metals immersed in an aqueous solution will be different from the value given by Table 18.1, and the order of the metals in a *galvanic series* may differ from that in the electromotive series. A galvanic series places metals in order such that each is anodic to those lower in the series under actual conditions, and will therefore differ for different solutions.

TABLE 18.1.—ELECTROMOTIVE SERIES OF THE METALS

Metal and ion considered	Normal electrode potential (volts)
Anodic (corroded) end	
Li, Li^+	$-3{\cdot}04$
K, K^+	$-2{\cdot}92$
Na, Na^+	$-2{\cdot}71$
Mg, Mg^{++}	$-2{\cdot}37$
Be, Be^{++}	$-1{\cdot}85$
Al, Al^{+++}	$-1{\cdot}66$
Zn, Zn^{++}	$-0{\cdot}76$
Cr, Cr^{++}	$-0{\cdot}74$
Fe, Fe^{++}	$-0{\cdot}44$
Cd, Cd^{++}	$-0{\cdot}40$
Co, Co^{++}	$-0{\cdot}28$
Ni, Ni^{++}	$-0{\cdot}25$
Sn, Sn^{++}	$-0{\cdot}14$
Pb, Pb^{++}	$-0{\cdot}13$
H_2 H^+	$0{\cdot}00$(Reference)
Cu, Cu^{++}	$+0{\cdot}34$
Hg, Hg^{++}	$+0{\cdot}79$
Ag, Ag^+	$+0{\cdot}80$
Pt, Pt^{++}	$+1{\cdot}2$
Au, Au^{+++}	$+1{\cdot}50$
Cathodic (protected) end	

Electrolytic corrosion can occur on a single piece of metal when in contact with an aqueous solution. Owing to slight inhomogeneities in the metal, or varying concentration of ions or molecules in the liquid, particularly dissolved oxygen, local anodic and cathodic regions will form with small potential differences between them, and localized corrosion may occur.

18.7. Anodic and cathodic reactions

Various reactions are possible at the anode and at the cathode. The most probable ones at the anode are:

(*a*) formation of metal ions in solution

$$M \rightarrow M^{++} + 2e^-$$

(*b*) oxide formation

$$M + 2OH^- \rightarrow MO + H_2O + 2e^-$$

(*c*) hydroxide formation

$$M + 2OH^- \rightarrow M(OH)_2 + 2e^-$$

(*d*) formation of an insoluble salt

$$M + 2X^- \rightarrow MX_2 + 2e^-$$

At the cathode the most probable reactions are:

(*e*) metal deposition

$$M^{++} + 2e^- \rightarrow M$$

(*f*) hydrogen evolution

$$2H^+ + 2e^- \rightarrow H_2$$

(*g*) reduction of anions or cations

$$2M^{+++} + 2e^- \rightarrow 2M^{++}$$

$$2X^{--} + 2e^- \rightarrow 2X^{---}$$

(*h*) molecular reduction, especially oxygen

$$\tfrac{1}{2}O_2 + H_2O + 2e^- \rightarrow 2OH^-$$

When metal ions form and pass into solution (reaction *a*), they may react with other ions present to form a precipitate, which by forming away from the anodic surface does not give protection. Little polarization occurs and the rate of reaction is controlled by the rate at which electrons can be absorbed in the cathodic reaction. The reaction is therefore said to be under cathodic control.

If the anodic reaction produces substances (reactions *b*, *c*, and *d*) which adhere to the surface as a film, protection is given against further corrosion, leading to *passivity*. Unless there is film breakdown, further corrosion is controlled by diffusion as in Section 18.4. The film may fail by dissolving, if soluble, or by undermining or cracking. Oxides are generally more protective than hydroxides.

Metal deposition (reaction *e*) will occur if the solution contains a sufficient concentration of ions of a more noble metal. Thus copper is deposited on the blade of a steel knife when placed in a solution of copper sulphate.

Hydrogen evolution can occur if the potential anywhere on the metal surface is below the value for equilibrium of the reaction as listed in Table 18.1.

The oxidation or reduction of ions in solution is determined by the potential necessary for their change-of-valency reaction. Examples that can occur are the oxidation of chromium Cr^{+++} ions to chromate ions in which chromium has a valency of six, and the reduction of ferric Fe^{+++} to ferrous Fe^{++} ions.

The reduction of oxygen (reaction *h*) can occur only if there is dissolved oxygen in the solution. Of all processes this is probably the most important. Absence of dissolved oxygen will stifle this reaction and, if potentials are not sufficient to allow one of the other possible cathodic reactions to occur, then, due to cathodic control, corrosion at the anodic area may not occur.

18.8. Pourbaix diagrams

In the previous section, various anodic reactions that can occur were listed. For multivalent metals, there will be similar reactions appropriate to each valency. Which reaction will occur and the rate at which it proceeds is dependent upon the concentration of the ions or other corrosion product in the solution, the pH value of the solution and the electrode potential of the metal. If a state of chemical equilibrium is reached, then further reaction and hence corrosion will cease. If, however, the ion or other product is removed without reaching equilibrium, then the reactions will continue.

As may be seen from the equation on p. 320 the equilibrium concentration of ions in the solution will vary with the electrode potential, the concentration decreasing with increasing negative potential. Thus for iron in an acid solution dissolving to give Fe^{++} ions, a concentration of $[Fe^{++}] = 10^{-6}$ kmol of ions m^{-3} is reached where the potential is $-0{\cdot}62$V. Such a concentration is so small that iron would corrode only to a negligible extent to maintain equilibrium. At a potential of, say, $-0{\cdot}4$ V the equilibrium concentration would be about 10^2 kmol of ions m^{-3}. However, before such a concentration would be reached, the Fe^{++} ions will react in the manner described in the next section and corrosion would be continuous. If an external potential is applied to give a negative potential to the iron of at least −0·62 V, then the corrosion rate is so low that the metal is said to be *immune*.

Also, as mentioned in the previous section, certain products of anodic reactions may form a surface layer which gives passivity. The formation of these are dependent upon electrode potential and pH value of the solution.

The information already discussed can be shown for any one metal on a diagram of electrode potential versus pH—a Pourbaix diagram*—which is an equilibrium diagram for electrochemical reactions. A simplified form of the diagram for pure iron in contact with an aqueous solution at 25 °C is shown in fig. 18.4. If the iron is exposed to conditions represented by a point in either of the regions labelled corrosion then corrosion can occur and will do so unless some special feature of the kinetics of the process prevents it.

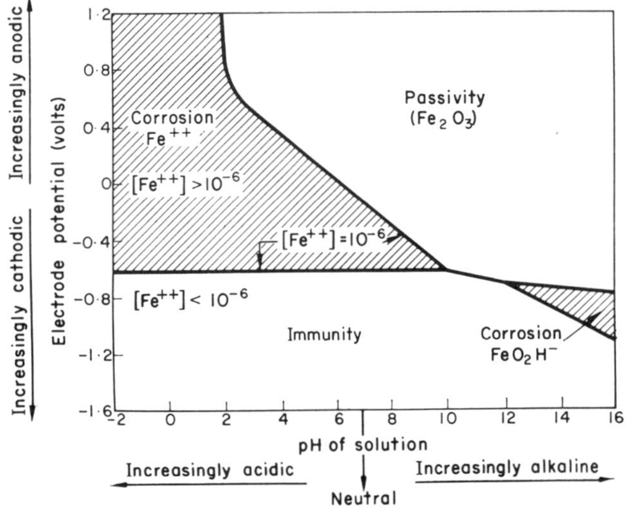

Fig. 18.4.—Simplified Pourbaix diagram for the iron-water system at 25 °C. [Fe^{++}] represents the equilibrium concentration of Fe^{++} ions in kmol of ions m^{-3}

The areas labelled *immunity* and *passivity* refer to the conditions described earlier in the section where corrosion is limited either by a very low equilibrium concentration of ions or by the formation of a protective layer of corrosion product. Pourbaix diagrams have been prepared for all the elements except the inert gases and are of help in forecasting conditions under which corrosion may or may not occur. Their use in selecting methods of corrosion prevention is described in Section 18.11.

* See *Atlas of Electrochemical Equilibria in Aqueous Solutions*, M. Pourbaix (translation by J. A. Franklin), Pergamon Press 1966.

18.9. Forms of metallic corrosion

The general corrosion mechanisms have been described, the most commonly experienced causes being dissimilar metals in contact and differential aeration, but each mechanism may make itself apparent in various forms.

Uniform attack is the term applied where all the surface is corroded to the same degree. Under such conditions, the useful life is easily estimated and unexpected failure should not occur.

Pitting and *intergranular corrosion* are non-uniform and can proceed in otherwise undamaged material. Their presence may not be known until failure occurs. Pitting may start as a localized anodic area due to an inhomogeneity. Once started, the solution around the anodic and cathodic areas will have different concentrations and form a concentration cell which accelerates the process. The inhomogeneities may occur in the metal or in the solution. For example, a drop of water on a steel surface will have a somewhat higher concentration of dissolved oxygen near the outside than at the centre. An anodic area would form at the centre. The Fe^{++} ions released would react with OH^- ions to form ferrous hydroxide, thereby decreasing the OH^- concentration, while the cathodic reaction at the edges would increase the OH^- concentration. These two effects will increase the potential difference and hence the corrosion current. The ferrous hydroxide forms a pale green precipitate away from the anodic area and then reacts with oxygen and more OH^- ions to form rust, which is a form of ferric hydroxide. The process is shown diagrammatically in fig. 18.5.

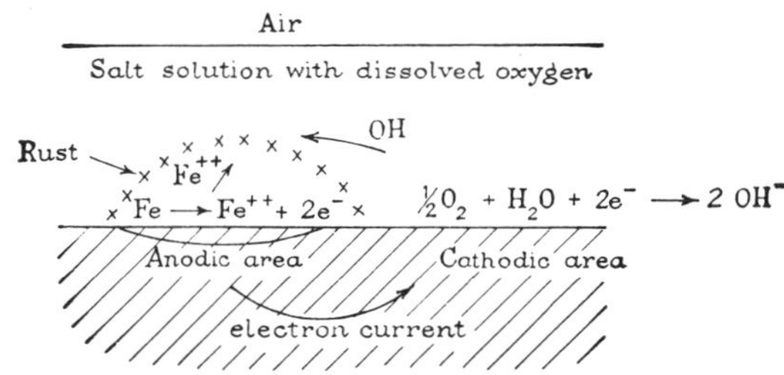

Fig. 18.5.—Processes in the rusting of iron

Surface inhomogeneities which cause localized anodic and cathodic areas are:

Breaks in a protective layer.

Deposits of foreign matter or loose corrosion products.

Local inhomogeneities in the metal, such as oxide particles.

Localized stressing of the metal, the stress causing a change in the electrochemical potential.

Where any corrosion products are removed mechanically by rapidly moving liquids or gases, corrosion will be accelerated. Local removal may cause pitting due to such corrosion-erosion or *impingement attack*.

Intergranular corrosion occurs where the grain boundaries are much more reactive than the remainder of the alloy. Weld decay in unstabilized stainless steels (Section 17.12) is an example of this.

Two-phase alloys are more prone to corrosion than are similar single-phase alloys, but even in the latter, preferential corrosion of one of the components may occur. *Dezincification* of brasses is an example of this effect. It is more likely to occur with higher zinc contents. In certain environments the brass goes into solution, a concentration cell is formed, and the copper is re-deposited as a spongy mass which has no strength. When this occurs generally over the surface, i.e., *uniform attack*, the effect is not serious. When, however, it is localized, giving *plug*-type dezincification, deep pits filled with spongy copper form and may even lead to perforation of pipe walls, etc.

Stress corrosion is a form of intergranular corrosion that is more pronounced when the material is subjected to a tensile stress. The destructive effect is then concentrated on a limited number of grain boundaries and travels in a general direction perpendicular to the applied stress. Its relation to fracture mechanics is discussed in Section 18.10.

*Corrosion fatigue** is the action of corrosion in the presence of repeated stresses and is far more serious than the action of either factor individually. The cracks that develop usually follow the slip bands set up by the plastic deformation.

Fretting corrosion is a particular form of corrosion fatigue that may occur in situations where there is slight relative movement of contacting surfaces due to the action of an alternating load. This may be found in bolted joints and other fitted assemblies. Surface contact at high spots results in localized plastic flow and cold welding. The welds subsequently rupture and loose metal particles are formed. In the presence of oxygen these particles oxidize and may become more abrasive. If protective oxide films are ruptured, thereby exposing bare metal, more oxidation can occur. The effect produces localized pitting which may become the source of fatigue cracks.

When two dissimilar metals are used in contact, the rate of corrosion of the anodic material is determined in part by the relative anodic and

* P. T. Gilbert, "Corrosion Fatigue," *Met. Rev.*, Vol. 1, 1956, p. 379.

cathodic areas. When the anodic area is relatively smaller, any cathodic control factors have little effect and corrosion is intense, whereas when the cathodic area is smaller, the corrosion of the anodic part is much less.

18.10. Stress corrosion cracking and fracture mechanics

Stress corrosion cracking can be considered in fracture mechanics terms (see Section 13.19) and studies made on conventional fracture toughness test specimens with the crack subjected to the corrosive environment. From the specimen geometry, initial crack depth, and applied load the stress intensity at the start of the test, K_{Ii}, can be calculated. If stress corrosion cracking occurs, the stress intensity increases as the crack depth increases until a critical value of K_I is reached (K_{ISC}) when the specimen fails by rapid fracture. It is found that K_{ISC} is constant for any one material. With a higher value of K_{Ii}, less time is needed for the critical value to be reached. This can be seen in the results of a typical series of tests shown in fig. 18.6. With an initial stress intensity corresponding to

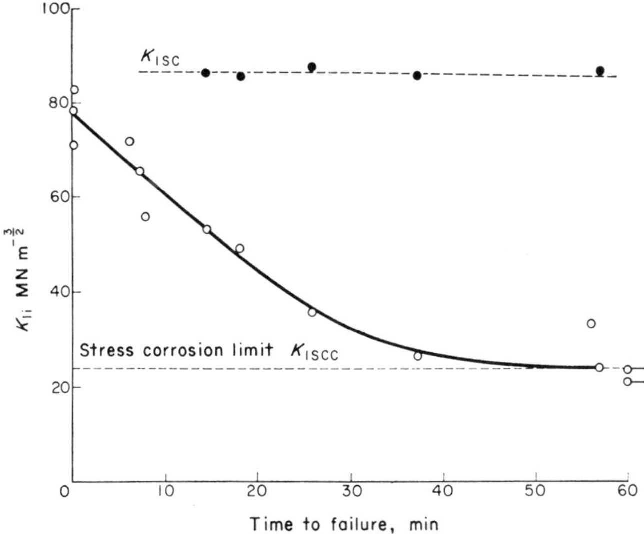

Fig. 18.6.—Effect of initial stress intensity on time to failure by stress corrosion for a $4\frac{1}{4}$% Ni-Cr-Mo steel.

○ Value of initial stress intensity factor K_{I_i}

● Value of stress intensity factor at failure, K_{ISC}

(after M. J. May, Fracture Toughness, Iron and Steel Institute publication No. 121, London, 1968, p. 95)

K_{ISC} of the material, the specimen fails immediately. With decreasing values of K_{Ii}, the time to failure increases until below a K_I level of 24 MN $m^{-3/2}$ stress corrosion cracking does not occur and the specimen does not break. This limit which can be considered as analogous to the fatigue limit (see Section 14.8) has been termed the stress corrosion limit, K_{ISCC}. In this case it is seen that K_{ISCC} is less than one third of K_{ISC}. If a high strength material is liable to be subjected to a corrosive environment, the possibility of stress corrosion cracking must not be overlooked in any fracture mechanics design considerations.

18.11. Prevention of corrosion

Various methods which can be used to reduce the extent or likelihood of corrosion are discussed in this section.* Some are applicable at the design stage and others during service.

A study of the Pourbaix diagram may indicate some methods which are illustrated here by considering the corrosion of iron in ordinary tap water. When a piece of iron is placed in a vessel of water of pH 7, an electrode potential of about − 0·45 V develops and the corresponding point is seen from fig. 18.7 to be in the corrosion region. Corrosion can be

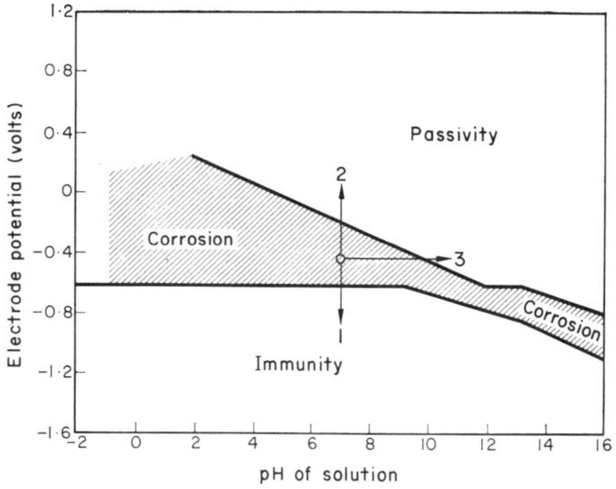

Fig. 18.7.—Pourbaix diagram for iron in contact with tap water showing three methods for controlling corrosion: 1, cathodic protection, 2, anodic protection, 3, protection by increase of alkalinity of solution

* W. D. Clark, "Design from the Viewpoint of Corrosion," *Met. Rev.*, Vol. 3, 1958, p. 279.

controlled by altering the conditions so that the point moves into the immunity or passivity region. Three possible methods are listed in the following sub-sections.

1. *Cathodic protection*

By the deliberate use of a metal that is anodic to the iron, connected to it by a conducting path and also exposed to the same aqueous solution, the potential of the iron will be lowered into the immunity region. The protection is gained at the expense of the *sacrificial metal* which is itself corroded.

Examples are the use of zinc plates in boilers and on steel hulls of ships near bronze propellors.

Alternatively, a potential difference from a D.C. supply may be connected between the metal to be protected and another electrode. This method is used, for example, in the protection of buried pipe lines. Though relatively simple to apply, there is a continual running cost due to the electricity used.

2. *Anodic protection*

A potential of opposite sign is applied so that the point representing the conditions moves into the passivity region. Two possible difficulties are: (a) the point is moved through a region of strong corrosion so that, if protection falls short, corrosion may be severe, (b) the film must be protective, which can be found only by experiment, the Pourbaix diagram merely indicating that the conditions are such that the solid product will form.

3. *Control of pH value*

The system point can also be moved into the passivity region by making the solution more alkaline. This may also be seen from fig. 18.8 which shows the variation of the aqueous corrosion of steel with pH value, it being least when the solution is sufficiently alkaline (pH 11–12).

Boiler water is made sufficiently alkaline by dosing the feed water with caustic soda. If, however, a high concentration of alkali is allowed to build up in a boiler, the steel may show a form of stress corrosion known as *caustic embrittlement.*

4. *Use of inhibitors*

Certain materials will contribute to the formation of more protective passive layers. Examples used with iron are chromates and phosphates, these being strong oxidizing agents. The influence of the former on the Pourbaix diagram is shown in fig. 18.9.

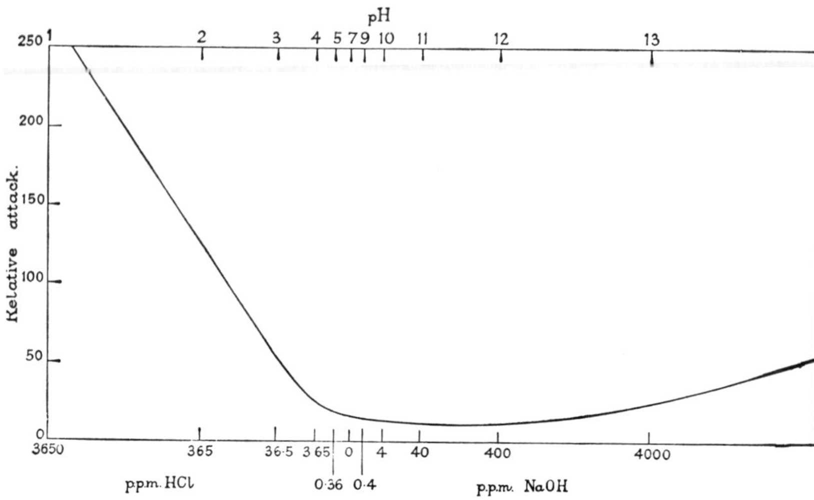

Fig. 18.8.—Attack on steel at 310 °C by water of varying degrees of acidity and alkalinity

[Redrawn from curve by Partridge and Hall—International Nickel]

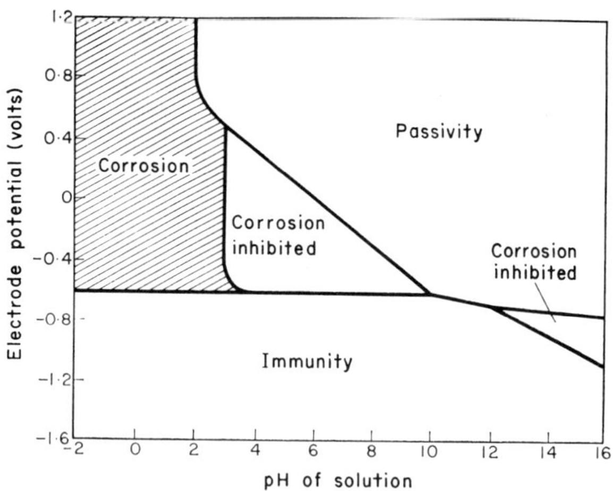

Fig. 18.9.—Inhibiting effect on the corrosion of iron due to the addition of a chromate to the aqueous solution. Compare with fig. 18.4

Paints containing inhibitors reduce the probability of film breakdown under the paint.

5. *Choice of metal*

The more noble metals offer better corrosion resistance, but the most inert ones are the most expensive. The most useful ones are those which form a stable protective film.

Stainless steels, containing at least 12% chromium and often some nickel, form a film which depends for its stability on Fe^{+++} and Cr^{+++} ions. Under reducing conditions (low potentials) the Fe^{+++} ions are reduced to soluble Fe^{++} ions and under oxidizing conditions (high potentials) the Cr^{+++} ions are oxidized to soluble Cr^{6+} ions, so that film breakdown can occur under extreme conditions.

Aluminium forms a very thin protective film which is resistant to solutions of pH 5–9. If, however, ions of noble metals are present even in extremely minute quantities, deposition of the metal occurs, followed by rapid corrosion of the aluminium. The Duralumin-type alloys which contain copper have a poor corrosion resistance. Also the film on pure aluminium is easily damaged and, if local rubbing continues so that a new protective layer is continuously removed, pits form.

Copper, being a noble metal, has good corrosion resistance and is almost completely passive in water and alkaline solutions.

6. *Construction*

Where different metals have to be used in the same structure and are liable to exposure to a corrosive medium, direct contact between the metals should be avoided by insulating layers, such as washers and sleeves at bolted joints, or suitable paints between contacting surfaces.

7. *Protective coatings*

The majority of corrosion protection is achieved by the application of additional surface layers.

Organic materials used are oils and greases for temporary use, and paints, varnishes, or lacquers for more permanence. The last three may also contain pigments for decorative purposes.

Inorganic materials used are the vitreous enamels which are basically silicate or borosilicate glasses (see Section 20.5) containing pigments to render the enamel opaque and of suitable colour. Chromates, phosphates, etc., which have been mentioned above as being corrosion inhibitors, act as oxidizing agents and promote the formation of oxide films which give anodic control.

Metallic coatings are widely used to give protection. If the base metal is isolated from the corrosive environment by a metal lower in the galvanic series, as for example by chromium, nickel, or tin on steel, the protection continues as long as the surface metal is self-protected. If the surface layer is scratched, however, the steel will be anodic and corrode where exposed. In the presence of an acid, tin becomes anodic to steel, hence giving protection when tinplate is used as a container for fruit, etc.

Galvanic protection is provided when the surface metal is higher in the galvanic series, e.g., cadmium and zinc on steel. If the surface is scratched, the surface metal is corroded, and a protective layer of cadmium or zinc oxide builds up over the scratch. Zinc and tin are usually applied by hot dipping, i.e., dipping the steel into the molten metal, and chromium, nickel, and cadmium by electroplating.

Thick diffusion coatings can be made by heating a metal in the presence of powdered aluminium (calorizing), chromium (chromizing), zinc (sherardizing), etc. Layers can also be applied by metal spraying, the covering metal being melted and blown on to the base metal as fine drops.

For metal cladding, a thin layer of corrosion-resistant material is made to adhere to sheets or plates of the base metal by rolling or some similar process. Alclad sheet is a Duralumin-type alloy with a thin surface layer of pure aluminium. Stainless steel is employed as a cladding for mild-steel sheets.

QUESTIONS

1. Discuss the role of oxygen in the corrosion of steel.

Describe in outline a method and the principle involved by which each of the following may be protected against corrosion:

(*a*) an exposed steel structure

(*b*) a steel boiler

(*c*) a ship's hull

(*d*) a buried steel pipe line. [MST]

2. The following crystallographic data refer to two metals and their oxides:

Material	Structure	No. of metal atoms in unit cell	Unit cell dimension nm *a*	*c*
Ca	fcc	—	0·5588	—
Cd	hcp	—	0·2979	0·5617
CcO	cubic	4	0·478	—
CdO	cubic	4	0·468	—

For each metal calculate the percentage change in volume when the metal changes to oxide. On the basis of the Pilling-Bedworth principle would these oxides be protective or non-protective against further oxidation? Give reasons why this principle does not hold in all cases.

3. Explain the mechanism of rust formation beneath a drop of water on a steel surface. [E]

4. Compare and contrast the principles involved in *cathodic protection* and *anodic protection* against corrosion.

Describe a Pourbaix diagram, and explain how it is used to show the conditions under which each protection method applies. [E]

5. Explain the terms *critical stress intensity factor*, K_{IC} and *critical strain energy release rate*, G_{IC}.

A thin-wall spherical pressure vessel 1·5 m in diameter with a wall thickness of 20 mm is manufactured from a high yield strength steel for which K_{IC} is 110 MN m$^{-3/2}$. Initial inspection shows that the most serious flaws in the vessel are short cracks in the inner surface lying with their planes normal to this surface and extending into the wall to a depth approximately equal to half their length at the inner surface. The stress intensity factor K_I for these cracks is given by

$$K_I = 2\sigma\sqrt{a}$$

where σ is the tensile hoop stress and $2a$ is the crack length at the inner surface.

In service the vessel contains a fluid at a steady working pressure of 18 MPa. Chemical action of the fluid on the steel causes cracks to grow slowly at a rate da/dt given by

$$da/dt = 6{\cdot}0 \times 10^{-13}\ K_I \text{ m s}^{-1}$$

where K_I is in MN m$^{-3/2}$. If in the initial proof test under inert conditions the vessel holds a pressure of 60 MPa, estimate the upper limit to the initial flaw size. What is the critical flaw size under working conditions? Hence comment on the probable mode of failure of the vessel in service assuming this to be due to slow crack growth. Estimate the life of the vessel in service.

It may be assumed that the cracks remain geometrically similar during growth and that general yielding does not occur. [E]

6. An investigation of the rate of crack growth in cold rolled brass when exposed to a solution of ammonium sulphate and subject to a static stress gave the following results:

Nominal stress σ (MPa)	Crack length c (mm)	Crack growth rate dc/dt (mm year^{-1})
4	0·25	0·3
4	0·50	0·6
8	0·25	1·2

Show that these results are consistent with the crack growth rate being a simple function of the stress intensity factor $K = \sigma\sqrt{(\pi c)}$.

The critical strain energy release rate, G_{IC}, for unstable fracture of this material in this environment is 55 kJ m^{-2} and $E = 110$ GPa It is proposed to use this brass for piping in an ammonium sulphate plant. The pipes must sustain a tensile hoop stress of 85 MPa and experience has shown that longitudinal scratches 0·02 mm deep are likely to occur on the inner surfaces of the pipes. Estimate the time that a pipe would last without fracturing once the ammonium sulphate solution started to flow through it. [E]

CHAPTER 19

Welding

19.1. Nomenclature

Welding is the joining of metal components by means of the inter-atomic forces—as distinct from joining methods like bolting, riveting, etc. A broad classification of welding methods would be into *plastic welding* and *fusion welding*. In *plastic-welding* processes, the parts are brought together under pressure, usually accompanied by heating to a temperature below the melting-point, so that plastic flow can occur and the parts become bonded together. The blacksmith welds steel parts together by the simple expedient of hammering. The main use of this method today is in resistance welding, the heating being due to the passage of a heavy electric current through a small region of contact where the components are brought together under pressure. These methods are *spot welding*, where joining is localized, and *seam welding* where the parts, usually sheet metals, pass between two pressure rollers while the current flows continuously.

Fusion welding is any process in which metal is made molten to make the joint. Either the parent metal or added filler metal is melted by an oxy-acetylene flame or electric arc, or some other means, and the molten metal allowed to solidify without the application of any pressure. Owing to the relative amounts of molten metal and unmelted metal, the weld metal is cooled quickly so that it solidifies in the manner of a chilled casting, that is, with columnar crystals and dendritic appearance.

The *heat-affected zone* is that metal which is not deformed plastically in plastic welding or melted in fusion welding, but which has its temperature raised sufficiently to cause metallurgical changes.

In the brief survey given here of some of the metallurgical changes that can occur due to welding, only arc welds will be discussed. Similar considerations will apply to welds made by other methods, with due allowance made for differences in temperatures and cooling rates.

19.2. Heat distribution in the vicinity of the weld

In successive regions of the heat-affected zone, metal will have reached all temperatures intermediate from room temperature to the temperature of the molten metal. The rate of cooling will depend upon the method of welding and the size of the specimen. In electric-arc welding, the heat is applied very locally, so that temperature gradients are steep and the heat-affected zone may extend only a fraction of an inch from the weld. In oxy-acetylene welding, on the other hand, the heat is not applied in so localized a manner, so that the heat-affected zone is larger, temperature gradients are not so steep, and the rate of cooling is generally less. Figure 19.1. shows possible temperature–time curves for various regions in the heat-affected zone of an electric-arc weld.

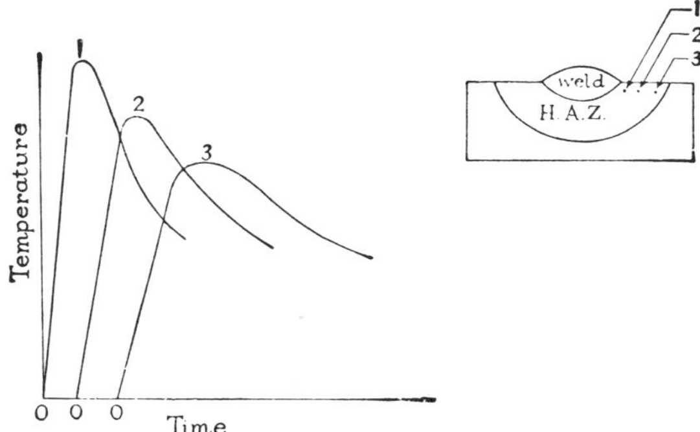

Fig. 19.1.—Typical heating and cooling curves for points in the heat-affected zone of a weld

19.3. Metallurgy of weld metal and heat-affected zone

The metallurgical changes that occur in the heat-affected zone will reflect the whole metallurgical behaviour of the particular alloy, and can be deduced from a knowledge of the equilibrium diagram and the laws of recrystallization and grain growth, together with the top temperature and rate of cooling for each point. Alternatively, the resultant microstructure would enable the heating and cooling history of any point to be deduced.

If the weld metal is not shielded from the atmosphere, then gases may dissolve in the liquid metal, especially oxygen, nitrogen, and hydrogen (the last is particularly likely if damp electrodes are used). On solidification, some will diffuse out but the rest will not escape and may form pockets of gas or compounds—oxides, etc.—which may have a serious effect upon mechanical properties.

The flux coating provided on most types of electrodes will help remove oxides that form and also provide a shielding layer to exclude the atmosphere during welding and the subsequent cooling. The layer of slag can also reduce heat losses and so control the rate of cooling.

Certain welding processes are carried out with an inert gas surrounding the arc to exclude all atmospheric gases.

In alloys there may be a loss of one or other of the elements due to volatilization. This can be compensated by using a filler rod of suitable composition which has an excess of the more volatile element.

Any effects due to previous heat treatment may be partially or completely destroyed, and the effects of previous cold work will be removed, by recovery, recrystallization, and grain growth.

The rate of cooling can influence the grain size and also, when phase changes occur on cooling, the final structure. Rapid cooling of steels may give a martensitic structure or a split transformation.

19.4. Welds in steel

A section of a fusion weld in mild steel is shown in fig. 19.2. Various zones are marked and their relationship to the iron-carbon equilibrium diagram shown. The zones are not sharply defined, but blend into one another. The unaffected zone, where the base metal has not reached the critical temperature, shows the typical grain structure of a normalized steel. In the transition zone, where the top temperature was between A_1 and A_3, the original pearlite and, towards the weld, increasing amounts of ferrite had changed to austenite on heating. The ferrite and pearlite have reformed on cooling with the pearlite progressively more dispersed towards the weld end. Where the top temperature was above A_3, the metal was entirely austenitic and grain growth occurred as the top temperature became higher. The final structure gives some indication of the size of the austenite grains and shows that there has been grain refining near the A_3-temperature, and a much larger grain size near the weld. The weld metal shows the typical cast-metal structure of columnar grains with a dendritic pattern.

Higher-carbon steels and low-alloy steels have slower critical cooling rates and hence are more prone to form martensite. Also, as stated in Section 16.10, a larger austenite grain size gives a higher hardenability, so that martensite is more likely to form in those parts of the heat-affected zone that reach the highest temperatures and where most austenite grain growth occurred. Figure 19.3 shows a weld in a medium-carbon steel, together with a graph of the hardness variation.

Temperatures are not uniform in the metal around a weld during both heating and cooling. As the weld metal and heat-affected zone cool, the hotter parts will still flow easily when the cool parts have become relatively rigid, but as the former parts cool, their thermal contraction will be prevented so that residual tensile stresses are set up.

Cracks may form in the weld metal or the adjacent hard metal due to the presence of

(i) residual stresses
(ii) martensitic structures
(iii) hydrogen.

Examples are shown in fig. 19.4. Such cracks may not cause immediate failure of a whole structure but may grow due to fatigue or other causes until complete fracture of the joint occurs. Hence precautions should be taken to avoid such cracking.

If cracking occurs, it is most likely to happen when the weld has cooled to below 150 °C, which may be some considerable time after the completion of welding.

During welding, hydrogen is absorbed by the molten weld metal from the arc atmosphere. While subsequently cooling, the solubility for hydrogen decreases rapidly. Much of the dissolved hydrogen escapes by diffusion into the atmosphere but some also diffuses into the HAZ and parent metal.

Control of the composition of electrode coverings and baking of electrodes before welding to drive off moisture keep the *potential hydrogen level* of the process low and so are beneficial. Also a slow cooling rate gives time for more hydrogen to diffuse out of the weld metal and escape. Processes such as metal–inert gas (MIG) welding, in which a consumable electrode in the form of a continuous wire is used and is surrounded by a stream of dry inert gas, are also effective in keeping the potential hydrogen level low.

A slower cooling rate is also beneficial in reducing the likelihood of martensite formation.

The rate of cooling of a weld and its HAZ depends on the heat input and also on the relative size of the structure. Thick plates when welded will provide a greater sink for heat and so cause more rapid cooling. Hence in such structures it is necessary to reduce cooling rates by preheating the whole structure in the vicinity of the joint before welding and sometimes postheating as well.

The extent of the precautions needed increases as the thickness of plates, etc. increases and as the hardenability of the steel increases.

A metal which when joined by welding will give more satisfactory service is said to possess good *weldability*. In the case of steels, lower carbon contents give better weldability and should be used when possible where welding is to be used as a joining method.

19.5. Welds in non-allotropic materials

Figure 19.5 shows the heat-affected zone of a weld in work-hardened copper. Progressing through the heat-affected zone towards the weld there are successively, recovery, recrystallization, and grain growth, the

Weld metal M.P. Heat-affected zone

Fig. 19.2.—Fusion weld

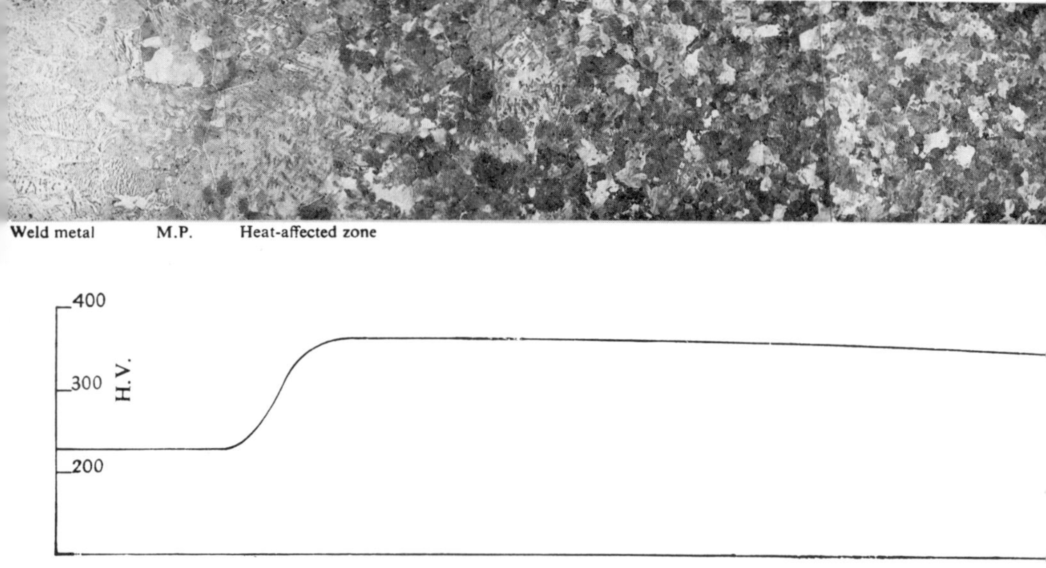

Fig. 19.3.—Fusion weld in a 0·5 % C steel

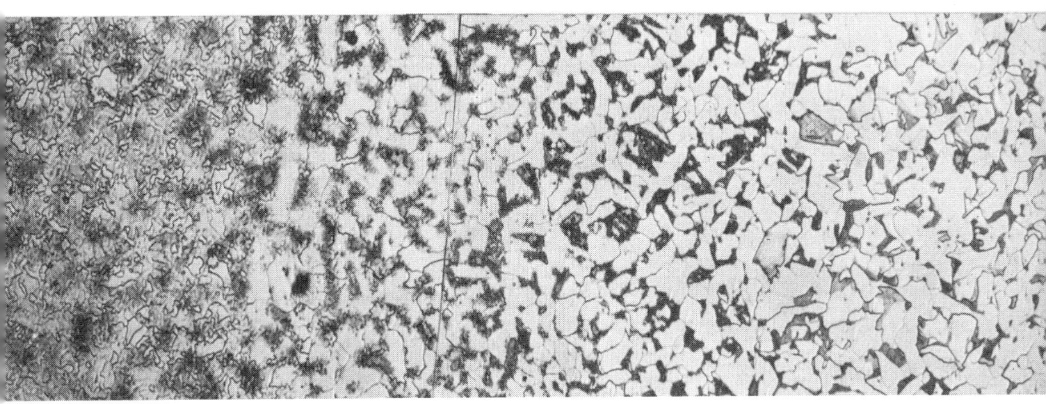

A3 A1 Unaffected metal

in a 0·15% C steel (×100)

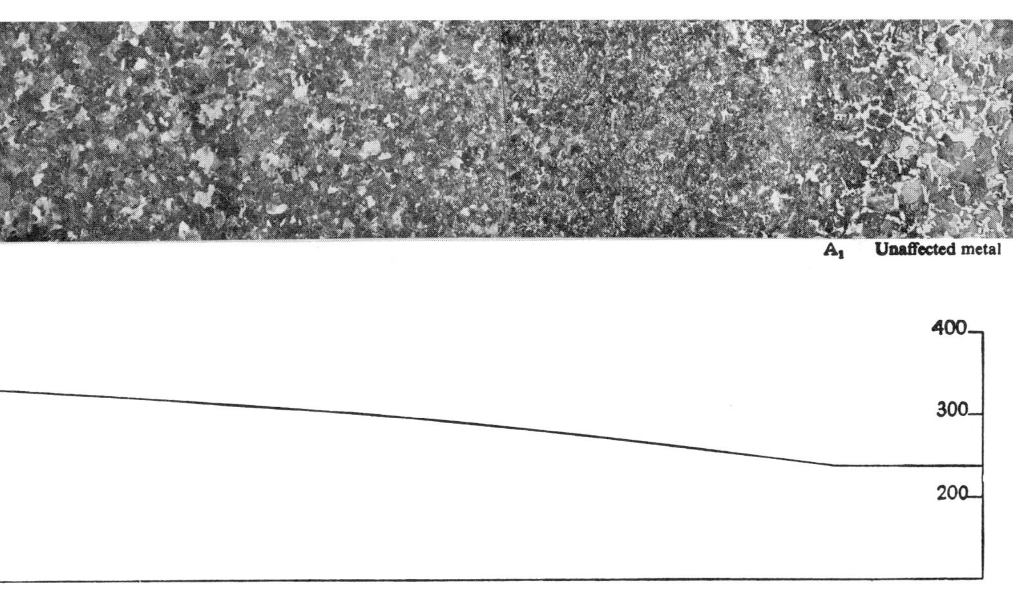

(×100). Variation of hardness is shown

Fig. 19.4.—Hydrogen-induced cracking in (*a*) weld metal

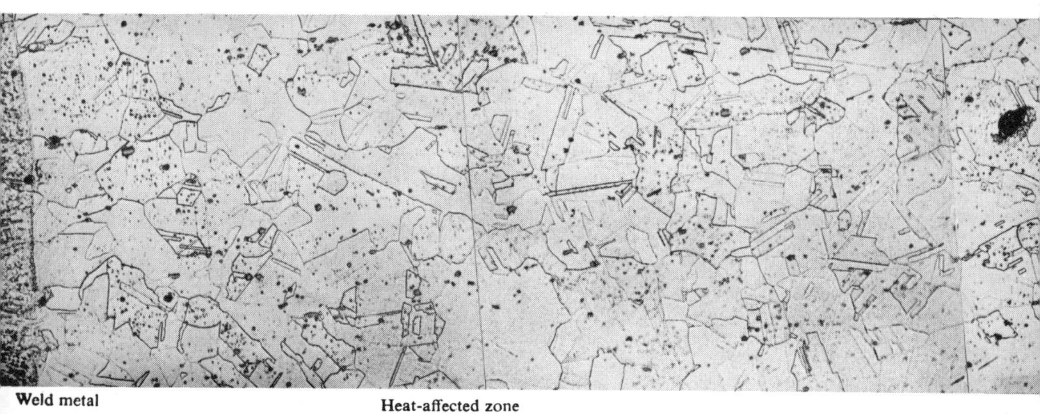

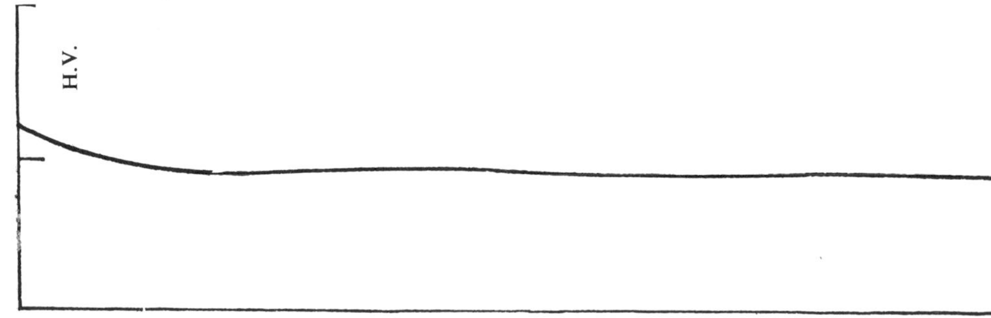

Fig. 19.5.—Fusion weld in work-hardened copper (×

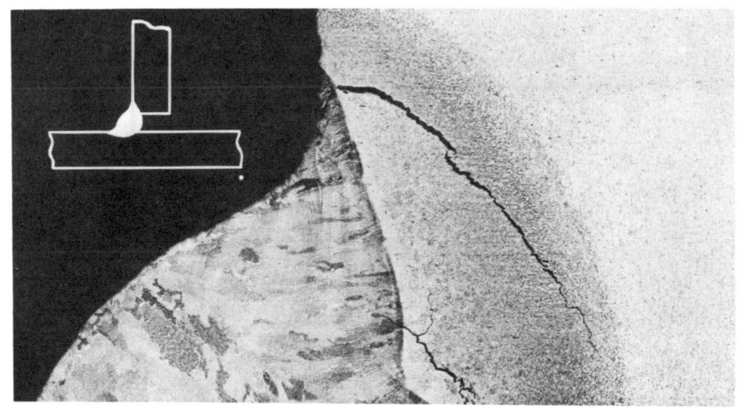

(*b*) heat affected zone.

[Micrographs by courtesy of the Welding Institute]

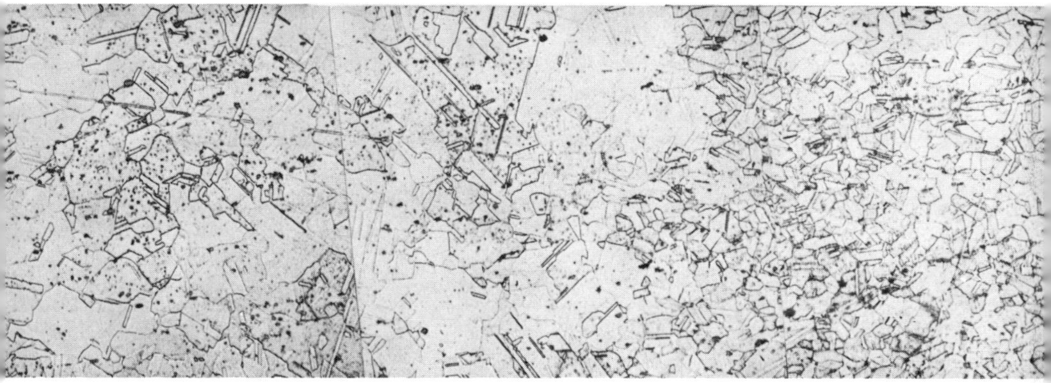

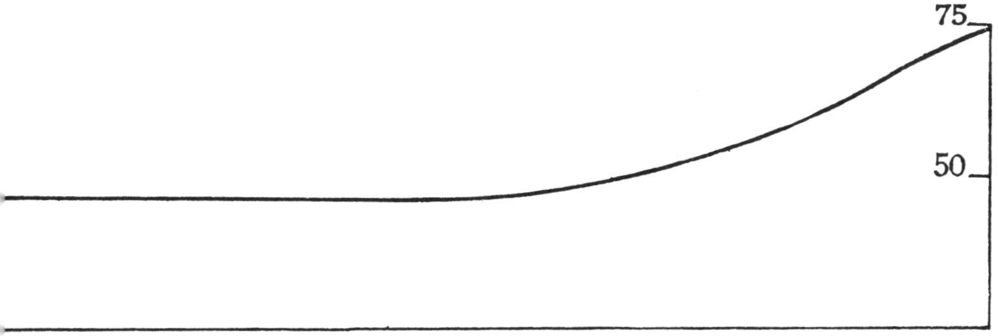

75

50

Decrease in hardness is due to recrystallization and grain growth effect of which on the mechanical properties is seen in the graph of hardness variation. Any calculation of strength of the article based on the properties of the work-hardened material would not be valid in the vicinity of the weld.

In a precipitation-hardened material, the effect of heating would be to solution-treat the metal and so remove the hardening effects. Depending on the rate of cooling, some subsequent age-hardening might or might not be possible. If the component is small enough to be placed inside any available heat-treatment furnace, the correct heat treatment could be applied to the weld metal and heat-affected zone. Otherwise, as in the case of the cold-worked material, the properties would not be homogeneous throughout the component.

QUESTIONS

1. A single-pass fusion weld is made in a specimen of 0·4% plain carbon steel. The weld and the parent metal are then sectioned and microscopical examination of the section reveals several zones of different grain size and structure. Describe these zones and explain how they arise.

What precautions should be taken to ensure a sound weld in such a material? [MST]

2. Discuss the metallurgical factors which are involved in the fusion welding of metals and comment on the limitations which these factors may place upon the industrial use of fusion welding. [MST]

3. Two sheets of hard-rolled aluminium are joined by a single-run arc-weld. Describe the resulting metallurgical structures that occur in the weld and in the heat-affected zone. Discuss the variation in mechanical properties across the weld zone.

How does the presence of the weld affect the overall strength? [E]

CHAPTER 20

Inorganic Non-metallic Materials

20.1. Introduction

The term *ceramics* is used to denote those products made from inorganic materials and which have non-metallic properties. Simple examples are ionically bonded magnesia, MgO, and covalently bonded silicon carbide, SiC, which crystallize in the cubic structures of sodium chloride (fig. 4.21) and diamond (fig. 4.20), respectively.

Some of these materials do not crystallize from the melt except at very slow rates of cooling, but instead increase steadily in viscosity as they cool and form *glasses*.

Traditional ceramic materials of interest to the engineer include stone, brick, concrete, clay, glass, vitreous enamel, and refractories. The majority of these are composed of silicates the structures of which are considered in Section 20.4.

New ceramic materials include oxides, carbides, borides, and other similar compounds which are being developed because they have properties of especial advantage in particular applications such as high temperature work, solid-state electronics, and in nuclear reactors.

20.2. Deformation of crystalline ceramics

In the oxides in which the bonding is mainly ionic, the oxygen ions, being larger than the metal ions, form the framework of the structure with the metal ions occupying interstitial spaces. In the non-oxides, the bonding is mainly covalent, especially in the carbides, nitrides, and borides, and atoms of the more electropositive elements, which are larger, form the framework.

Ductility is possible in single crystals of simpler ionic materials, the operative slip mechanisms being determined by both geometrical and electrical considerations. Sodium chloride and magnesia slip on $\{110\}$ planes in $\langle 110\rangle$ directions (see fig. 4.22), and at high temperatures may also slip on $\{100\}$ planes, again in $\langle 110\rangle$ directions.

In such ionic materials, grown-in dislocations are effectively locked by small amounts of impurity. Hence, the dislocations that are responsible for slip must be nucleated after the crystal is grown. This requires higher stresses than those necessary to cause movement of the dislocation, so that such materials have well-defined yield stresses.

For a polycrystalline aggregate to show extensive ductility, each crystal must have at least five independent slip systems. Sodium chloride does not satisfy this requirement at low temperatures and so is a brittle material.

Silver chloride is exceptional in that slip can occur on both {110} and {100} planes at room temperature. It therefore has a larger number of slip systems than sodium chloride and the polycrystalline material is ductile.

Alumina, Al_2O_3, has a more complicated hexagonal structure comprising close-packed layers of O^{--} ions with incomplete layers of Al^{+++} ions between (see fig. 20.1). The minimum amount of slip of one close-packed layer necessary to move the aluminium ions to identical positions, i.e., the Burgers vector, is large. As for metals, the energy of a dislocation is proportion to $|\mathbf{b}|^2$ and so dislocations are not easily formed in Al_2O_3 except at high temperatures (> 1000 °C).

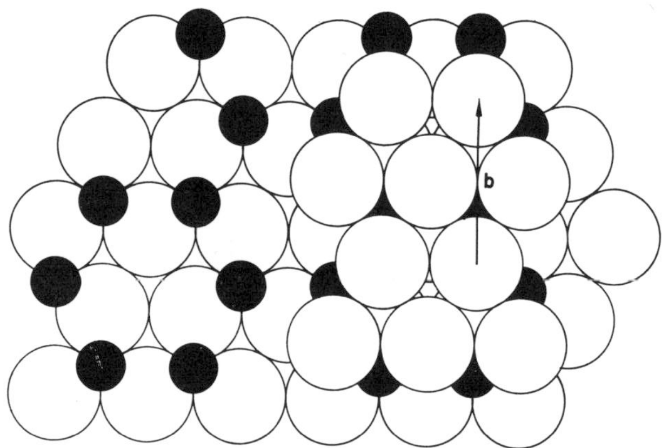

Fig. 20.1.—Structure of Al_2O_3 showing Burgers vector **b**

Most ceramic materials are more complex so that dislocation motion is even more difficult and the mode of failure is usually one of brittle fracture. If the brittle fracture can be suppressed by applying very large hydrostatic compressive stresses, then slip systems can be activated by very high shear stresses, and plastic deformation ensues.

20.3. Properties and uses of some engineering ceramics

Some physical and mechanical properties of a few simple ceramic materials are listed in Table 20.1.

TABLE 20.1—MECHANICAL AND PHYSICAL PROPERTIES OF SOME CERAMICS

		Magnesia MgO	Sintered alumina Al_2O_3	Vitreous silica SiO_2	Hot pressed SiC	Reaction bonded SiC	Hot pressed Si_3N_4	Self bonded Si_3N_4	Commercial graphite C
Melting point	°C	2800	2040	1710	decomposes at 2300		sublimes above 1900		sublimes ~3800
Density	ρ kg m^{-3} 10^3 ×	3·6	3·9	2·5	3·2	2·7	3·2	3·2	1·6
Young's modulus	E GPa	210–310	350–380	70	350–470	210	150–320	170	7–14
Coefficient of thermal expansion	α 10^{-6} ×	13·5	7–9	0·55	4·5	4·5	2·9	2·9	5
Thermal conductivity	k W $m^{-1}K^{-1}$	36–45	12–32	1–2	100	42	10–16	4	130–160
Tensile strength	σ_f MPa	100	400–500	80–160	350–800	200	500–900	200–300	17
Thermal shock resistance $R = \frac{k\sigma f}{\alpha E}$	W m^{-1} 10^3 × (mean value)	1·2	3·4	3·2	31	8·9	13·4	2	47

The values quoted are typical but subject to variation depending upon degree of compaction, etc.

In ceramics with entirely strong primary bonding, the melting points will be high, and higher where only di- and trivalent ions or covalent bonds are present. Materials, such as magnesia and alumina, which have melting points of 2800 °C and 2040 °C respectively, make excellent refractory materials for lining high temperature furnaces, etc.

Several ceramics, which have high compressive strengths and resistance to abrasion, are excellent as abrasive powders and for incorporating in grinding wheels. Magnesia, alumina, and diamond powders are used for metallurgical polishing (see Section 5.1). Silica particles are used in sandstone and glasspaper. Corundum (alumina contaminated with iron oxide) is used in emery paper and grinding wheels, but silicon carbide (carborundum) and diamond are much harder and superior for these purposes. Alumina is used for tips of lathe tools, etc., for high speed machining in which application it can operate successfully when red hot and also cut hard materials, such as white cast iron, which rapidly blunt more conventional tools.

Because there are no free conducting electrons in most ceramics, thermal conductivity takes place only by transfer of vibrational energy from atom to atom and may be very small compared with the conductivity by free electrons in metals. Hence ceramics are generally good thermal insulators. Also crystal boundaries and any pores may decrease the conductivity. However, it may increase considerably at high temperatures due to radiation across pores and for other reasons.

In these ceramics with low thermal conductivity, any localized change in the temperature will not be followed rapidly by the remainder, so that temperature gradients are set up which can cause thermal stresses, possibly leading to failure.

Diamond and graphite are exceptional in having high thermal conductivities. Graphite is unique in being highly anisotropic in this respect. Its structure is one of parallel hexagonal layers (see fig. 20.2) having strong bonding within each layer and only weak van der Waals forces between layers. The thermal conductivity in the plane of the layers is 100 times that perpendicular to the layers. Hence it is used in applications such as the surface lining of rocket motors where it is deposited with the high conductivity layers parallel to the tube surface. Thus thermal insulation is provided between the gases and the rocket tube, and the temperature over the lining surface is fairly constant even if heating is localized.

Both silicon carbide and silicon nitride, Si_3N_4, are now receiving much attention for use in high temperature applications. They both have high mechanical strengths, moderately low coefficients of thermal expansion and high thermal conductivities, which give them excellent thermal shock resistance, silicon carbide being better in that respect than any other

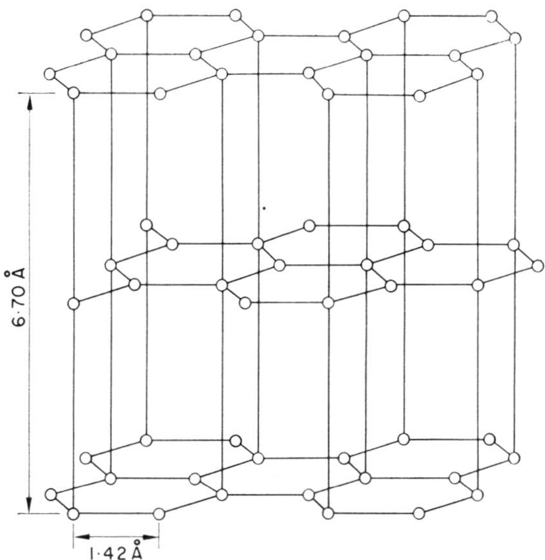

Fig. 20.2.—Structure of graphite

material except graphite. They retain high strengths at temperatures well above 1000 °C. They are being used experimentally in gas turbines for the blades and most other hot parts. Because of their low friction and wear characteristics they are used for ball and roller bearings where they operate satisfactorily without lubrication.

In an oxidizing atmosphere, a thin adherent film of silica forms on the surface of silicon carbide, but silicon nitride oxidizes internally. In both cases, further oxidation of the bulk is negligible. Pyrolytic coatings of silicon carbide improve the oxidation resistance of graphite in nuclear and rocket motor applications. Pistons and piston rings made from silicon nitride have been used successfully in diesel engines.

The densities of ceramics are low compared with those of most metals, which may give them weight-saving advantages. However, they are brittle by nature and so are sensitive to all forms of stress concentrations and flaws. There is a wide scatter of strength values between nominally identical specimens due to the variation of internal pores, etc. Hence, in the design of ceramic components, in addition to paying attention to the avoidance of stress concentrations as far as possible, a statistical approach to the material strength is employed.

With the exception of graphite, which has a high and highly anisotropic electrical conductivity like its thermal conductivity, and a few semiconductors like silicon carbide, most ceramics have extremely high electrical resistivity at temperatures below 300 °C. Hence they are used for electrical insulators and give best results when completely vitrified to be as free as possible of any pore spaces and with a glazed surface to prevent moisture absorption. Porcelain (see Section 20.4) is most widely used for this purpose, but is replaced by steatite and other ceramics for specialized uses such as high frequency applications.

Alumina is used for sparking plug insulators. These articles have to withstand rapid fluctuations of temperature and pressure, the maxima being about 850 °C and 6 MPa, and voltages up to 12 000 V and also maintain gas-tight joints with the metal conductor and base.

Ceramics are very durable, being chemically resistant to most acids, alkalis, and organic solvents and are unaffected by oxygen.

20.4. Silicate structures

Silicon combines with oxygen by bonds that are intermediate between ionic and covalent. Each silicon atom combines with four oxygen atoms, which are arranged in a regular tetrahedron with the silicon at the centre as in fig. 20.3. Each oxygen atom requires a further electron to complete a stable shell of eight, which it may acquire in one of two ways:

1. It can receive an electron from a metal atom so that a negative silicate ion and a positive metal ion are formed.

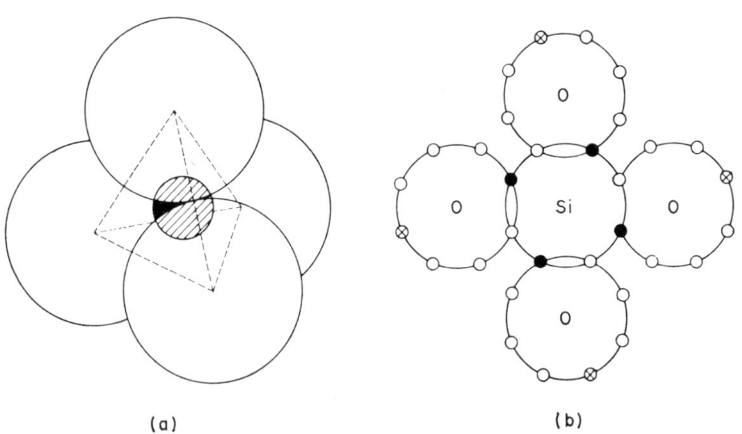

Fig. 20.3.—(*a*) SiO_4 tetrahedron. The silicon is small compared with the oxygen atom. (*b*) two-dimensional representation of electron structure. ● Electron from silicon, O electron from oxygen, ⊕ electron from positive ion

2. It can share an electron pair with another silicon atom forming a multiple group: that is, two tetrahedra can share a corner. Two silicon atoms rarely have more than one oxygen atom in common, i.e., two tetrahedra can share a corner. Two silicon atoms rarely have more than one oxygen atom in common, i.e., two tetrahedra hardly ever share more than a corner.

These SiO_4 tetrahedra can thus combine with each other and with metal ions in a multitude of ways and so give a great diversity of silicate structures. The various types of structure will be considered in order of increasing complexity.

20.4.1. *Monosilicates*

These are the simplest, being composed of $SiO_4{}^{4-}$ ions and metal ions. The silicate ion can be represented two-dimensionally as

O^-

O^- — Si — O^-

O^-

An example is forsterite or magnesium silicate, Mg_2SiO_4, in which each magnesium ion has provided two electrons so that two magnesium ions are needed for each silicate ion.

It is a refractory with a melting point of 2124 °C. On melting, the liquid comprises Mg^{++} and $SiO_4{}^{4-}$ ions.

20.4.2. *Disilicates*

Two tetrahedra sharing one oxygen atom give hexavalent disilicate ions $Si_2O_7{}^{6-}$ which can be represented by

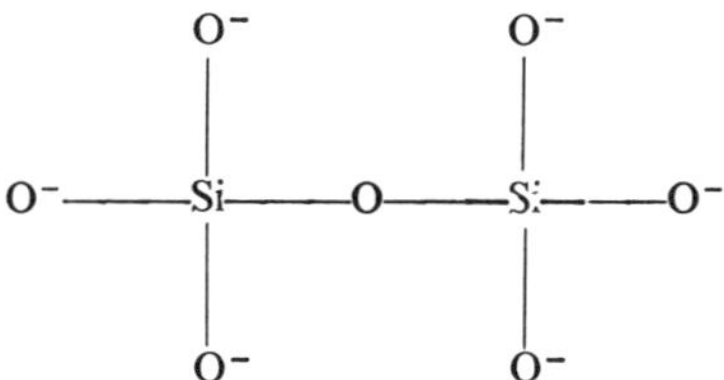

The disilicates are rare and unimportant.

20.4.3. *Chain Structures*

Sharing of two corners of each tetrahedron with other tetrahedra will give a chain structure with the empirical formula $(SiO_3)^{--}$. Each silicon atom is connected to two others by oxygen bridges and to two charged oxygen atoms:

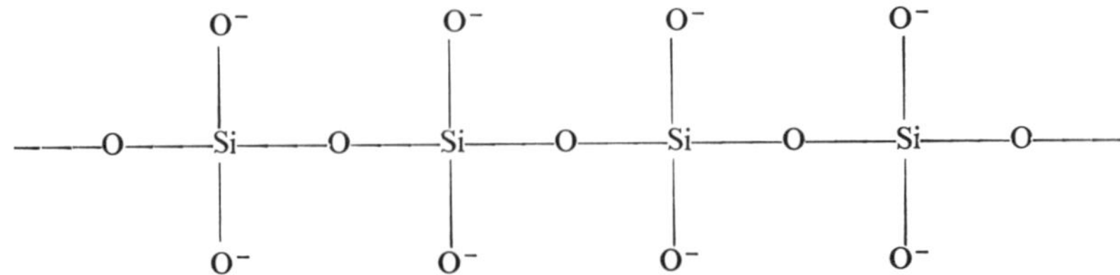

In a crystal these chains lie parallel to one another and are crosslinked by ionic bonding to the metal ions.

A slightly greater degree of sharing gives a double chain with the empirical formula $(Si_4O_{11})^{6-}$. The arrangement of the oxygen atoms in each type of chain is shown in fig. 20.4.

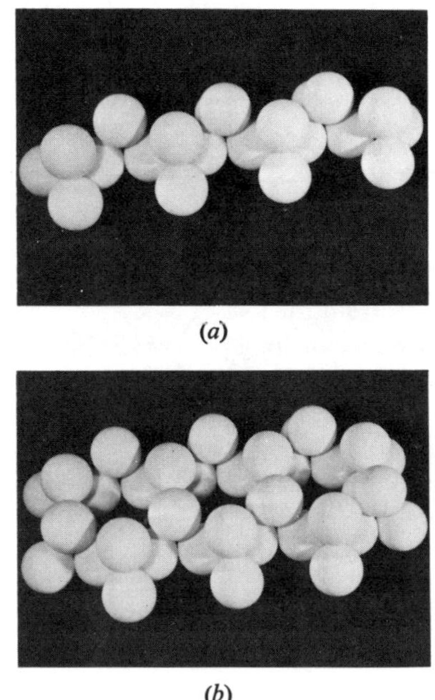

(b)

Fig. 20.4.—Models of (*a*) single chain and (*b*) double chain structures of oxygen atoms in silicates

In materials containing these chain structures, the Si—O bonds are stronger than the ionic bonds so that cleavage planes are the ones crossed only by the latter. The chains being stronger, such materials have a fibre structure like asbestos (see Section 20.10).

20.4.4. *Sheet structures*

Sharing of three corners of each tetrahedron gives sheet structures with the empirical formula $(Si_2O_5)^{--}$. On one side of the sheets each oxygen atom is bonded to two silicon atoms, while on the other every oxygen atom is available to form an ionic bond (fig. 20.5). The sheet structure may join up with other layers in many ways. Some typical ones are shown by kaolin, talc, and mica.

Fig. 20.5.—Model of silicate sheet structure

Aluminium hydroxide, $Al(OH)_3$ is also a layer structure with $(OH)^-$ ions on each side of the aluminium ions. Replacement of some of the OH^- ions on one side by the unsatisfied oxygen atoms of the silicate sheet gives a clay, kaolin, $Al_2(OH)_4Si_2O_5$.

A model and diagrammatic cross-section of kaolin are shown in fig. 20.6. There are no primary bonds between successive layers, but because the structure is non-symmetrical, strong polarization exists and hence there is strong secondary bonding between the layers in the crystals. Also due to this high degree of polarization, molecules of a polar liquid, such as water, are easily adsorbed between layers and not readily removed. The water between layers permits easy slip which gives wet kaolin a high degree of plasticity typical of clays.

Talc, $Mg_3(OH)_2(Si_2O_5)_2$, on the other hand, is a symmetrical structure with silicate sheets on both sides of a central sheet of Mg^{++} and OH^- ions (fig. 20.7). Due to the symmetry, there is no polarization, which is why talc does not absorb water. Also the van der Waals forces between layers will be small so that crystals slide easily over one another—providing a lubricating action.

The micas are more complicated, one silicon atom in every four being replaced by an aluminium atom. This results in an excess negative charge in the alumino-silicate sheet which is balanced by alkali metal ions which

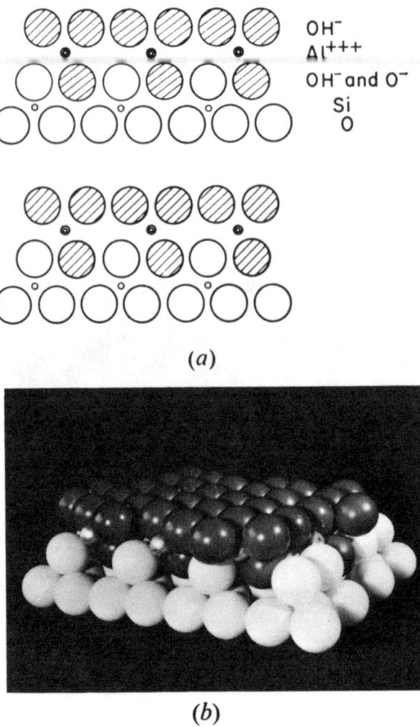

Fig. 20.6.—Structure of kaolin. (*a*) sequence of layers, (*b*) model

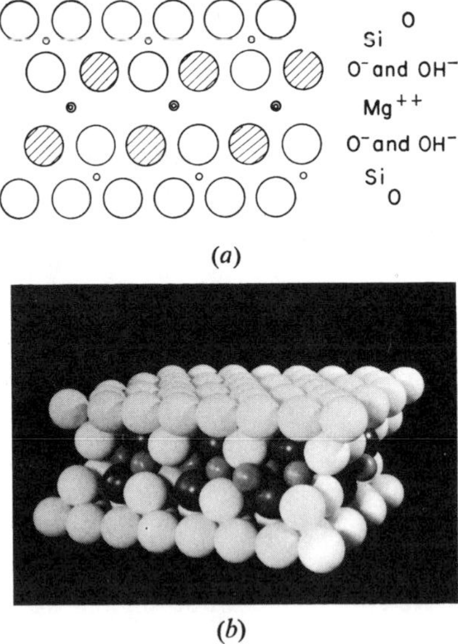

Fig. 20.7.—Structure of talc. (*a*) sequence of layers, (*b*) model

hold the sheets together with ionic bonds (fig. 20.8). The micas are relatively strong, but cleave easily on the weakest planes which are those containing the alkali atoms.

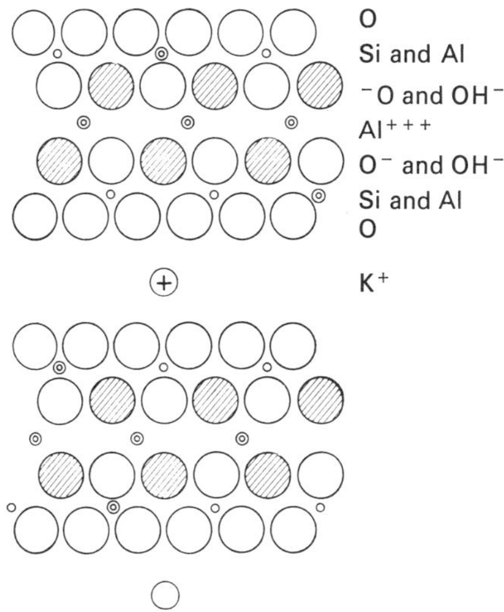

Fig. 20.8.—Structure of mica. Sequence of layers in muscovite, $KAl_2AlSi_3O_{10}(OH)_2$

20.4.5. *Network structures*

When each corner of every tetrahedron is shared, a three-dimensional structure results. SiO_2 has three allotropic forms, each stable over a different temperature range. However, a change from one form to another involves the breaking of bonds and making of fresh ones so that the reactions are sluggish and each phase tends to exist in a metastable state over a wide temperature range. Of more importance is that, on heating, there is a temperature for each phase at which the bonds across each oxygen atom straighten causing a sudden expansion or dilatation. This means that crystalline silica would not be suitable for use as a refractory where there were rapid changes of temperature.

Substitution of aluminium atoms for some of the silicon in SiO_2 with additional alkali atoms to compensate for charge deficiencies give the feldspars. Silica and feldspar are the commonest minerals.

20.5. Vitreous structures

Certain inorganic materials, notably silica (SiO_2), boric oxide (B_2O_3) and phosphorous pentoxide (P_2O_5), have slow rates of nucleation of crystals from the melt and will, in general, on cooling, form glasses. The atoms in a glass are linked in a three-dimensional manner as in a crystal but the structure is not regular, only short-range order being present (fig. 20.9). Pure silica glass has a low coefficient of thermal expansion and a high softening temperature. Also the dilatation found in the crystalline forms does not occur, all of which give it good thermal shock resistance (see Table 20.1).

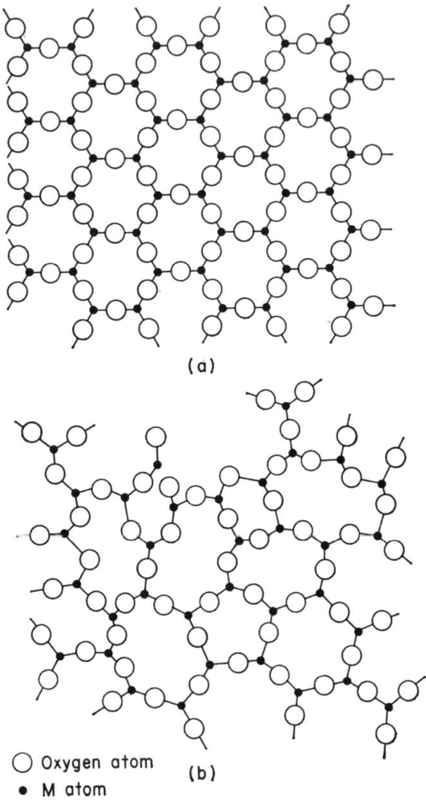

Fig. 20.9.—Two-dimensional representation of an oxide M_2O_3 in (*a*) the crystalline, (*b*) the glassy form

The oxides of the alkaline metals, which are incapable of forming glasses on their own, will combine with silica in the vitreous state. Some of the bridging oxygen bonds between silica atoms are broken and replaced by two non-bridging oxygen atoms which carry negative charges, electrical neutrality being maintained by the metal ions.

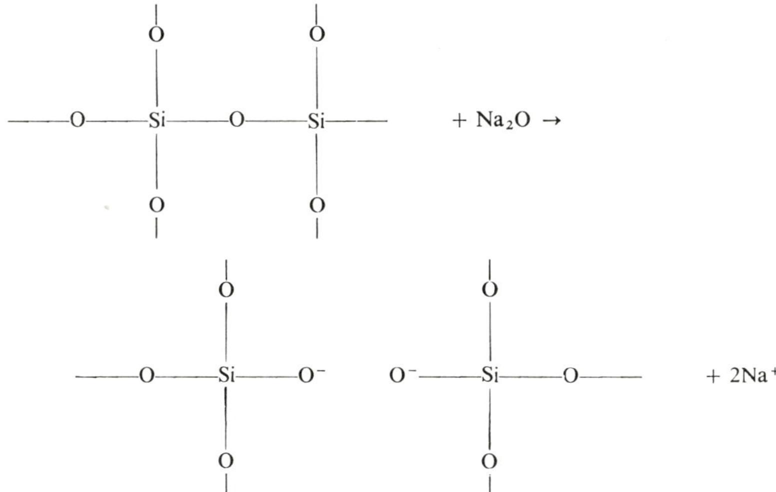

The introduction of such breaks in the structure weakens it so that the softening temperature is lowered and the thermal expansion is increased.

20.6. Mechanical behaviour of glasses

Because glasses lack the long-range order of crystals, dislocation motion is impossible. At high temperatures, glasses deform by viscous flow, the coefficient of viscosity η varying approximately linearly with log $1/T$ which suggests a thermally activated process.

At high temperatures, near the melting point, glass can be moulded to shape easily. Also any locked-up stresses can be removed by annealing at a somewhat lower temperature.

At temperatures sufficiently low for viscous flow to be negligible, glasses behave elastically and fracture in a brittle manner, being extremely notch sensitive (see Section 13.6). Hence, unless care is taken during and after manufacture to remove locked-up stresses and microcracks, the tensile strength of glass is very low.

20.7. Glass-ceramics

By the introduction of suitable catalysts, some materials formed in the glassy state can by subsequent heat treatment be devitrified, i.e., converted to the crystalline state. If the catalyst is highly dispersed in a very finely divided state, a high density of nuclei can be formed, giving a fine-grain microstructure. Such materials are known as *glass-ceramics*.

The mechanical strength of a glass-ceramic is considerably greater than that of glass of the same material in the vitreous state, or of a ceramic article made by pressing and sintering. Many physical properties are also superior and some can be varied over wide ranges by change of composition. Materials with coefficients of thermal expansion from small negative to high positive values are possible.

Due to the scattering of light at the crystal interfaces, glass-ceramics are usually opaque.

A great advantage of glass-ceramic articles is that they can be shaped by the conventional processes used in glass manufacture and do not change dimensions or surface finish appreciably during devitrification.

Very high strength glass products can be made by producing prestressed articles which have high surface compressive stresses balanced by moderate tensile stresses in the interior. Higher tensile loads can then be applied before any surface flaws are subject to a net tensile stress and become liable to propagate.

One method of making such toughened glass is by rapid cooling of the hot surfaces during manufacture so that they become rigid while the centre is still in a viscous state. As the centre cools more slowly, it undergoes thermal contraction which is resisted by the already cold outer layers, giving a residual stress distribution as shown in fig. 20.10.

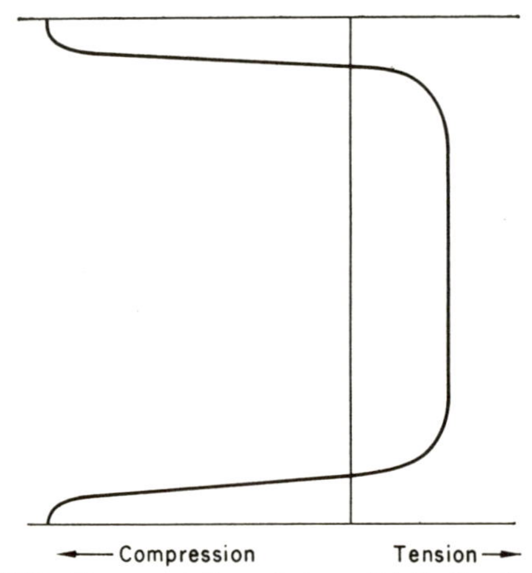

Fig. 20.10.—Representation of stress distribution through a sheet of toughened glass

In another method, by chemical treatment of the surface layers devitrification will occur there while the bulk remains glassy. If the surface layer has a lower coefficient of thermal expansion than the interior, then on cooling high surface compressive stresses will be developed. Glass with a strength of 750 MPa can be made in this way.

20.8. Cermets

High-temperature ceramics are more refractory and have greater chemical stability and abrasion resistance than metals, but have lower resistance to thermal and mechanical shock. A cermet is a combination of a ceramic and metal which exhibits the desirable characteristics of both.

A common example is the use of tungsten carbide cemented with cobalt for the tips of tools for machining hard materials, e.g., in masonry drills. Alumina with iron or chromium is used in the high-temperature parts of flame holders and nozzles for jet propulsion units, for thermocouple protection tubes, etc. There are numerous uses in nuclear reactor technology.

20.9. Fabrication of ceramic articles

Ceramics in general cannot be deformed by hot or cold working and so must be fabricated by some other method. Machining is usually not suitable because of the hardness and danger of brittle fracture.

Casting from the melt may give a partly or entirely glass article. If a crystalline material is necessary and a glass-ceramic process (see preceding section) is not practicable, then articles may be made to shape by compacting powdered material in a mould by mechanical pressure and then *sintering*. This comprises raising the temperature to a value at which solid state diffusion occurs and the particles become firmly joined. Pores will be present and although porosity may be reduced by hot pressing it cannot be eliminated unless the temperature is high enough for some of the powder to melt. If the sintering has to be continued for a long time, uneven grain growth may occur, which has a deleterious effect upon the strength.

Self-bonded silicon carbide is made by bonding SiC powder and graphite with an organic binder and shaping by pressing, extruding or machining. The green body is then heated in contact with silicon, which melts at 1420 °C, impregnates the pores by capillary action and reacts with the graphite to form more SiC which bonds with the original SiC. The product shows no dimensional change from the green body and so no machining is necessary, but it can be lapped to a fine finish.

To make reaction-bonded silicon nitride, silicon powder is compressed in a die or slip cast (see below) and the shaped piece is subsequently heated in nitrogen at a temperature above 1200 °C. It is much more porous than hot-pressed silicon nitride, but nevertheless has high strength and articles can be produced to high tolerances without machining.

In slip casting, a suspension of the powdered material in water is poured into a mould. Removal of the water gives a solid article which can be sintered.

Bricks and earthenware articles such as clay tiles and pipes are made by moulding wet clay to the desired shapes and then heating in kilns. The first action is to drive the water off, after which the temperature is raised to a sufficiently high value for long enough to give complete bonding. If the articles are required to have an impermeable surface they are glazed. Glazes are glasses of low fluidity containing silica and other oxides. The raw materials may be applied to the fired or unfired clay articles by dipping the latter in a liquid mixture known as slop glaze. The articles are then fired to a sufficient temperature for the glaze to vitrify.

Porcelain is a high quality ceramic product usually white, always translucent and of very low porosity. A typical composition of the raw material is 50 % kaolin, 25 % quartz, and 25 % feldspar.

Steatite, used as an insulator for high frequencies and manufactured from talc mixed with a suitable fluxing material, consists of crystals of magnesium metasilicate and free silica in a glassy matrix.

20.10 Asbestos

Asbestos is the name conventionally used for a number of naturally-occurring fibrous materials. These silicate minerals are in two groups.

(i) The *amphiboles*, with the general formula $M_7Si_8O_{22}(OH)_2$ where M can be sodium, iron, calcium, or magnesium or a mixture of them, are chain silicates and occur naturally in fibrous form.

(ii) *Crysotile*, $Mg_3(Si_2O_5)(OH)_4$, is a sheet structure of pairs of sheets. There is a mis-match between the magnesium hydroxide and the silicate sheets so that the pair of sheets takes a rolled form. The least strain configuration occurs for diameters in the range 5–50 nm and the material consists of fibrils which are either a series of concentric tubes or spirals with inside diameters 5–10 nm and outside diamters 10–80 nm. The submicroscopic fibrils, which may be 0·1–10 mm in length, are bunched together into fibres which may be 3–50 mm in length. Crysotile occurs in layers in serpentine rock, which has the same chemical formula, and from which it is formed by solution and recrystallization.

Asbestos is a good thermal insulator and is stable at temperatures up to 500 °C above which temperature it decomposes to an amorphous phase. The fibres can be woven into string and cloth. It is also used as a reinforcing material in cement for a wide range of asbestos cement products—roof sheeting, guttering, water tanks, etc. However, its use poses a health hazard to the extent that it is now being replaced by other materials for many applications.

20.11. Cement and concrete

Cement, made by heating a mixture of calcareous (lime-containing) and argillaceous (clay-containing) materials in kilns to about 1500 °C and grinding the mixture to a very fine powder, contains four principal constituents:

tricalcium silicate	$(CaO)_3 . SiO_2$
dicalcium silicate	$(CaO)_2 . SiO_2$
tricalcium aluminate	$(CaO)_3 . Al_2O_3$
tetracalcium aluminoferrite	$(CaO)_4 . Al_2O_3 . Fe_2O_3$

to which is added a small amount of

gypsum	$CaSO_4 . 2H_2O$

The proportions of the various constituents can be varied by using different proportions of the raw materials.*

When mixed with about half its weight of water, the cement forms a plastic mass. Hydration reactions then occur and the cement sets to a hard mass. Hydrated crystals of colloidal dimensions (1–100 nm) form at the surfaces of the cement particles. They form plates and hollow fibres which grow into a network as hydration proceeds. The continuous, porous, solid mass that results is called the *cement gel* (fig. 20.11). The *gel pores* have an average distance across of about 2 nm.

The tricalcium aluminate reacts rapidly giving the initial set but little contribution to the final strength. The gypsum is added to delay this reaction. The calcium silicates are the effective cementing compounds. Tricalcium silicate reacts rapidly producing most of its strength in a few days, whereas dicalcium silicate is much slower in hydrating, taking as much as a year for most of it to react. The fourth constituent is of secondary importance, contributing little to the strength. As with all chemical reactions, the rate of reaction increases rapidly with temperature. At 0 °C the hydration is extremely sluggish. The hydration of cement is exothermic, a great deal of heat being evolved in the early stages of the reactions. This can be beneficial in cold weather, as the temperature of the mixture is raised, maintaining a reasonable rate of setting. Conversely it is a disadvantage in a large mass of concrete where the heat produced would not be lost rapidly and the temperature might rise so high as to set up undesirable thermal stresses. For these reasons, special cements have been developed. The temperature rise can be reduced by the use of *low-heat cement* which contains smaller proportions of tricalcium aluminate and

* The composition limits for Portland cement are given in B.S. 12. Part 2: 1971.

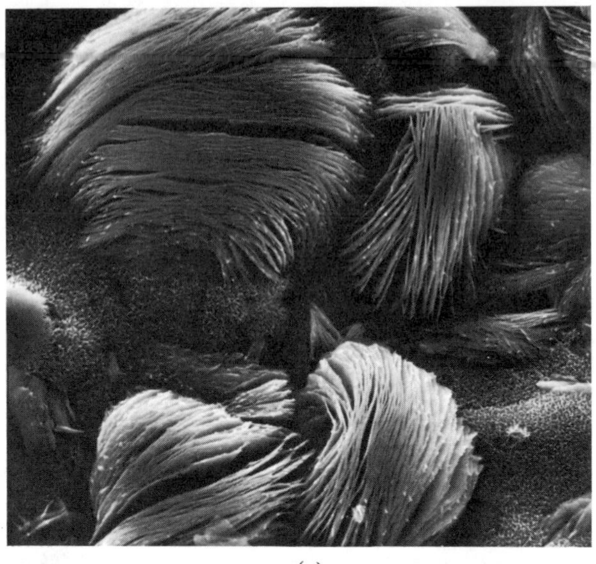

(*a*)

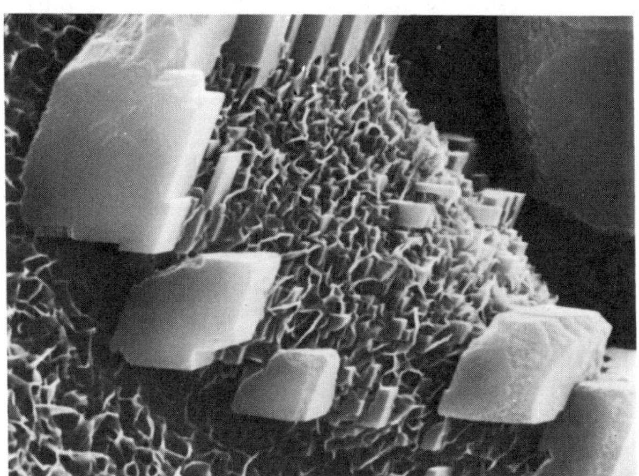

(*b*)

Fig. 20.11 (*a*) Hydration products of Portland Cement showing typical clusters of buckled sheets against a background of porous calcium silicate hydrate. The structures shown were formed on the surface of cement which was exposed to an excess of water. In conventional concrete the structural elements would form in a restricted space and would therefore be far more compact than this micrograph would suggest (×8000). (*b*) Porous calcium silicate hydrate with relatively massive crystals of Portlandite (×9000)

(Micrographs by courtesy of Professor R. H. Mills)

tricalcium silicate. For cold weather use or where high early strength is required, *rapid hardening cement* which contains a higher proportion of the early setting constituents should be used (see fig. 20.12).

Water in the cement paste which is in excess of that needed for hydration will leave *capillary pores* which are sources of weakness and cause liability to frost damage, as any trapped water will expand on freezing. Also if insufficient water is used, the hydration will not be complete, again causing weakness. The variation of strength with water/cement ratio is shown in fig. 20.13.

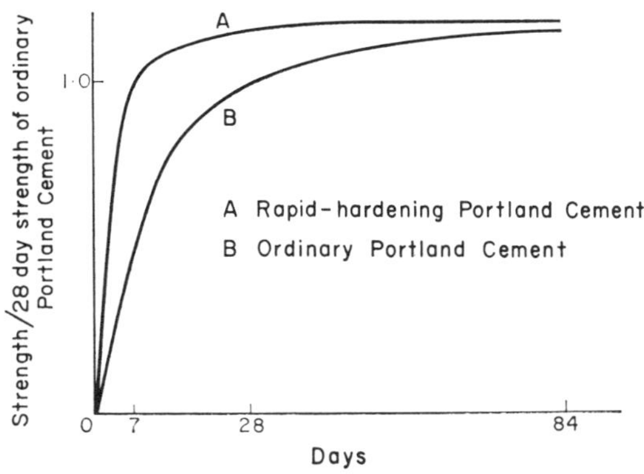

Fig. 20.12.—Increase of strength with time for concretes of same water/cement ratio

For economy in the use of cement, aggregates comprising sand and gravel are mixed with it in making concrete. A range of size distribution in the right proportions (i.e., a suitable grading) is necessary, so that the aggregate packs densely and there should be sufficient cement paste to coat all the aggregate particles and fill all voids. Water is needed to produce hydration and also to make the mix workable. Entrapped air or excess water will give porosity with a loss of strength. Water in excess of that necessary for complete hydration may be needed to give sufficient workability to the concrete to permit proper compaction. The aggregate used must have sufficiently high compressive strength, so that it is not the weakest part of the final product, and also must be free of any chemicals (particularly organic compounds) which would hinder setting. Provided that these conditions are satisfied, the strength of the concrete depends upon the strength of its cement paste.

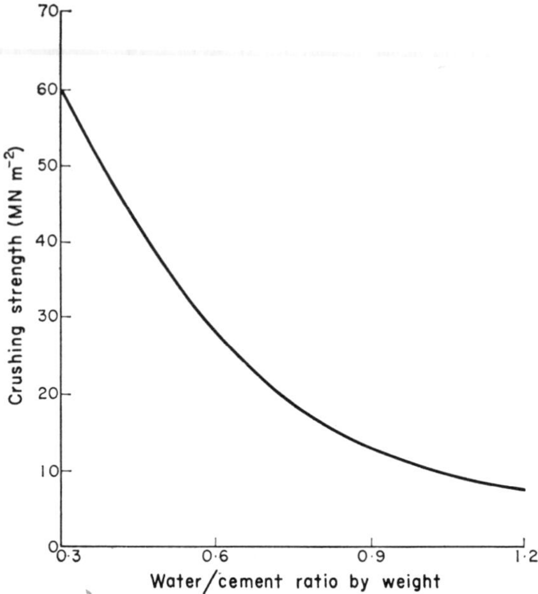

Fig. 20.13.—Variation of 28 day strength of cement and concrete with water/cement ratio

QUESTIONS

1. The caesium chloride structure consists of two interpenetrating simple cubic lattices, one of anions and the other of cations, with the cations at the body centres of the anion unit cells. The sodium chloride structure consists of two interpenetrating face-centred cubic lattices, with cations at the mid-edge positions of anion unit cells Each structure is stable only if anions can touch cations.

Assuming that the cations and anions behave as rigid spheres with radii r^+ and r^- respectively ($r^+ < r^-$), find the lowest limit of the radius ratio (r^+/r^-) for each structure to be stable.

Using ionic radii $Rb^+ = 0{\cdot}149$ nm and $Cl^- = 0{\cdot}181$ nm show that RbCl can crystallize with the CsCl structure and find the unit cell dimensions. [E]

2. Describe the structure, interatomic bonding, and valency characteristics of the silicate ion $(SiO_4)^{4-}$.

3. Describe how the arrangements of atoms in asbestos, kaolin, mica, and talc affect their physical and mechanical properties. [MST]

4. In choosing a material to give minimum weight for the following two cases, the relevant material parameters to be considered are E/ρ and $E^{1/2}/\rho$ respectively, (i) a

tensile member of specified length which must have a limited extension under a prescribed load, (ii) a uniform strut of specified length which must not buckle under a prescribed compressive load.

Rank the following materials in order of merit for these cases and state briefly why the materials with the better values of the parameters are not always used.

	Young's modulus E MPa	Density ρ kg m^{-3}
Alumina Al_2O_3	530 000	4000
Aluminium	72 000	2700
Balsa wood	3200	160
Iron	210 000	7870
Nylon	3000	1140
Spruce (parallel to grain)	13 000	500

[MST]

5. The equilibrium diagram for mixtures of the refractory oxides alumina (Al_2O_3) and silica (SiO_2) at temperatures above 1200 °C is shown in Fig 20.14 The phases are identified by the names given to the natural minerals. Silica undergoes an allotropic change from tridymite to crystobalite on heating through 1470 °C.

Determine the composition of the intermediate compound, mullite.

Describe the changes that occur in a mixture containing 65% alumina by weight as

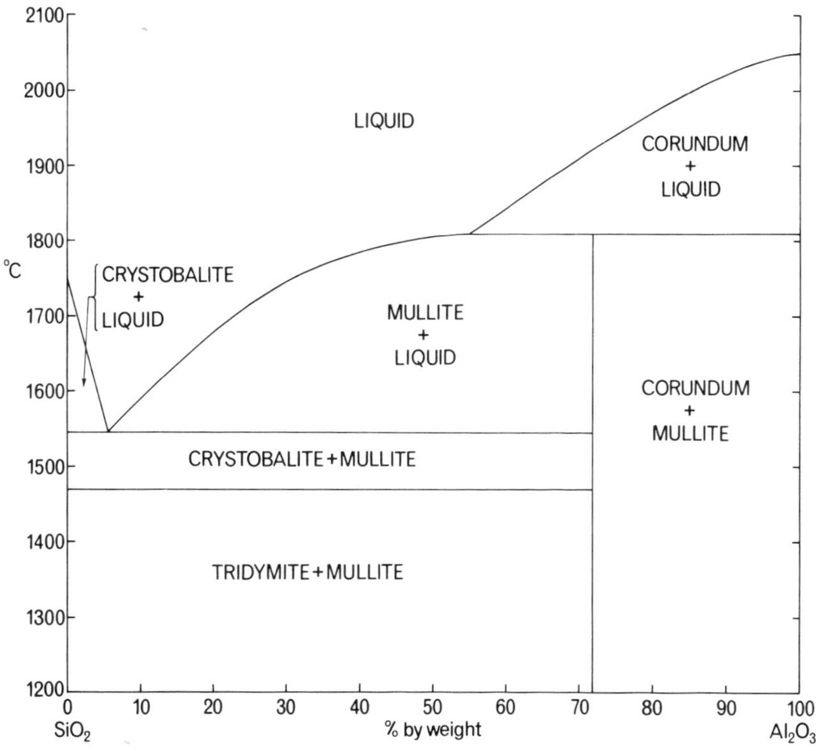

Fig. 20.14

it is slowly cooled from the liquid state to 1200 °C, giving detailed compositions at each change point.

Depending upon the temperature of operation, what compositions should be particularly avoided for refractory bricks made from alumina-silica mixtures? [E]

6 Sketch the structure of silica when in the form of
(i) a crystalline solid,
(ii) a glass.
How is the latter structure modified by the addition of sodium?

The variation of viscosity with decreasing temperature of a sodium silicate glass was determined experimentally. The results were as follows:

Temperature (°C)	1394	1156	977	838	727	636	560
Viscosity ($N\ s\ m^{-2}$)	5·02	$3{\cdot}1 \times 10$	$3{\cdot}80 \times 10^2$	$7{\cdot}08 \times 10^3$	2×10^5	10^7	$9{\cdot}12 \times 10^8$

Determine the value of the activation energy for viscous flow in this sodium silicate glass at 636 °C and 977 ° C and give reasons why the activation energy is not constant with temperature. [E]

7. When ceramics are subject to thermal shock, the heated surfaces may flake off because of the thermal stresses produced. This is known as spalling. The tendency of ceramics to spall is a function of the following properties: thermal conductivity, k; tensile strength, σ; coefficient of thermal expancion, α; Young's modulus, E; specific heat capacity, C_p; density, ρ. Explain briefly whether an increase or a decrease in each property is required to reduce the tendency to spall.

Some quoted values of these quantities for three ceramic materials are

	alumina (Al_2O_3)	magnesia (MgO)	silica (SiO_2)
k $J\ m^{-1}\ s^{-1}\ K^{-1}$	30	32	7
σ MPa	280	100	100
α K^{-1}	$8{\cdot}8 \times 10^{-6}$	$13{\cdot}5 \times 10^{-6}$	$0{\cdot}5 \times 10^{-6}$
E GPa	530	280	69
C_p $kJ\ kg^{-1}K^{-1}$	1·0	1·2	1·1
ρ $kg\ m^{-3}$	$4{\cdot}0 \times 10^3$	$3{\cdot}6 \times 10^3$	$2{\cdot}5 \times 10^3$

On the basis of these values, rank in order of resistance to spalling three types of refractory brick each made principally of one of these ceramics.

If T is the step change in surface temperature and x a coordinate distance measured normal to the surface, form non-dimensional groups which will appear in the expression for the thermal stress, $\bar{\sigma}$, at a time t after the change. [E]

CHAPTER 21

Organic Materials

21.1. The nature of organic compounds

Organic compounds are those that contain carbon and usually one or more of the elements hydrogen, oxygen, and nitrogen, together sometimes with halogens and other non-metallic elements. Metallic carbides and carbonates are conventionally classed as inorganic compounds, whereas compounds of metals with organic radicals are distinguished by the name of metallo-organic compounds. The covalent bonding of carbon atoms with four bonds arranged in a tetrahedral manner has been discussed in Section 4.12. Carbon atoms can link to one another in long chains and in ring structures, so that the number of different compounds which can be made is limitless. Organic solids consists of molecules which are held to each other by weak secondary or van der Waals bonds. In compounds with larger molecules the total force between any two molecules and hence the overall strength is larger.

Naturally-occurring organic materials have been of service to the engineer since the days of the earliest technician, e.g., timber, rope made from animal or vegetable fibres, felt packing, cork insulation, rubber, fuels, etc. In variety these have been augmented and superseded by *plastics*—synthetic organic materials containing large molecules. In the natural materials such as timber and rope the large molecules are ready-made whereas in the synthetic materials they are built up by joining simpler molecules by chemical means.

In most of these natural and synthetic materials, the molecules are very large so that there are fewer secondary bonds and hence greater strength. The properties can only be properly understood in terms of the molecular structure which will be discussed in the following sections.

21.2. Saturated hydrocarbons

In the saturated hydrocarbons, the carbon atoms present in each molecule are joined by single covalent bonds and any further carbon bonds are completed by hydrogen atoms.

The simplest member of the series is *methane* (CH_4), each molecule of which contains one carbon atom joined to four hydrogen atoms, these

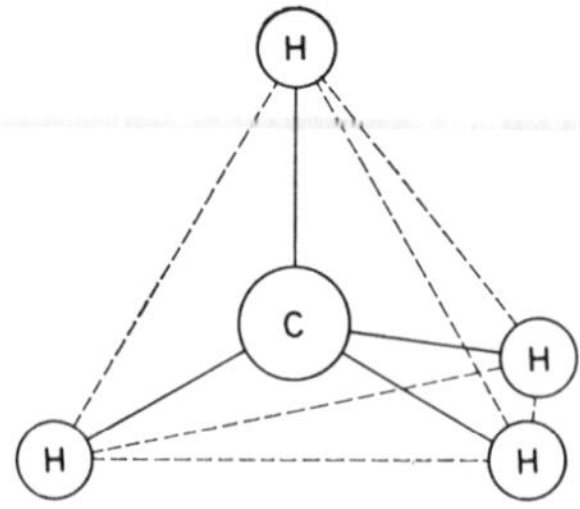

Fig. 21.1.—Model of methane molecule

being arranged symmetrically to give a regular tetrahedron as shown in fig. 21.1.

The next in the series is *ethane* (C_2H_6) in which the two carbon atoms are linked by one covalent bond. Although this and all other organic structures are three-dimensional, it is convenient to represent them by two-dimensional structural formulae. Thus ethane is

```
    H   H
    |   |
H — C — C — H
    |   |
    H   H
```

and the next member of the series, *propane* (C_3H_8) is

```
    H   H   H
    |   |   |
H — C — C — C — H
    |   |   |
    H   H   H
```

Each member has been derived from the previous one by removing a hydrogen atom and replacing it with a *methyl* group (—CH_3). If this process is repeated each time to a carbon atom at the end of the chain, we get a homologous series with the general formula C_nH_{2n+2} which may be written as $CH_3 \, . \, (CH_2)_{n-2} \, . \, CH_3$ or

```
    H   [ H ]       H
    |   [ | ]       |
H — C — [ C ] ————— C — H
    |   [ | ]       |
    H   [ H ]n−2    H
```

The names of the next few members of this *alkane* series are

C_4H_{10}	butane
C_5H_{12}	pentane
C_6H_{14}	hexane
C_7H_{16}	heptane
C_8H_{18}	octane, etc.

As n is increased, so also is the ratio of primary to secondary bonds. The molecules are larger, giving larger van der Waals forces, so that the strength increases and the melting and boiling points are raised (see fig. 21.2.). The lowest members of the series are gases at room temperature, the medium ones are oils, and the highest ones are solids, paraffin waxes.

As an alternative to deriving butane from propane as described above, the methyl group could have replaced a hydrogen atom on the central carbon atom.

Hence there are two alternative structures for butane:

```
    H  H  H  H              H     H     H
    |  |  |  |              |     |     |
 H—C—C—C—C—H             H—C—————C—————C—H
    |  |  |  |              |     |     |
    H  H  H  H              H     |     H
                               H—C—H
                                  |
                                  H
```

normal butane　　　　isobutane

Because of the structural difference, there are differences in the properties, isobutane having somewhat lower melting and boiling points. The different compounds with the same molecular formula are *isomers.* For any higher member of the series, the numbers of possible isomers will be greater.

There are other homologous series of compounds in which hydrocarbon groups are linked to typical organic radicals. Some of the radicals and the type of compound formed from each are:

—OH	alcohol	>CO	ketone
—CHO	aldehyde	$—NH_2$	amine
—COOH	acid		

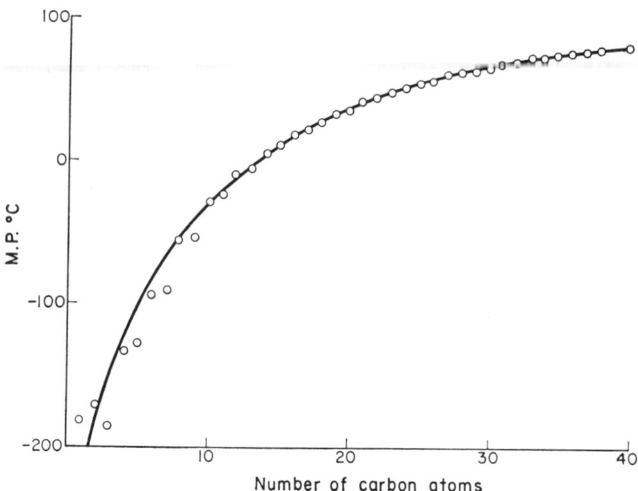

Fig. 21.2.—Melting point as a function of molecular size for *n*-alkanes

The alkanes do not react easily, the only notable reactions being (*a*) combination with oxygen to form carbon dioxide and water, which is strongly exothermic, and (*b*) when chlorine is present in diffused light, the substitution of chlorine for hydrogen. This can happen to varying degrees; thus methane reacts with chlorine to give a mixture of methyl chloride CH_3Cl, methylene chloride CH_2Cl_2, chloroform $CHCl_3$, and carbon tetrachloride CCl_4.

21.3. Unsaturated hydrocarbons

Double covalent bonds can form between two carbon atoms, the simplest compound of this type being ethylene C_2H_4. The two dimensional formula is written as

```
H   H
|   |
C = C
|   |
H   H
```

The next homologues in the series are propylene or propene

```
    H   H   H
    |   |   |
H — C — C = C
    |       |
    H       H
```

and the three isomers of butylene

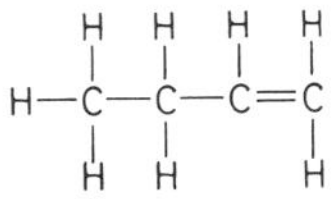

1 butene 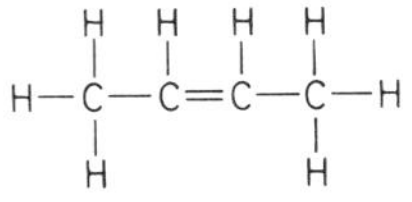　　2 butene 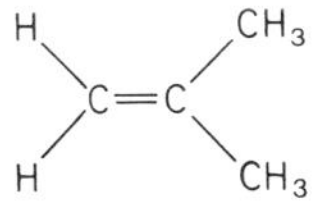　　isobutene

This series with the general formula C_nH_{2n} is known as the *olefines*.

The double bonds are relatively weak, and revert easily to single bonds, at the same time taking part in additive reactions, for example the reaction of ethylene with chlorine to give ethylene chloride.

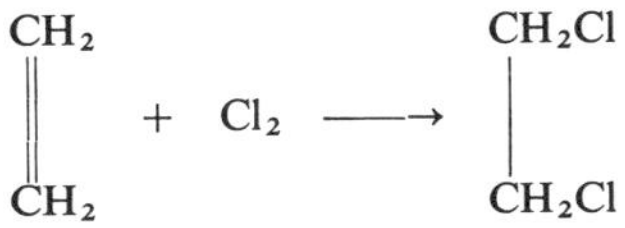

$$CH_2{=}CH_2 + Cl_2 \longrightarrow CH_2Cl{-}CH_2Cl$$

21.4. Aromatic hydrocarbons

The alkane-type hydrocarbons and the wide range of compounds that are derived from them by substitution of various groups for the hydrogen atoms are known as *aliphatic* compounds. The *aromatic* compounds are those based on the structure of benzene (C_6H_6) in which the carbon atoms form a ring with one hydrogen atom attached to each. The structure is thought to be a resonance between

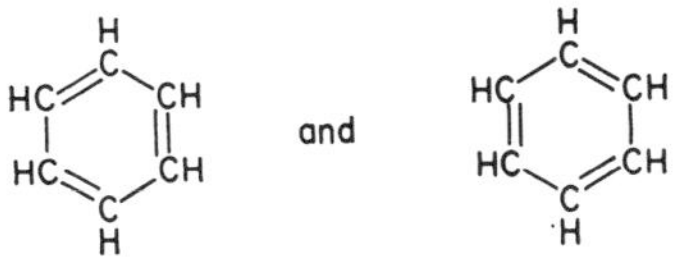

and

In structural formulae, the benzene ring is usually drawn as a simple hexagon, the carbon and hydrogen atoms not being shown. Any groups substituted for hydrogen atoms are indicated adjacent to the appropriate corner of the hexagon. Some examples are:

toluene 　　phenol 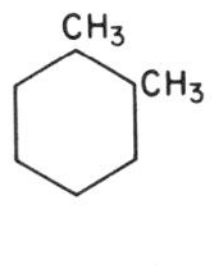　　orthoxylol 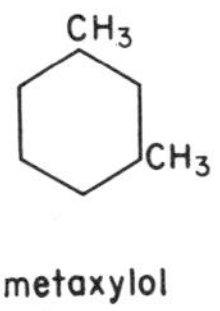　　metaxylol 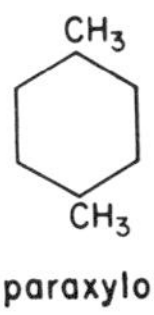　　paraxylol

The benzene ring is much stronger than the ethylene double bond and persists unchanged in many chemical reactions.

21.5. Addition polymerization

Under certain conditions (high pressure and temperature in the presence of a catalyst) ethylene molecules can react with one another to form a chain of carbon atoms—each being linked also to two hydrogen atoms. This and similar processes are known as *addition polymerization.* The units of the *monomer*, in this case ethylene, have joined to form a *polymer* (fig. 21.3)—this one is known as *polyethylene,* or more familiarly as *polythene.* Apart from the terminal groups, the structure is identical to that of an alkane and it behaves in a similar manner, being fairly inert chemically. The long-chain molecules are linked to each other by van der Waals forces and so the melting points will be low. Except when highly crystalline (see Section 21.9), it does not have a sharp melting point, but softens in the temperature range 100–150 °C.

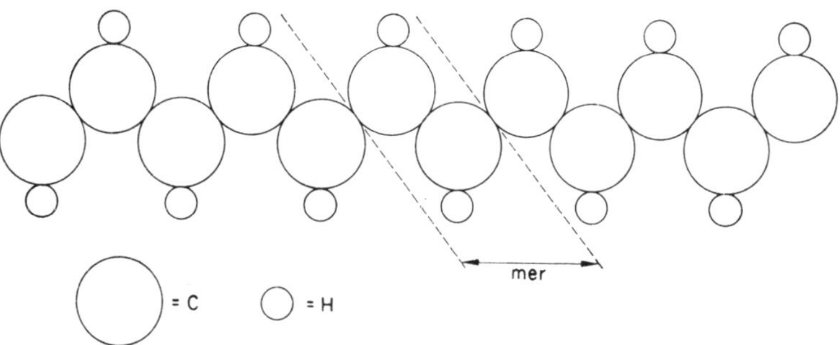

Fig. 21.3.—Polythene molecule showing unit or mer

The *vinyl* compounds are obtained by the substitution of different groups for one of the hydrogen atoms in ethylene. They and other substitution products of ethylene all have the double bond between the carbon atoms and hence may be polymerized in a similar manner. One of the commonest is vinyl chloride which polymerizes to polyvinyl chloride (p.v.c.).

$$\begin{array}{cc} H & H \\ | & | \\ C & = C \\ | & | \\ H & Cl \end{array} \longrightarrow \begin{array}{ccccccccc} & H & & H & & H & & H & \\ & | & & | & & | & & | & \\ - & C & - & C & - & C & - & C & - \\ & | & & | & & | & & | & \\ & H & & Cl & & H & & Cl & \end{array}$$

Some other monomers which also form commercially important polymers by addition polymerization are:

propylene ($CH_2{=}CH\,.\,CH_3$)

vinyl acetate $\left[\begin{array}{c} CH_3{-}\overset{\displaystyle O}{\overset{\|}{C}}{-}O{-}CH{=}CH_2 \end{array}\right]$

styrene ($CH_2{=}CH{-}C_6H_5$)

vinylidene chloride ($CH_2{=}CCl_2$)

methyl acrylate $\left[\begin{array}{c} CH_2{=}CH{-}\overset{\displaystyle O}{\overset{\|}{C}}{-}O{-}CH_3 \end{array}\right]$

methyl methacrylate $\left[\begin{array}{l} \qquad\quad CH_3 \\ \quad\;\; / \\ CH_2{=}C \\ \quad\;\; \backslash \\ \qquad\quad \overset{\displaystyle O}{\overset{\|}{C}}{-}O{-}CH_3 \end{array}\right]$

acrylonitrile ($CH_2{=}CH{-}CN$)

All these addition polymerization reactions require heat, light, pressure, or a catalyst.

Molecules of the monomer must be available in the region of the end of a chain if continued polymerization is to occur. Hence polymerization can proceed easily until most of the molecules have been used up, after which unpolymerized molecules must diffuse to the regions where further reaction can take place. Terminal radicals are necessary to produce stability, for example hydrogen peroxide is used to provide terminal —OH groups, thus:

$$\begin{array}{ccccccccc} & & H & & H & & & & H & & \\ & & | & & | & & & & | & & \\ HO & - & C & - & C & - & - - - & - & C & - & OH \\ & & | & & | & & & & | & & \\ & & H & & H & & & & H & & \end{array}$$

The *degree of polymerization* is the number of *mers* or repeating monomer units in the chain, i.e., it is the ratio of the molecular weight of the polymer to the molecular weight of the monomer. A polymer will contain a mixture of molecular sizes, so that the degree of polymerization represents an average value.

By using two or more types of monomer, *copolymers* can be produced. Vinyl chloride and vinyl acetate, for example, give

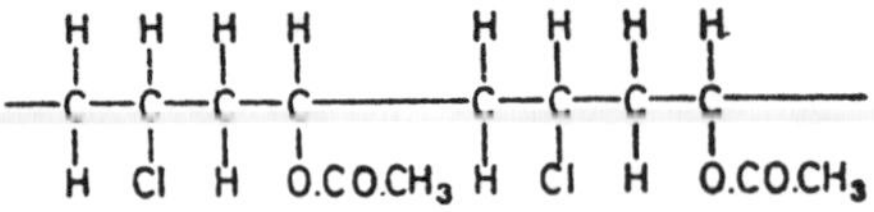

$$
\begin{array}{cccccccccccccccc}
 & \text{H} & & \text{H} & & \text{H} & & \text{H} & & \text{H} & & \text{H} & & \text{H} & & \text{H} \\
 & | & & | & & | & & | & & | & & | & & | & & | \\
— & \text{C} & — & \text{C} & — & \text{C} & — & \text{C} & —— & \text{C} & — & \text{C} & — & \text{C} & — & \text{C} \;—— \\
 & | & & | & & | & & | & & | & & | & & | & & | \\
 & \text{H} & & \text{Cl} & & \text{H} & & \text{O.CO.CH}_3 & & \text{H} & & \text{Cl} & & \text{H} & & \text{O.CO.CH}_3
\end{array}
$$

21.6. Condensation polymerization

This occurs between molecules with functional groups which can react so that they lead to polymer formation accompanied by the elimination of small molecules, such as water.

The acid–base reaction of organic chemistry is an example of a condensation reaction, an alcohol (R_1OH) and an acid ($R_2 . COOH$) condensing to form an *ester* with elimination of a water molecule:

$$R_1OH + R_2 . \underset{\underset{O}{\|}}{C}—OH \longrightarrow R_1—O—\underset{\underset{O}{\|}}{C}—R_2 + H_2O$$

where R_1 and R_2 denote saturated hydrocarbon radicals.

Some other reactions used in the manufacture of commercial polymers are:
an amine reacting with an acid to form an *amide*

$$R_1—NH_2 + R_2 . \underset{\underset{O}{\|}}{C}—OH \longrightarrow R_2 . \underset{\underset{O}{\|}}{C}—O—NH . R_1 + H_2O$$

and an isocyanate condensing with an alcohol to form a *urethane* or with an amine to form a *urea*.

$$R_1OH + R_2 . NCO \longrightarrow R_2—NH—\underset{\underset{O}{\|}}{C}—O—R_1$$

$$R_1NH_2 + R_2 . NCO \longrightarrow R_2—NH—\underset{\underset{O}{\|}}{C}—NH—R_1$$

If bifunctional molecules are used for the starting materials then chain polymers can result. For example, a double alcohol or glycol reacting with a double acid will give a polyester:

$$
\begin{array}{l} \;\;CH_2OH \\ \diagup \\ R_1 \\ \diagdown \\ \;\;CH_2OH \end{array}
+
\begin{array}{l} \;\;COOH \\ \diagup \\ R_2 \\ \diagdown \\ \;\;COOH \end{array}
\longrightarrow HO\!-\!\left[CH_2—R_1—CH_2—O—\underset{\underset{O}{\|}}{C}—R_2—\underset{\underset{O}{\|}}{C}—O \right]_n\!-\!H + H_2O
$$

Similar reactions with the appropriate materials will give *polyamides*, *polyurethanes*, and *polyureas*.

Chain polymers can also be formed by condensation reactions using a single type of unsymmetrical bifunctional molecule instead of two types of symmetrical molecules. An example is an amino acid, which contains an amine group on one carbon atom and an acid group on another and which polymerizes to a polyamide.

$$NH_2R.COOH \longrightarrow H\text{-}\left[\text{-}NHR.CO\text{-}\right]_n\text{-}OH + H_2O$$

The textile fibres, terylene and nylon are a polyester and a polyamide, respectively.

With polyfunctional molecules that can link to three or more other molecules in polymerization, three-dimensional network polymers can result. One example of this is found in the reaction between phenol (C_6H_5OH) and formaldehyde (CH_2O). A formaldehyde molecule can link two phenol molecules with elimination of a water molecule.

OH + H—C(=O)—H + OH ⟶ OH—CH₂—OH + H_2O

Any of the hydrogen atoms on a phenol molecule (other than that on the —OH group) can react in this manner so that one phenol molecule can be linked to one, two, or three other molecules. Spatial reasons would probably prevent more linkages per molecule.

Similar reactions occur in the formation of urea formaldehyde and melamine formaldehyde polymers.

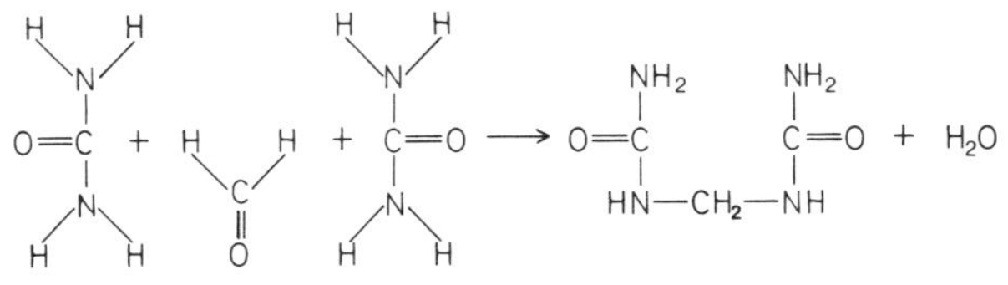

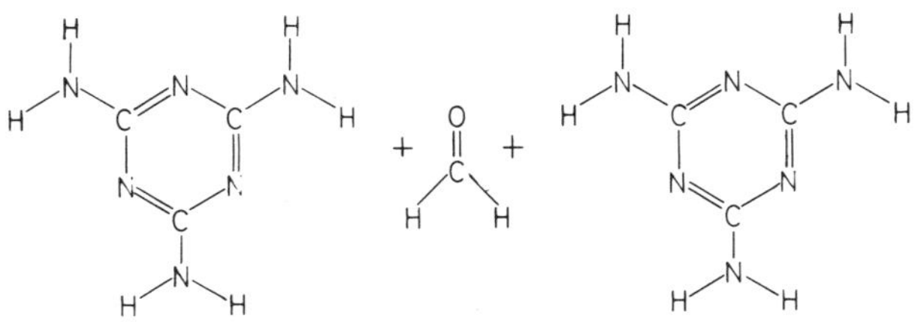

$$H_2N\text{-}CO\text{-}NH_2 + HCHO + H_2N\text{-}CO\text{-}NH_2 \longrightarrow O{=}C(NH_2)\text{-}HN\text{-}CH_2\text{-}NH\text{-}C(NH_2){=}O + H_2O$$

$$C_3N_3(NH_2)_3 + HCHO + C_3N_3(NH_2)_3$$

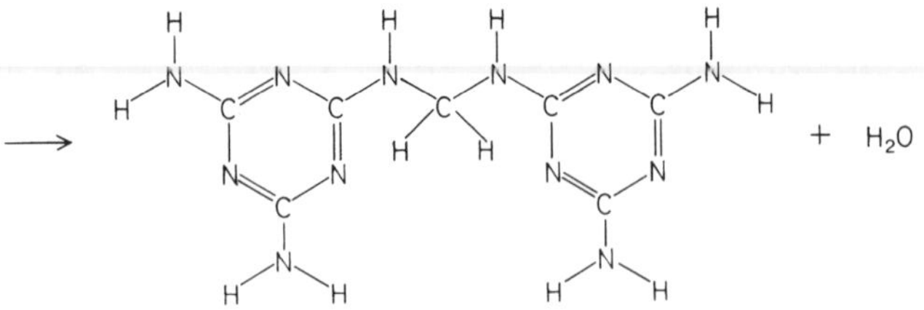

$\longrightarrow$ $+ \; H_2O$

21.7. Relation of mechanical properties to polymer shape

The products of addition polymerization so far discussed and some products of condensation polymerization are linear or chain molecules, and hence will be held together only by van der Waals forces. The exact nature of these forces depends upon the side groups present. In chains which are symmetrical, like polyethylene, the molecules are non-polar, whereas unsymmetrical chains, like polyvinyl chloride, are strongly polar. Hence the force of attraction between chains of the latter will be considerably greater than between chains of the former. Also, with more complicated shapes of side groups, it is less easy for molecular chains, which are intertwined, to straighten or for molecules to slip past one another, thus increasing the resistance to plastic deformation.

Linear polymers exhibit visco-elastic behaviour which can be represented approximately by a mechanical analogue consisting of springs and dashpots. The simplest analogue is shown in fig. 21.4. The deformation can be split into three parts:

1. An immediate elastic strain due mainly to straightening of the polymer chains to their limit before they have to slip past one another. This is reversible and occurs at all temperatures.

2. A visco-elastic strain which is due to the straightening of the twisted and curved polymer chains and which is impeded by internal viscosity. It is time-dependent and is recoverable again after a time when the load is removed.

3. Irrecoverable viscous flow due to the slipping of chains past each other.

The viscosity associated with the two latter parts decreases with increasing temperature. At very low temperatures, only mechanism 1 plays a part so that the polymer is a hard, rigid, glassy material. As the temperature is raised, mechanism 2 becomes important, the temperature below which it is insignificant being known as the *glass transition* tem-

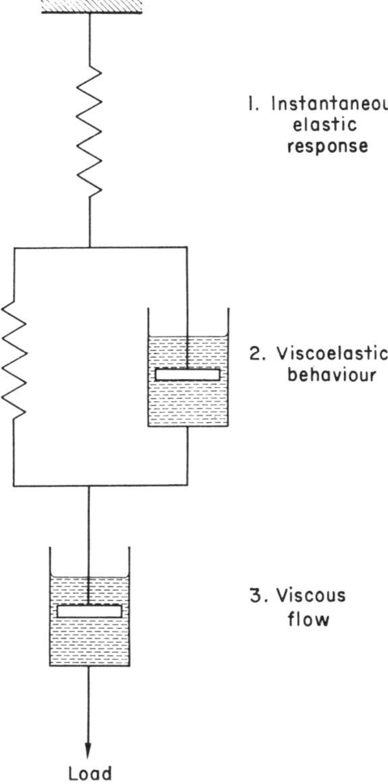

Fig. 21.4.—Mechanical analogue (comprising springs and dashpots) which simulates response of a polymer to an applied stress. Mechanism 1 dominates at low temperatures, 2 at intermediate temperatures and 3 at high temperatures

perature. At a still higher temperature, the *flow* temperature, mechanism 3 becomes important. A more complicated system of springs and dashpots is needed to represent more exactly the actual behaviour of most polymers, but the three basic mechanisms are common to all.

Polyethylene is well above its glass transition temperature at room temperature and behaves in a rubbery manner on stressing with a delayed recovery on unloading. Because, like all alkanes, it is inert to most chemicals, it is widely used for containers and water piping. Because the flow temperature is below 100 °C, its use for piping is restricted to cold water supplies.

Polypropylene, with the same chemical inertness, has methyl groups attached to the main chain of carbon atoms, thereby increasing the van der Waals bonding between chains and raising the flow temperature. It can be used satisfactorily for hot water piping.

Polymers with large side groups such as polystyrene and the acrylic plastics (polymethyl acrylate and polymethyl methacrylate—more commonly known by the trade name of Perspex) are below their glass transition temperatures at room temperature and so are hard and brittle. They are transparent and colourless. Polystyrene is unaffected by water, strong acids, and alkalis and has good electrical properties. The acrylic plastics are mostly used in sheet form and as protective coverings.

Textile fibres from polyesters and polyamides are made by extrusion of filaments through tiny dies or spinnerettes in a molten state (melt spinning). The fibres thus produced are twisted into yarns for weaving into fabrics. By stretching melt-spun fibres several hundred percent, the molecular chains become aligned and the strength is increased (see Section 21.9).

Polycarbonates, such as

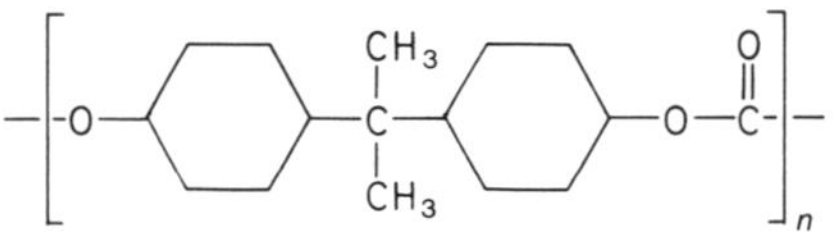

have exceptional strength and rigidity due to the benzene ring structure, flexibility due to the single bonds and toughness due to the side groups. This polymer, which is as transparent as glass and is nearly as tough as steel, is used for vandal-proof glazing.

Polyimides, such as

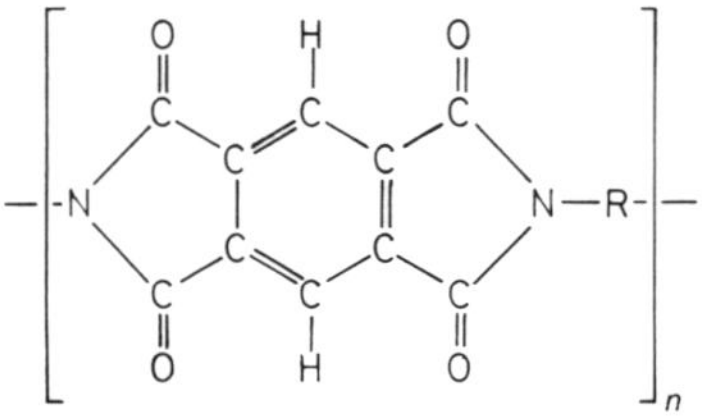

have high stiffness and thermal resistance due to the triple ring structure. This particular material can be used successfully from 4 K up to 400 °C.

Polytetrafluoroethylene (p.t.f.e.) is a white translucent plastic of stiff rubbery consistency at room temperature. It has an extremely low coefficient of friction and is used for unlubricated bearings. Also it has a low value of dielectric constant which does not vary with frequency, so it is useful as an insulating material for high-frequency applications.

These materials which are visco-elastic fluids at higher temperatures are called thermoplastic resins. Fabrication methods with these materials follow the general sequence of

(i) heating to soften,

(ii) mechanically deforming, and

(iii) cooling to harden.

Apart from melt-spinning mentioned above, the processes used include extrusion to make lengths of constant cross-section, injection moulding whereby the heated material is forced into a split mould, and the moulding of sheet or tubular material. In sheet forming, a sheet is clamped at the edges and heated. Then it is placed in contact with a mould of the desired shape and the region between sheet and mould evacuated, so that the heat-softened material is forced by atmospheric pressure to take the shape of the mould. Alternatively, the heated sheet may be pressed between a pair of matching moulds. Plastic bottles are formed by applying pressure to the inside of a thermoplastic tube which has had the bottom pinched off to close it. The tube is blown in a split mould of the correct shape.

Thermoplastics can be joined by welding, i.e., heating the contact area and then mechanically pressing them together while cooling.

Polymers such as phenol formaldehyde (bakelite) and urea formaldehyde, can have three or more linkages to each group and so build up three-dimensional networks which are continuous structures of covalent bonds. When completely polymerized, such materials will not undergo any irrecoverable plastic deformation and will fracture only in a brittle manner. They are usually stronger than thermoplastic resins and can be used at higher temperatures. Articles of these resins are usually manufactured from partially polymerized material which is pressed into moulds and then polymerization is completed by heating. They are known as *thermosetting* resins.

Some linear polymers can be built into rigid networks by cross-linking reactions. For example, bombardment of polythene by neutrons from a nuclear reactor will break some C—H bonds and form a new C—C bond between separate chains.

The shape of chain molecules can also be made more complicated by branching, i.e., deliberately adding side chains, for example by condensation polymerization.

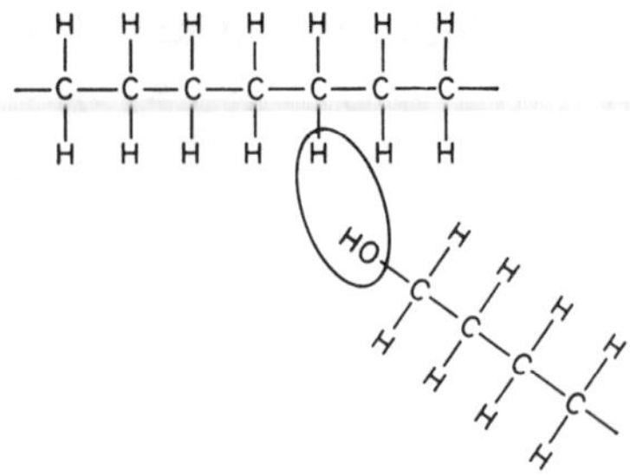

$$
\begin{array}{ccccccc}
H & H & H & H & H & H & H \\
| & | & | & | & | & | & | \\
-C- & C- & C- & C- & C- & C- & C- \\
| & | & | & | & | & | & | \\
H & H & H & H & H & H & H
\end{array}
$$

HO, C, C, C, C, H

With so many variations possible in the shape and size of polymer molecules, a wide range of properties are obtainable. The science of designing polymer molecules to give desired properties is known as *molecular architecture.*

Polymers can also be fabricated into expanded structures or foams. A hot molten thermoplastic containing a volatile solvent dissolved under pressure can foam when the pressure is released. A thermosetting resin foam can be made by polymerization of liquid raw material into which air or carbon dioxide has been whipped to form a froth. The elastic moduli and strengths of these expanded polymers are roughly proportional to density and also depend to some extent upon the cell size. These materials find uses for cushioning, protective packing, lightweight rigidity, and for lightweight thermal insulation.

21.8. Elastomers

Natural rubber is a polymer of *cis*-isoprene with the unit structure

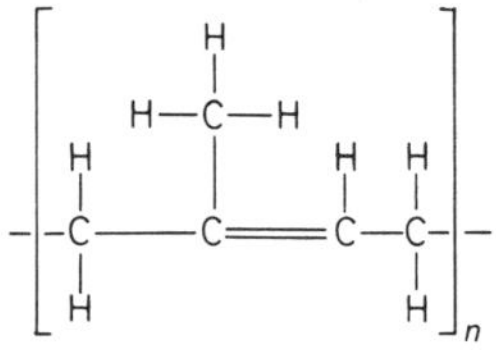

$$
\left[
\begin{array}{ccccc}
 & & H & & \\
 & & | & & \\
 & & H-C-H & & \\
H & & | & H & H \\
| & & | & | & | \\
-C- & & C= & =C- & -C- \\
| & & & & | \\
H & & & & H
\end{array}
\right]_n
$$

The methyl group and hydrogen atom attached to the unsaturated carbon atoms both lie on the same side of the chain. The double bond does not permit rotation and so the chain locally has the shape shown in fig. 21.5a. Rotation is possible at every single C—C bond and so a chain molecule can adopt many different configurations from fairly straight to highly twisted. Such rotation and different configurations would also be possible with any of the chain polymers discussed earlier in this Chapter.

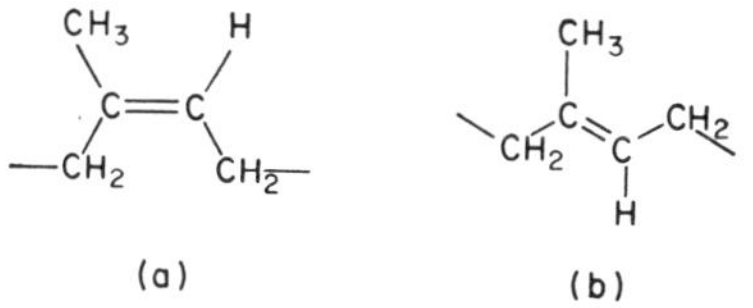

$$\begin{array}{c} CH_3 \quad\quad H \\ \diagdown \quad \diagup \\ C{=}C \\ \diagup \quad \diagdown \\ -CH_2 \quad\quad CH_2- \end{array}$$

(a)

$$\begin{array}{c} CH_3 \\ | \\ \diagdown CH_2 - C = C - CH_2 \diagdown \\ | \\ H \end{array}$$

(b)

Fig. 21.5.—Unit of (*a*) rubber and (*b*) gutta-percha

Due to thermal agitation, the relative orientation of adjacent bonds will be constantly changing by virtue of the possible rotation, and there will be a statistical distribution of the configurations dependent upon their relative energies. In many polymers the energy is less when the chains are stretched and lie more nearly parallel to one another. Hence for these materials the ordered (or stretched and parallel) state will predominate. In the rubbers, the difference in energy between the stretched and the tangled forms of the molecular chain is small. Therefore all configurations are almost equally likely to occur. Because there are more possible tangled configurations than stretched ones, the former will predominate. Hence in rubbers the tendency to maximum entropy dominates whereas in other classes of polymers this is overcome by the tendency to minimum energy.

When a tensile stress is applied to a rod of rubber, the chains will stretch, so that a considerable elongation can occur. On removing the stress, thermal agitation will return the chains to the tangled form causing the rod to return to its original length.

If the temperature of a piece of rubber is increased while under a constant stress, there will be a contraction, i.e., a negative coefficient of thermal expansion. Between −20 °C and 70 °C the elastic modulus is proportional to the absolute temperature. The glass transition temperature is in the range −30 °C to −50 °C, below which natural rubber is a rigid brittle solid.

Polymers of this type which have a large elastic range are known as *elastomers*. The various synthetic rubbers have a similar double bond structure, as in polychloroprene or neoprene, which is formed by the polymerization of chloroprene

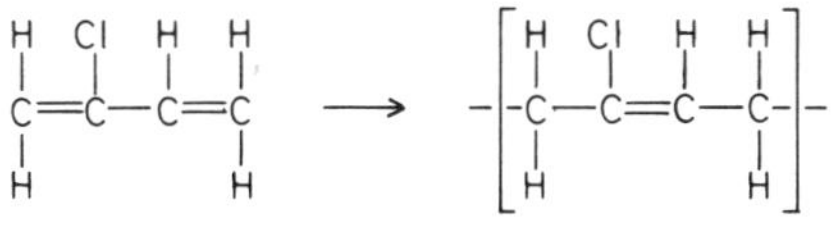

$$\begin{array}{cccc} H & Cl & H & H \\ | & | & | & | \\ C = & C - & C = & C \\ | & & & | \\ H & & & H \end{array} \longrightarrow -\left[\begin{array}{cccc} H & Cl & H & H \\ | & | & | & | \\ -C - & C = & C - & C- \\ | & & & | \\ H & & & H \end{array}\right]-$$

The structure isomorphous with rubber, in which the groups attached to the unsaturated carbon atoms lie on opposite sides of the chain, does not have the rubberlike property. *Trans*-isoprene is the isomer of isoprene with this structure, and when polymerized gives a much more rigid material called gutta-percha (fig. 21.5*b*).

Plain chains of polyisoprene will slide fairly easily past one another. With a small amount of anchoring by cross-linking the plastic flow can be suppressed while still permitting large elastic extensions. This is achieved by *vulcanizing* with sulphur. A few of the unsaturated bonds are broken and sulphur atoms used to bridge the gaps.

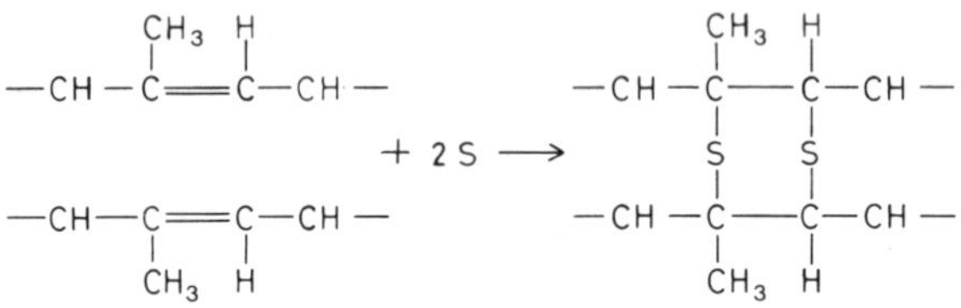

$$
\begin{array}{cccc}
 & CH_3 & H & \\
 & | & | & \\
-CH- & C= & C- & CH- \\
 & & & \\
-CH- & C= & C- & CH- \\
 & | & | & \\
 & CH_3 & H &
\end{array}
\; + 2S \longrightarrow \;
\begin{array}{cccc}
 & CH_3 & H & \\
 & | & | & \\
-CH- & C- & C- & CH- \\
 & | & | & \\
 & S & S & \\
 & | & | & \\
-CH- & C- & C- & CH- \\
 & | & | & \\
 & CH_3 & H &
\end{array}
$$

With higher sulphur contents, more cross-linking occurs which restricts the range of elastic stretching. When all the unsaturated bonds have been used up by full vulcanization, as in ebonite, the rigidity reaches a maximum.

The hardening of rubber which occurs slowly when it is exposed to light is due to a similar cross-linking process, oxygen being the bridging atoms. This oxidation reaction is catalysed by the action of light.

21.9. Crystallinity in polymers

Polymers which are completely crystalline are rare but some have regions of about 10 nm in size which have an ordered three-dimensional structure. This is possible only in cases where the chain has a regular structure. Polythene and polytetrafluoroethylene have such a structure, but in the vinyl polymers the side groups may be randomly arranged on one side or other of the chain (*atactic*). Only when they are all arranged on one side (*isotatic*) or alternately on the two sides (*syndiotactic*) is crystallinity possible. In copolymers, units of the two or more monomers do not necessarily occur in a regular sequence, so that these are more likely to be non-crystalline.

Crystalline polymers consist of crystalline regions with amorphous regions between, each individual chain passing through several regions of each sort, or else the chain passes through one crystalline region several times, being folded back on itself.

A crystalline polymer does not show a steady decrease in viscosity with increasing temperature, but maintains an almost constant value to a temperature at which there is a sharp drop, this corresponding to the melting or break-up of the crystal structure. The behaviour of partially crystalline polymers is intermediate between that for crystalline and that for amorphous material. Where flexibility is required of a material, crystallinity is undesirable.

When polymer material comprising linear chains is elongated, there is an initial, almost linear, elastic region followed by yielding and elonga-

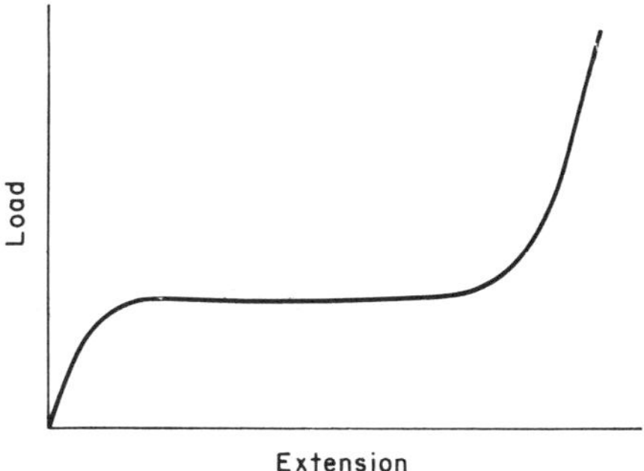

Fig. 21.6.—Typical load–extension curve for a linear polymer

tion which may be several hundred percent. Then the load–extension curve (fig. 21.6) rises sharply. In the extension or " drawing " the molecular chains have become oriented in the direction of drawing and this alignment gives a crystalline structure. The longitudinal strength of fibres is increased many times by the drawing.

21.10. Wood

The primary structural material of trees and plants is a natural polymer, *cellulose*. This is a long chain material, the polymer having the structure

```
   ┌                               ┐
   |        H    OH                |
   |        |    |                 |
   |       C ─── C                 |
   |      /|     |\                |
   |     / OH    H \               |
 ─ | ─CH             CH─O─ | ─
   |     \       H   /             |
   |      \      |  /              |
   |       O ─── C                 |
   |             |                 |
   |            CH2OH              |
   └                               ┘n
```

It is therefore a regular chain and the presence of the hydroxyl groups makes it highly polar, so that the secondary forces between chains are high. In wood the cellulose chains are aligned and the material is highly crystalline. An amorphous carbohydrate material, lignin, is the next important constituent. This and other materials bond the cellulose crystals together and form tubular cells through which moisture and sap pass.

Growth is most rapid in spring, giving thin open cells, while that in summer is slower producing denser and stronger cells. This annual cycle produces the rings and grain structure of wood.

The majority of cellulose chains run parallel to the grain giving greater strength in that direction than in the transverse direction.

The strong polar nature of the cellulose chains provides the strong affinity of wood for moisture, which causes swelling. On drying cut timber, the water is lost more readily from surfaces perpendicular to the grain than from those parallel to the grain and this tends to cause warping. Also in larger pieces the moisture is lost more readily from the surfaces than from the bulk so that surface shrinking occurs leading to cracking.

These difficulties can be overcome by laminating, i.e., preparing thin sheets which are easily dried, and sticking them together with adhesives, the run of the grain being arranged to give strength in the desired directions. Also by impregnating wood with suitable plastics, moisture absorption can be prevented.

On heating in air, wood decomposes before it would soften, but by heating in steam, which prevents drying out, it softens and can be readily shaped.

The strong polar forces between chains make wood insoluble in water. By treatment with carbon disulphide and sodium hydroxide which reduce the chain lengths and neutralize the polar influence of the hydroxyl groups, a water-soluble derivative is formed. A solution of this is forced through minute holes or narrow slots into an acid hardening bath to form cellulose fibres or sheets, known as rayon and cellophane, respectively. These are stretched to orient the chains.

21.11. Epoxy resins

The polymers of this series (Araldite, etc.) are formed from materials containing an epoxy group

```
         O
        / \
   —C———————C—
```

which can act as a polymerizing point. For example, an alcohol would react to give an ether and hydroxyl group

$$R.OH + \underset{\diagdown\ O\ \diagup}{CH_2\text{———}CH}\text{—} \longrightarrow R.O\text{—}CH_2\text{—}\underset{\displaystyle OH}{\underset{|}{CH}}\text{—}$$

Then the hydroxyl group would react with another epoxy group to form a side linkage

$$\underset{\diagdown\ O\ \diagup}{CH_2\text{———}CH}\text{—} + R.O\text{—}CH_2\text{—}\underset{\displaystyle OH}{\underset{|}{CH}}\text{—}$$

$$\longrightarrow RO\text{—}CH_2\text{—}\underset{O\text{—}CH_2\text{—}\underset{\displaystyle OH}{\underset{|}{CH}}\text{—}}{\underset{|}{CH}}\text{—}$$

Various other agents can also be used to give polymerization. While these resins are thermosetting, some will cure at room temperature once the monomer and curing agent are mixed. The mixture passes from the liquid to the solid state with very little shrinkage, so that the bonds initially established are preserved and there are no high internal stresses. Also they differ from most thermosetting resins in that there are no by-products.

The resulting materials have high adhesive strength due to the polarity of the hydroxyl (—OH) and ether (—C—O—C—) groups. They will also form chemical bonds with surfaces, e.g., metals, where active hydrogen may be found.

They are very inert chemically, being unaffected by caustic chemicals, and are extremely resistant to acids. In this respect they are much superior to phenolic and polyester resins which are attacked by caustic soda.

A typical polymer is

$$\text{—}\underset{\underset{|}{O}}{\underset{|}{CH}}\text{—}CH_2\left[\text{—}O\text{—}C_6H_4\text{—}\underset{CH_3}{\overset{CH_3}{\overset{|}{\underset{|}{C}}}}\text{—}C_6H_4\text{—}O\text{—}CH_2\text{—}\underset{\underset{|}{O}}{\underset{|}{CH}}\text{—}CH_2\text{—}\right]$$

Due to the greater distance between cross-linking points, these materials have greater flexibility than phenol formaldehyde and similar thermosetting resins, and this gives them much greater toughness.

The epoxy resins find many uses as adhesives, coatings, and cast articles.

21.12. Fillers

These are usually non plastic and insoluble compounds, added to change the properties of the pure polymer in some definite manner. Thus a reinforcing filler may be used to improve the mechanical strength. Up to 50% by weight of cellulose fibre materials, in the form of rags, paper, cloth, or wood, are commonly used in the thermosetting resins, the resultant strength depending upon the state of subdivision. Wood flour is often used as a filler in phenol formaldehyde (bakelite) mouldings for the purposes of reducing cost and keeping the density low while not having much effect upon the strength. Paper and linen are used mainly in laminated structures, which are manufactured as sheets, tubes and blocks, and which have greater strength and toughness.

Urea formaldehyde and melamine plastics are frequently used because of their translucence, permitting them to be coloured easily. If the filler is not to destroy this translucence, it must not impart any colouring, and so highly purified cellulose must be used.

Asbestos and mica can be used to increase the heat resistance and also have good electrical properties. Powdered minerals may be used to impart hardness, but, not having a fibre structure, will not increase the tensile strength.

Carbon black in the form of particles with diameters in the range 20–200 nm are used as fillers in rubbers, especially for vehicle tyres. They improve the stiffness, the strength by a factor as great as 10, and abrasion resistance by as much as five times. The finer the particles, the more effective they are.

21.13. Plasticizers

These are compounds added either to aid the compounding or moulding process or to modify the final properties. They partly separate the long chain molecules so that the forces between them are decreased. The effect on the thermoplastic materials is to give them a more flexible or rubber-like nature, i.e., they lower the glass-transition temperature. Thus pure polyvinyl chloride is a hard substance at room temperature, but when plasticized with, for example, tricresyl phosphate, it becomes flexible.

21.14. Silicones

Polymers with backbone chains of elements other than carbon have been synthesized. The most notable are the *silicones* in which the backbone chain is composed of silicon and oxygen atoms—*the siloxane link*

```
 |    |    |    |    |    |    |
—Si—O—Si—O—Si—O—Si—
 |    |    |    |    |    |    |
```

The most common silicones have simple hydrocarbon radicals attached, e.g., methyl ($—CH_3$) and phenyl ($—C_6H_5$). Cross-linked structures can also be synthesized, such as:

```
       CH3    CH3
        |      |
 —O—Si—O—Si—O—
        |      |
       CH3     O
               |
          —O—Si—O—Si—O—
               |      |
              CH3    CH3
```

These materials are noted for their temperature stability over a wider range than the carbon-based polymers. Silicone rubbers, consisting of methyl siloxane chains have good temperature resistance up to 150 °C, good flexibility at temperatures as low as −90 °C and are resistant to oils.

QUESTIONS

1. Define the terms *polymerization, copolymerization.*
Discuss the effect of the symmetry of the mer molecule on the properties of long-chain polymers. [MST]

2. Describe the necessary structural features of compounds which can be formed into thermoplastic polymers.
Discuss, with examples, the influence of the type of monomer upon the mechanical and physical properties of the product. [MST]

3. What is meant by the terms *addition polymerization*, *copolymerization*, and *degree of polymerization?*
The mer weight of a certain polymer is 75 kg/kmer. In a series of tests on various mixtures the following data were obtained:

Average molecular weight (kg/kmol)	Tensile strength (MPa)	Solution viscosity (s)
6000	17	25
8000	52	30
10 000	61	40
12 000	66	80
14 000	69	220

Considering the strength and viscosity of the material as functions of the degree of polymerization, comment on the significance of the results. [MST]

4. What is meant by the terms *addition polymerization*, *degree of polymerization*, and *cross-linking* as used in connection with the structure of polyisoprene?
The creep behaviour in tension of an amorphous polymer can be modelled by a spring-and-dashpot series unit in series with a spring-and-dashpot parallel unit. Representing the spring and dashpot constants by E_1, η_1 and E_2, η_2 respectively, derive an expression for the creep strain as a function of time after the sudden application of a

constant tensile stress σ. Sketch the corresponding creep-strain–time curve.

Propose a molecular interpretation for the mechanical behaviour predicted by the components of the model. [E]

5. (*a*) On rapidly loading an elastomer which has no cross-linking it behaves in a linear elastic manner with a Young's modulus of 10 kPa. When held in the stretched state, the stress relaxes to half its value in one hour. On the assumption that the elastomer can be represented by a spring-and-dashpot series model, calculate the rate of increase of tensile strain with time if a specimen is subjected to a constant tensile stress of 400 Pa.

(*b*) An elastomer containing 48% butadiene, 40% isoprene, 10% sulphur, and 2% carbon black by weight is vulcanized so that all the sulphur is used in cross-linking. What fraction of the possible cross-links are joined by vulcanization? [E]

6. A thermoplastic material yields at a tensile stress σ_y given as a function of the absolute temperature T by

$$\sigma_y = A \exp\left(\frac{T_1}{T-T_0}\right)$$

where $T_0 = 273$ K, $T_1 = 23{\cdot}03$ K, and A is a constant of the material

The same material fractures at a tensile stress σ_f given by

$$\sigma_f = A\left(20-\frac{T}{T_2}\right)$$

where $T_2 = 28{\cdot}3$ K

By considering a graph of $\log_e(\sigma/A)$ against temperature

(*a*) determine the brittle–ductile transition temperature, and

(*b*) sketch and compare stress–strain curves for the material at 280 K, 286 K, and 330 K assuming that the yield strain of the material is constant. [E]

CHAPTER 22

Composites

22.1. Introduction

A composite material is one composed of two (or more) distinct phases, one of which is a matrix surrounding fibres or particles of the other, there being some form of bonding at the interface between them. It is designed to provide a combination of desired properties that is not exhibited by either material on its own.

Usually one of the materials has distinct advantages, for example it may have high strength and low density, but because it is brittle or not formable in large sections cannot be used on its own.

Precipitation-hardened alloys (Section 11.9), dispersed-oxide hardened metals (Section 11.10), and elastomers with fillers such as carbon black (Section 21.12) are all examples of particulate composites but will not receive further consideration here. Particulate composites which use rubbery particles to increase the toughness of a brittle matrix will be considered in Section 22.8.

Most of this chapter will deal with the manufacture and properties of fibrous composites.

22.2. Reinforcing materials for fibrous composites

Ceramics have high values of Young's modulus and because they are non-ductile can have high tensile strengths in the absence of flaws. Also the specific gravities are low compared with almost all metals so that in terms of strength/weight and stiffness/weight ratios they have distinct advantages. The main drawback, however, is in fabricating large components from these materials in which the high tensile strength can be maintained.

These materials can, however, be manufactured as fibres or whiskers (see Section 11.1) with high strengths, some selected examples being given in Table 22.1. Any of these can then be incorporated into a ductile matrix to form a composite that has mechanical properties intermediate between those of the ceramic and those of the matrix. By suitably orienting the fibres, components with highly directional strength or stiffness can be made.

Glass fibre is made by drawing out fine threads from molten glass and is done as a continuous process. The freshly drawn glass has a high

TABLE 22.1.—STRENGTH AND STIFFNESS DATA FOR VARIOUS CERAMIC FIBRES AND WHISKERS COMPARED WITH MILD STEEL

			E glass fibre	Silica fibre	Boron fibre	Alumina whisker	Carbon whisker	Carbon fibre		Silicon nitride whisker	Mild steel
								High modulus	High breaking strain		
Fibre diameter,	μm		8–10	50–80	100	3–10	0·5–5·0	7·5	8·0	5	
Density, ρ,	kg m^{-3}	10^3 x	2·54	2·2	2·5	3·96	2·2	2·0	1·74	3·2	7·87
Tensile strength, σ_m,	GPa		1·8	3·5	2·7	21	20	2·0	3·0	7·0	0·5
Young's modulus *E* parallel to fibre axis,	GPa		70	77	420	430	700	560	230	380	210
σ_m/ρ,	$m^2 s^{-2}$	10^6 x	0·7	1·6	1·1	5·3	9·1	1·0	1·7	2·2	0·06
E/ρ,	$m^2 s^{-2}$	10^9 x	27·5	35	170	110	320	280	130	120	27

strength due to freedom from surface imperfections but this deteriorates with time due to the corrosive effect of the atmosphere. By encasing the fibres in a polymeric material such as an epoxy resin (see Section 21.11) a now familiar composite—glass-fibre reinforced plastic (GRP)—is produced. A special alkali-resistant glass fibre has been developed for use in making glass-fibre reinforced cement (GRC).

Whiskers are grown by deposition of material from vapour on to a suitable substrate. The whiskers, each of which is a single crystal, grow at points where screw dislocations emerge on the surface of the substrate. In the furnace, the temperature and pressure are controlled to give gas supersaturations lower than those necessary for three-dimensional growth so that the crystals continue to build up at the tips and not at the sides. The present stage reached in silicon nitride whisker manufacture is the production of batches of a few kilogrammes each, the total process time being 40 hours. Whiskers are used in a polymer matrix or more usually in a metal matrix. For example, an alumina-nickel composite is used for high temperature applications.

Boron filaments can be made by drawing from the melt, though there are serious limitations on this method. They are usually made by vapour deposition on to continuous filaments of a metal or glass. Boron fibres are used in boron-polymer and boron-aluminium composites.

Carbon fibres are produced by controlled pyrolysis of organic textile fibres. The fibres used consist of long-chain molecules which have a preferred orientation along the fibre axis. During heating, certain chemical reactions occur between the polymer chains so that six-membered ring structures are formed parallel to the fibre axis. One fibre source is polyacetonitrile which initially reacts as follows:

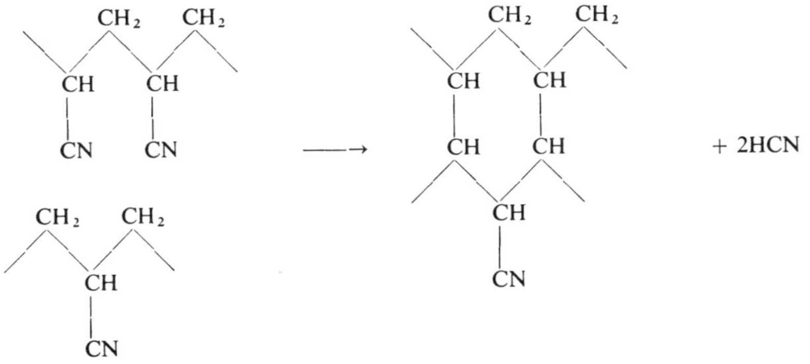

```
      CH2   CH2                          CH2   CH2
  \  /   \ /   \                      \  /   \ /   \
   CH     CH                           CH     CH
   |      |                            |      |
   CN     CN          ——→              CH     CH          + 2HCN
                                      /  \   /  \
   CH2   CH2                              CH
  /   \ /   \                             |
       CH                                 CN
       |
       CN
```

The reaction depends upon the rotation of the reacting groups to favourable positions which is very temperature-dependent. After complete pyrolysis, the resultant carbon fibres consist of long *fibrils* about 10 nm wide and within these fibrils are oriented graphite crystals. The hexagonal layers of carbon atoms which contain the high modulus and high strength direction lie along the fibril axes. Two grades of carbon fibre are manufactured. One, of higher density, has a much higher elastic modulus and the other, less dense, a higher tensile strength.

Carbon fibres are used with epoxy resins or metals such as aluminium as matrix for the manufacture of articles where strength/weight ratio is highly important, e.g., compressor blades in gas turbines.

Metals are also used for reinforcement either as fine wires (e.g., stainless steel) or as whiskers (e.g., tungsten). These are usually used in metal matrices.

22.3. Manufacture of fibre composites

The reinforcing material is manufactured either as whiskers, or as continuous fibres which may be single fibres in the case of boron or bundles of filaments in the cases of glass and carbon. Whiskers can be used only by randomly mixing them into the matrix material (unpolymerized plastic or molten metal) which is moulded or extruded to shape before the matrix hardens. Fibres and filaments can be laid down in definite orientations giving anisotropic properties, e.g., highest strength in a specific direction. Glass and carbon fibres are supplied as unidirectional rovings or *tows*, assembled as flat tapes and sheets of aligned fibres loosely stitched together or as woven sheets. Also they can be cut into short lengths, known as *chopped strand*, either used randomly or as chopped strand mat. Hence products can be made with one-, two- or, to a limited extent, three-dimensional reinforcement (see fig. 22.1).

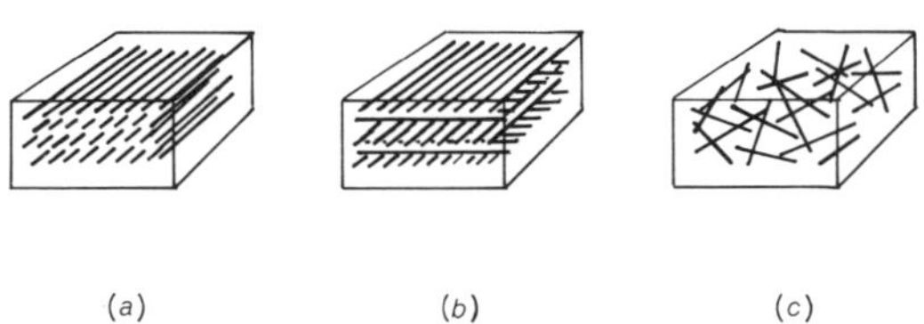

(*a*) (*b*) (*c*)

Fig. 22.1.—One-, two- and three-dimensional reinforcement in fibre composites

The more commonly used methods of manufacture of products with a resin matrix, usually a polyester or an epoxy, are:

(i) Hand lay-up. Layers of aligned sheet, woven fabric, or chopped strand mat are cut to size and laid on forms of the right shape. Liquid resin is applied to the successive layers as required.

(ii) Preform. Complex shapes are made by depositing chopped fibres on a screen of appropriate shape and then moulded to final form in a press.

(iii) Spray-up. Bundles of rovings are fed through a chopper and the chopped fibres blown simultaneously with a spray of resin on to a mould. Compaction is usually completed by the use of rubber rollers.

(iv) Filament winding. For axially symmetrical objects, a continuous filament is wound on to a mandrell or appropriate shape, the pattern of winding being regulated to give directional strength as required. Resin is also applied continuously as needed.

For metal matrix composites, the fibres are first combined with the matrix, consolidated to form a dense material, and then, if necessary, further worked to form the required shape.

GRC products are mostly manufactured by the spray-up method, but filament winding has been adopted for pipes.

22.4. Elastic properties of a composite

If each component behaves in a linear elastic manner, the composite will also behave in a linear elastic manner, the Young's modulus, E_c, having a value intermediate between those of the separate composites. Consider a composite that contains a volume fraction f of a material with Young's modulus E_f in a matrix with Young's modulus E_m Bounds on the value of E_c can be calculated for two extreme cases.

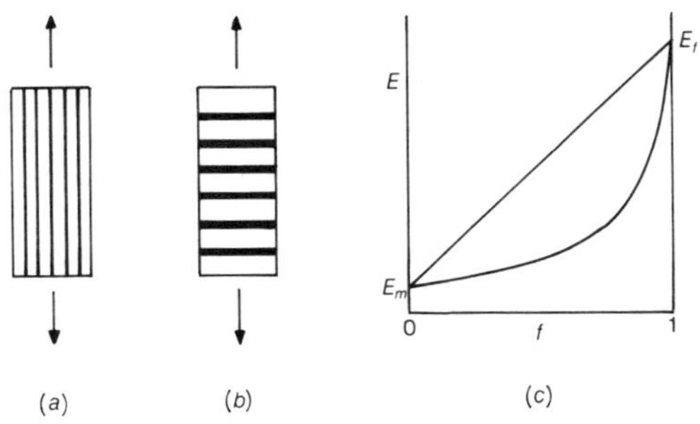

Fig. 22.2.—(*a*) Aligned fibre composite. (*b*) Lamellar composite. (*c*) Variation of modulus with volume fraction for $E_f = 10\ E_m$

(i) With continuous fibres aligned in the direction of stress (fig. 22.2*a*) and considering unit area of cross-section of the composite, the mean stress σ_c is given by

$$\sigma_c = f\sigma_f + (1 - f)\sigma_m \tag{22.1}$$

where σ_f and σ_m are the stresses in the fibre and in the matrix, respectively. We can assume that the strains in the fibres and in the matrix will be equal so that if ε is the strain

$$E_c\varepsilon = fE_f\varepsilon + (1 - f)E_m\varepsilon$$

or

$$E_c = fE_f + (1 - f)E_m$$

(ii) If all the reinforcing material is in flat sheets perpendicular to the direction of stress (fig. 22.2*b*) then the stress σ in the two materials will be equal but the strains ε_f and ε_m will differ. The total strain ε is

$$\varepsilon = f\varepsilon_f + (1 - f)\varepsilon_m$$

so that

$$\frac{\sigma}{E_c} = f\frac{\sigma}{E_f} + (1 - f)\frac{\sigma}{E_m}$$

or

$$E_c = \frac{E_f E_m}{fE_m + (1 - f)E_f}$$

The resulting variations of modulus with volume fraction for these two cases are shown in fig. 22.2*c*.

Although resin matrices do not show strictly linear elasticity, their behaviour is often sufficiently close to linear for the above considerations to apply.

22.5. Strength of a fibre composite

For a composite with continuous fibres all aligned in the stressing direction, the stress can be predicted by the law of mixtures, as in equation (22.1), even when both constituents are not behaving elastically.

Thus if we have a brittle fibre in a ductile matrix which is an ideal elastic-plastic material (see Section 12.10) and the failure strain of the fibre is greater than the yield strain of the matrix, the stress–strain curve for the composite will be of the form shown in fig. 22.3.

In a more general case of a brittle fibre in a ductile matrix which shows some strain hardening (fig. 22.4*a*), let σ_{fu} be the breaking stress of the fibres at a strain ε_f, σ_m be the stress in the matrix at the strain at which the fibres fail and σ_{mu} be the tensile strength of the matrix. When the composite is strained to ε_f, the stress will be given by the law of mixtures as

$$\sigma_c = f\sigma_{fu} + (1 - f)\sigma_m \qquad (22.2)$$

and is represented by line AB on fig. 22.4*b*. When the fibres have all broken and the load is taken entirely by the matrix, then at the point of matrix

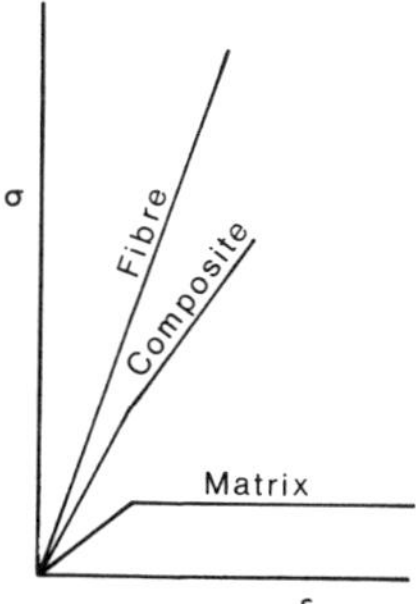

Fig. 22.3.—Stress-strain curves of fibre, matrix and 50% volume fraction composite

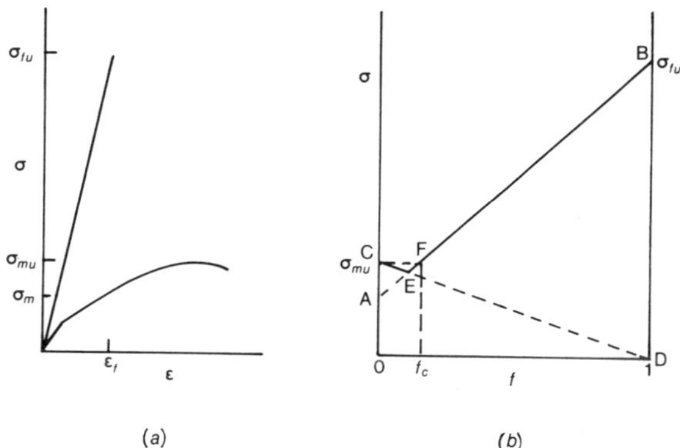

Fig. 22.4.—(*a*) Stress–strain curves of brittle fibre and weak ductile matrix. (*b*) Strength of composite as function of composition

failure the stress calculated on the total cross-sectional area of the composite will be

$$\sigma_c = (1 - f)\sigma_{mu}$$

which is represented by line CD on fig. 22.4*b*. These lines intersect at E and show that for small volume fractions of the fibre the strength is controlled by the strength of the matrix, while at larger volume fractions the strength is controlled by the fibres because once they have broken the load that can be supported by the matrix is less. The strength of the composite thus exceeds the strength of the matrix only if the volume fraction of fibres exceeds a critical value f_c given by point F, i.e.,

$$\sigma_c = f_c\sigma_{fu} + (1 - f_c)\sigma_m = \sigma_{mu}$$

or

$$f_c = \frac{\sigma_{mu} - \sigma_m}{\sigma_{fu} - \sigma_m}$$

There is also obviously a maximum volume fraction that is possible if all interstices between fibres are to be filled and if all the fibres are to be separated by matrix material. In the case of fibres with cylindrical cross-sections and all of the same diameter, the maximum value of f is just over 0·9.

If the fibres are discontinuous, the stress will not be uniform along the length of each fibre. The tensile stress builds up from the fibre ends, the force being transmitted to the fibre by interfacial shear from the matrix in the manner shown in fig. 22.5*a*. The interfacial shear will at first increase as the strain in the composite increases, but will be limited either by the yield strength in shear of the matrix material or by the bond strength between fibre and matrix. This bond strength will be high only if the matrix wets the fibre during manufacture. If the combination is non-wetting, coatings are applied to the fibres to promote wetting.

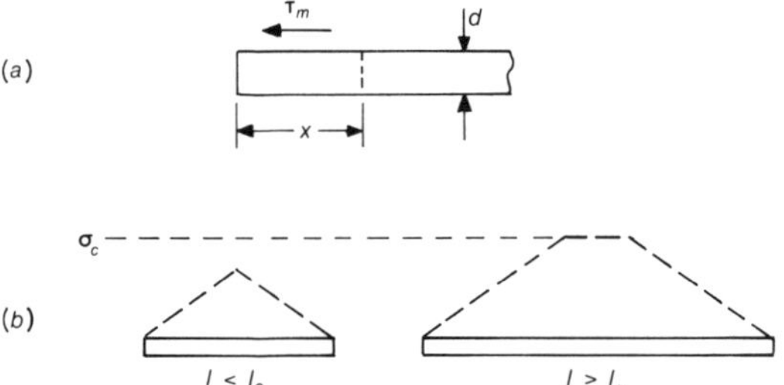

Fig. 22.5.—(*a*) Transfer of load from matrix to fibre by interfacial shear. (*b*) Distribution of longitudinal stress in fibre

If τ_m is the maximum interfacial shear stress, then at a distance x from the end of a fibre of circular cross-section and diameter d the total shear force that can be transmitted to the fibre is $\pi d \tau_m x$ and the force in the fibre is $\pi d^2 \sigma_f/4$. Hence

$$\sigma_f = \frac{4x\tau_m}{d}$$

If the fibre is to reach its fracture stress σ_{fu}, then the minimum length of fibre is

$$l_c = 2x = \frac{\sigma_{fu} d}{2\tau_m}$$

Hence the aspect ratio (l/d) of the individual fibres must be not less than the critical value $\sigma_{fu}/2\tau_m$ if they are to develop their full strength. Otherwise the matrix will fail without breaking the fibres.

The average stress in a fibre of length l ($>l_c$) when the stress in mid-length is equal to the breaking stress is given by

$$\sigma_f = \sigma_{fu}(1 - \frac{l_c}{2l})$$

and σ_f can be substituted for σ_{fu} in equation (22.2) to give the strength of the composite. We see that discontinuous fibres produce less strengthening than continuous ones.

For fibres with a brittle matrix, the fracture strain of which is less than the failure strain of the fibres, the matrix will crack and transmit all the load to the fibres. These can stretch, with failure of the interfacial bond until they also reach their failure strain and may break at some distance from the plane of the matrix crack, so exhibiting pull-out. Where fibres run at right-angles to the direction of stretching, there may be debonding which initiates matrix cracking.

In compression, the strength of the composite depends on the ability of the matrix to constrain the fibres from buckling. Failure is usually preceded by delamination of individual fibres in unidirectional composites or sheets of fibres in bi-directional composites.

22.6. Specific stiffness and specific strength

Composite materials are frequently of advantage where mass or weight saving is a prime requirement and their usefulness can be assessed in terms of their specific stiffnesses and specific strengths, that is, the ratios of Young's modulus and tensile strength respectively to the density. Figure 22.6. shows values of these parameters for some representative composites compared with various other materials in common use.

22.7. Toughness of fibre composites

The fracture of a fibre composite may involve plastic deformation of the matrix, fracture of the matrix, plastic deformation of the fibres, pull-out of the fibres, and fracture of fibres, all of which contribute to the energy required. A fractured surface of a carbon-fibre-reinforced composite where pull-out was a significant feature is shown in fig. 22.7.

In a composite of brittle fibres in a brittle matrix in which both matrix and fibres break along the same plane, as in fig. 22.8*a*, the toughness may be assumed to be given by the law of mixtures

$$G_c = fG_{cf} + (1 - f)G_{cm}$$

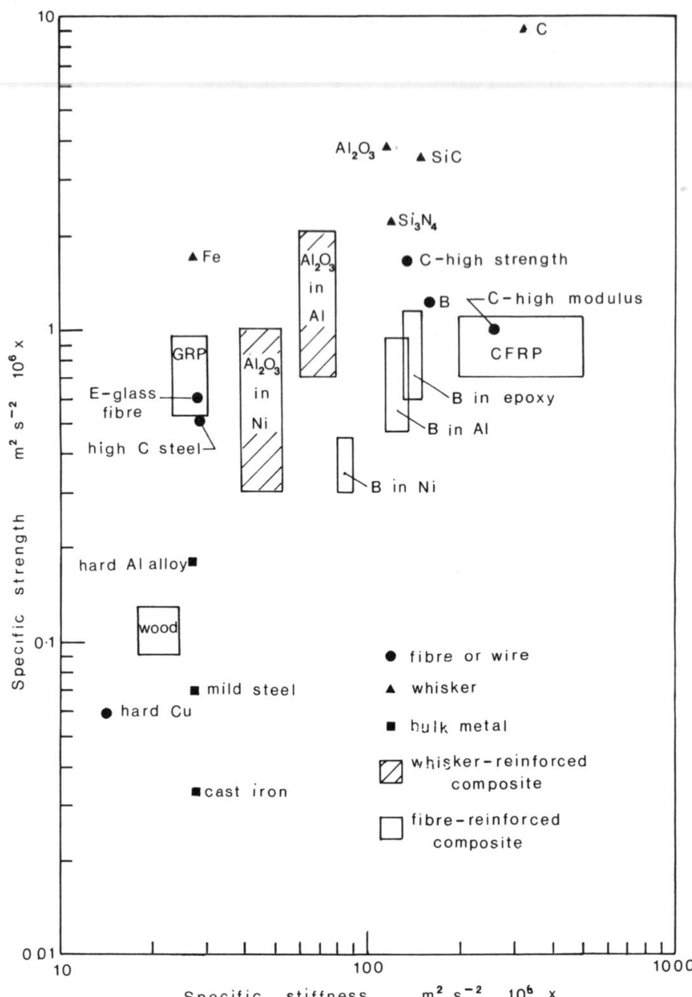

Fig. 22.6.—Specific strengths and specific stiffnesses of various fibre-reinforced composites compared with those of the reinforcing fibres and some common materials

where G_{cf} and G_{cm} are the toughnesses of fibre and matrix, respectively.

Composites of brittle fibres in a tough matrix in which the fibres break ahead of an advancing crack (fig. 22.8*b*) will exhibit heavy deformation of the remaining matrix which bridges the crack. These bridged regions will neck down to fracture in a ductile manner. Most of the work done will be associated with this plastic deformation. Hence, a smaller volume

Fig. 22.7.—Fracture surface of carbon fibre reinforced composite showing pull-out of fibres (×2000)
[By courtesy of Dr. P. W. R. Beaumont]

fraction of fibre would give a larger G_c and, according to Section 13.20, higher strength in the presence of a notch, even though it would give a lower un-notched strength.

Ductile fibres in brittle matrices will also give high toughness associated with the necking of the fibres after the matrix fractures (fig. 22.8*c*). Here increased volume fracture of fibres would increase both toughness and unnotched strength.

In composites with a weak fibre–matrix interface or with low matrix shear strength, the transverse tensile and shear stresses in the vicinity of an advancing crack tip may cause longitudinal cracks to form along the plane of the fibre as shown in fig. 22.8*d*. This debonding blunts the crack and temporarily stops its further advance. After travelling a small distance, this secondary crack may initiate further cracks perpendicular to the tensile stress at each of its ends, the whole process repeating when these cracks meet other fibres as in fig. 22.8*e*. This process will increase the toughness G_c.

With discontinuous fibres, the ends of which lie within $l_c/2$ of the plane of the advancing crack or where brittle fibres break away from the plane, the fibres will pull out of the matrix (fig. 22.8*f*). The work to pull out a length l of fibre would be

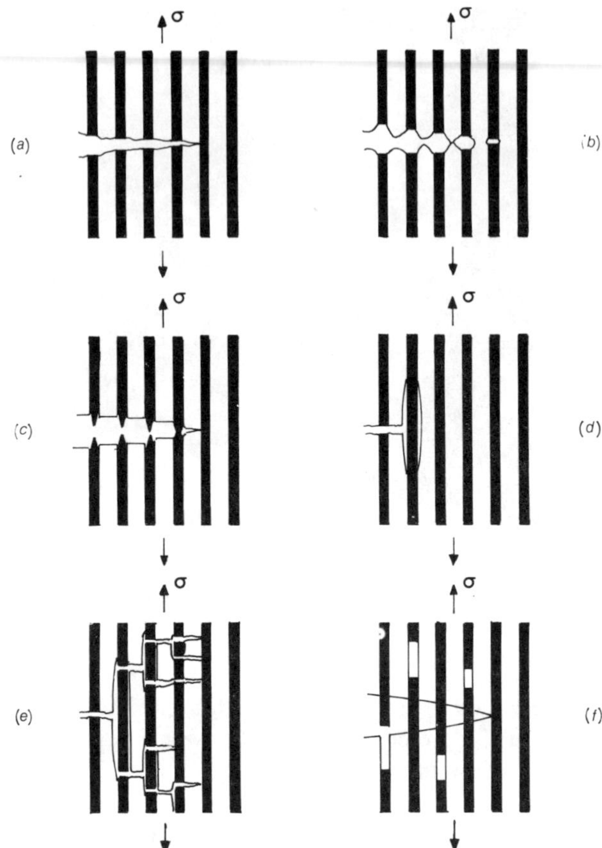

Fig. 22.8.—Fracture modes in fibre composites: (*a*) brittle fibres and brittle matrix with strong interface, (*b*) brittle fibres and ductile matrix, (*c*) ductile fibres and brittle matrix, (*d*) (*e*) debonding along fibres leading to multiple cracking, (*f*) pull-out of fibres

$$\int_0^l \pi d\tau_m x \, dx = \pi d\tau_m \frac{l^2}{2}$$

where the symbols have the meanings assigned to them in Section 22.5. The number of fibres per unit area is

$$N = \frac{f}{\pi d^2/4}$$

Also, if we assume that the pull-out lengths are distributed uniformly between 0 and $l_c/2$, the total pull-out work is

$$\frac{\dfrac{4f}{\pi d^2}\displaystyle\int_0^{l_c/2} \pi d\,\tau_m \frac{l^2}{2}\,dl}{\displaystyle\int_0^{l_c/2} dl} = \frac{f\tau_m l_c^{\,2}}{6d}$$

$$= \frac{fd\sigma_{fu}^{\,2}}{24\,\tau_m}$$

which is an additional contribution to the toughness.

It is therefore seen that in most cases the toughness of a composite will be greater than the value that would be given by applying the law of mixtures.

22.8. Fracture toughness of polyblends

Some glassy polymers are used in conjunction with rubbery polymers to produce tough materials. Examples are high impact polystyrene (HIPS) and acrylonitrile butadiene styrene (ABS). In HIPS the rubbery phase is introduced as particles of diameter 1–10μm and in volume fractions of 5–20%. ABS is a copolymer of rubbery and glassy components.

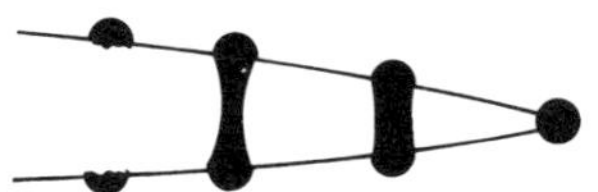

Fig. 22.9.—Rubbery particles bridging crack in a polyblend

As a crack spreads through the glassy matrix (fig. 22.9), the rubbery particles act as bridges and stretch to a considerable extent before breaking. The work to fracture is therefore greatly increased.

QUESTIONS

1. A composite material consists of particles of A in a matrix of B. Discuss the manner in which the Young's modulus of the composite material will vary with the properties of the components and with the shape of particles of A.

If the stress–strain relationships for A and B are as shown in Fig. 22.10, determine and draw on graph paper the stress–strain curves, up to 2% strain, for a composite containing 50% by volume of each constituent (*a*) assuming A to be in continuous fibres lying in the direction of loading and (*b*) assuming A to be in the form of lamellae lying normal to the direction of loading.

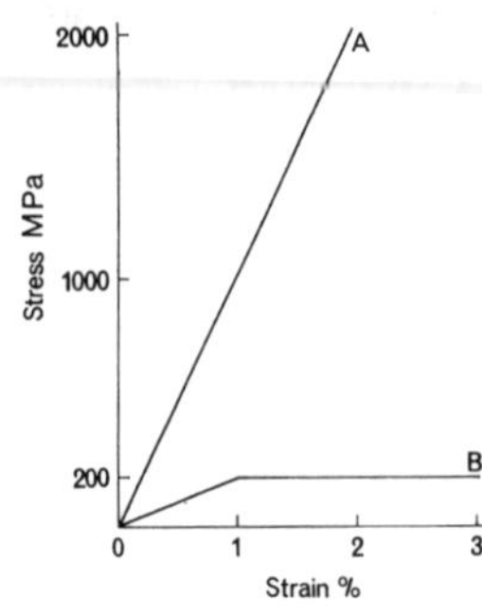

Fig. 22.10

2. A composite material contains 60% by weight of carbon fibres in epoxy resin, the fibres running parallel to the loading direction. The effective Young's modulus of the fibres in the composite can be taken as 27% of that of graphite loaded in the basal plane and the density as 80% of that of graphite. Calculate the effective Young's modulus and the specific stiffness of the composite in the direction of the fibres.

What are the advantages of using carbon fibres in composite materials?
(For graphite: density = 2·2 Mg m^{-3}, Young's modulus (basal plane) = 10^3 GPa; for epoxy: density = 1·12 Mg m^{-3}, Young's modulus = 4·5 GPa.) [E]

3. It is required that a flat uniform plank should have a specified bending stiffness per unit width. Show that the minimum weight of the plank is achieved by using a material with a maximum value of $E^{1/3}\rho^{-1}$, where E is Young's modulus of elasticity and ρ is the density of the plank material.

Such a plank is to be made from a composite material which consists of a polyester matrix that surrounds uniformly distributed continuous glass fibres. The fibres are aligned in the direction of maximum tensile stress. Given the following values of material properties, determine the optimum volume fraction of glass fibre and the corresponding equivalent value of $E^{1/3}\rho^{-1}$ of the composite plank.

	E (GPa)	ρ (kg m^{-3})
Glass	70	2400
Polyester	3·5	1200

[E]

CHAPTER 23

The Nucleus

23.1. The structure of the nucleus

The nucleus of an atom is very small compared with the overall size of the atom (see p. 11). It has a charge $+Ze$ and a mass which is almost an integral multiple of the mass of the hydrogen atom. This integer is known as the mass number A.

Experimental work has shown that the diameter of a nucleus is given by $d = 3\,A^{1/3} \times 10^{-15}$ m, i.e., nuclear diameters lie between about 3×10^{-15} m and 19×10^{-15} m.

At one time it was thought that there were only two fundamental particles—the electron and the proton—the proton being the nucleus of the simple hydrogen atom. A nucleus would then be composed of A protons to give the necessary mass and $(A-Z)$ electrons to reduce the total charge to the correct value.

Consideration of the electron in terms of wave mechanics shows that it could exist in the nucleus only if its wavelength were of the same order of magnitude as the nuclear diameter. The energy corresponding to such wavelengths is of the order of hundreds of MeV,* whereas the energies of particles in the nucleus are of a much lower order of magnitude; thus it is reasonable to assume that a nucleus cannot contain an electron and remain stable. The neutron was postulated, being a particle of the same mass as the proton and having zero charge. The existence of the neutron has since been confirmed experimentally. The mass number A is the sum of the number of protons Z and the number of neutrons N.

For the light elements, $N \approx Z$, but as the atomic number increases, the ratio N/Z increases to about 1·5.

23.2. Isotopes

Since the chemical properties of an element depend upon the number of electrons which atoms of that element have, and hence on Z, the value of N does not affect the chemical properties. If two nuclei with the same Z but different N existed, then there would be two chemically

* MeV = million electron volts.

identical atoms with different atomic weights. Such atoms do occur, and the two sorts of atoms with the same Z are known as *isotopes*.

Isotopes may be identified and distinguished by writing the Z and A-values together with the symbol for the element as follows: ${}_Z X^A$. Thus ${}_1H^1$ is a hydrogen atom. ${}_2He^4$ is the most common isotope of helium, etc.

Many elements have more than one naturally occurring isotope. The atomic weight of chlorine as determined by chemical methods is 35·5. However, physical tests show that there are two isotopes, ${}_{17}Cl^{35}$ and ${}_{17}Cl^{37}$, and there are about three times as many of the former, giving a weighted mean of 35·5. Also many isotopes have been produced in the laboratory by various nuclear reactions.

Figure 23.1 shows a plot of all the *nuclides*, or types of nuclei that had been reported up to the year 1956. The solid circles represent isotopes with stable nuclei and the crosses are the ones with unstable nuclei (see p. 361). The two terms *isotones* and *isobars* which appear in the diagram refer to nuclei with the same N and same A, respectively.

23.3. Mass defect

Atomic weights and nuclide masses are now expressed in terms of the *unified atomic mass unit* (u) (see Table 2.1) which is one twelfth of the mass of the ${}_6C^{12}$ atom. The masses of a hydrogen atom, ${}_1H^1$, a neutron, ${}_0n^1$, and a helium atom, ${}_2He^4$ are, respectively, 1·007 825 u, 1·008 665 u, and 4·002 604 u. The helium atom is made up of two protons, two neutrons, and two electrons which have a total mass of 2 × (1.007 825 + 1·008 665) u, i.e., 4·032 980 u which is 0·030 376 u more than the actual mass of the helium atom. This apparent loss of mass is referred to as the *mass defect*.

It has been shown, as a consequence of Einstein's relativity theory, that mass and energy are equivalent, the relation between energy E and mass m being $E = mc^2$, where c is the velocity of light.*

When protons and neutrons are combined to form a helium nucleus, the loss of mass is converted to energy according to this relationship. The mass defect per *nucleon* (proton or neutron) in this case is 0·007 59 u and is equivalent to 7·07 MeV. This is called the *binding energy* per nucleon and is a measure of the stability of the nucleus. If the atom were formed from the elementary particles, this energy would be released either as kinetic energy of particles or as electromagnetic radiation.

The binding energy per nucleon is

$$931{\cdot}5\,\frac{ZM_H+(A-Z)M_N-M}{A}\ \text{MeV}$$

* Hence 1 u = 931·5 MeV.

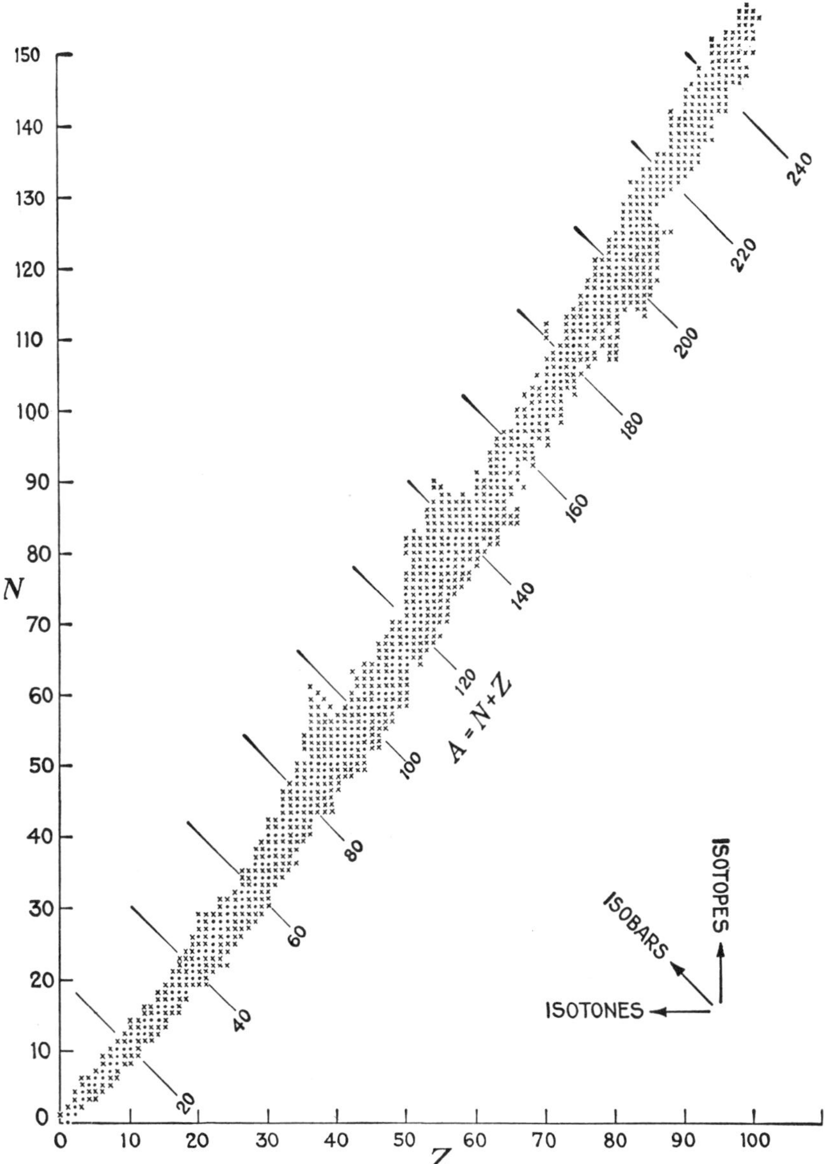

Fig. 23.1.—Plot of all isotopes reported up to 1956
● naturally-occurring or stable isotopes. × unstable isotopes

where M_H = mass of ${}_1H^1$ atom, M_N = mass of neutron, M = actual mass of atom in u.

The manner in which the binding energy per nucleon varies with atomic mass is shown in fig. 23.2. From the curve it may be seen that the nuclei of elements with mass numbers of about 80 are the most stable, as energy would have to be supplied to change them into lighter elements.

If nuclei in this middle mass range could be produced either by combining light nuclei or by splitting heavy nuclei, a great deal of energy would be released.

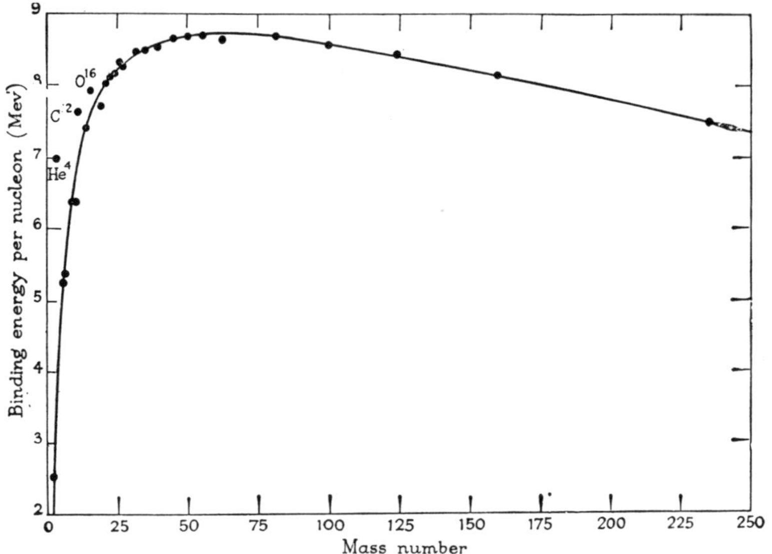

Fig. 23.2.—Variation of binding energy per nucleon with atomic weight. [Glasstone and Edlund]

The combination of light elements to form heavier ones is known as *nuclear fusion*. It is the source of energy in the sun, and fusion reactions have been initiated by man, but it has not yet been utilized as a practical source of power.

The splitting of heavy nuclei is known as *nuclear fission*. It may occur spontaneously, but only at a slow rate. The probability of fission is vastly increased if the nucleus is excited, i.e., its energy increased, by some 5 or 6 MeV. In nuclear reactors this excitation is provided by neutron bombardment (see Section 23.7). The excited nucleus may then break down to give two nuclei of mass number about 70–160 and 2 or 3 neutrons. As can be seen from fig. 23.2, the process releases about 1 MeV of energy per nucleon, or approximately 200 MeV per fission. The energy appears mainly as kinetic energy of the fission-product nuclei, some of the remainder being kinetic energy of the neutrons and energy of radioactive emission. Most of the energy is eventually converted to heat in the reactor, and this heat is used to provide power. The neutrons present among the fission products can make these processes self-sustaining.

23.4. Nuclear reactions

So far reference has been made primarily to elements whose nuclei do not change spontaneously. These are the ones marked as solid circles in fig. 23.1 and which lie along the mid-region of the band. Other nuclei are unstable and will tend to change into stable nuclei by one of the several processes listed in Table 23.1. These changes can occur only if the total mass of the products is less than the initial mass, i.e., if there is an increase in the mass defect, which is released as energy. These processes are known as *radioactive decay* processes.

TABLE 23.1.—DECAY PROCESSES OF UNSTABLE NUCLEI

Decay mode	Description	Change in atomic number	Change in mass number
α	α-particle ($_2He^4$ nucleus) emitted from nucleus	−2	−4
β^-	electron emitted	+1	0
β^+	positron (positive electron) emitted from nucleus	−1	0
K-capture	capture of an orbital electron from the K-shell by the nucleus	−1	0
isomeric transition	nucleus changes from excited to stable or ground state by emission of energy as γ-rays	0	0

α- and β-decay and K-capture may also be accompanied by the emission of energy in the form of γ-rays.

All nuclei with Z values greater than that of lead ($Z = 92$) appear to be unstable, many dissociating by α-decay. When the neutron/proton ratio lies outside the stable range, the nuclide will be radioactive and change in the direction of increased stability. If the nucleus has an excess of neutrons, it undergoes β^- decay, a neutron changing into a proton and an electron.

$$_0n^1 \rightarrow p + e^-$$

If the nuclide is on the other side of the stable band, the spontaneous reaction is the conversion of a proton to a neutron and a positron or β^+ ray.

$$p \rightarrow {}_0n^1 + e^+$$

A positron is a particle of the same mass and magnitude of charge as an electron but of opposite sign. This particle is short-lived, becoming annihilated when it meets an electron, the whole of the mass and kinetic energy then being converted to the energy of a γ-ray photon.

The same change, increase of N by 1 and decrease of Z by 1, is also brought about by K-capture. An electron from the innermost electron orbit of the atom is absorbed into the nucleus and combined with a proton to form a neutron.

After a β^- or β^+ decay, the nucleus may be left in an excited state and within a short time ($\sim 10^{-14}$ s) reverts to the ground state by emission of a γ-ray photon. γ-rays are electromagnetic radiations of very short wave-length.

23.5. Rate of radioactive decay

Every radioactive isotope has a probability of a nucleus decaying in a given time. This decay probability is characteristic of the isotope and is constant irrespective of the chemical or physical state of the atom. Hence, in a given specimen, the rate of decay at any instant is proportional to the number of radioactive atoms of the isotope present at that instant.

Thus if N atoms are present at time t

$$\frac{dN}{dt} = -\lambda N$$

where λ is the *decay constant*.

By integration, this gives,

$$N = N_0 e^{-\lambda t}$$

where N_0 is the number of atoms at $t = 0$.

The probability of decay is usually expressed in terms of the *half-life*, that is, the time taken for any given quantity of the particular isotope to decrease to half the original quantity. Hence, if τ is the half-life,

$$e^{-\lambda\tau} = 1/2$$

or

$$\tau = (\log_e 2)/\lambda$$

high, sufficient to cause a considerable temperature rise if it were released. In graphite, for example, it could raise the temperature by 200 K.

Atomic rearrangements are also possible which can lead to various effects, such as formation of a ferromagnetic phase in austenitic stainless steel.

23.9. Radiation damage due to absorption

The various reactions possible when a neutron is absorbed by a nucleus have been discussed in Section 23.7.

The new elements formed may alloy with or combine chemically with the original element giving dimensional changes and hence internal stressing, among other effects, or may set up conditions where corrosion is possible, whereas the original material was corrosion-resistant in its surroundings.

The transmutation products may be inert gases, such as krypton or xenon which do not combine chemically with the parent material. Initially the gas molecules will migrate to vacancies, which will then diffuse together so that pockets of gas are formed, the pressure becoming so great that the metal swells.

23.10. Conclusion

It is obvious from this extremely brief survey that in the application of materials in the nuclear-energy field a number of new factors appear, and the problem of selection of materials is no longer governed solely by the consideration of the straightforward factors such as strength, corrosion resistance, and weldability.

Features such as adequately large or small cross-sections and the effects of irradiation may severely restrict the choice for certain applications, and adequate corrosion resistance, etc., may then reduce the possibilities still further. Many of the rarer elements, which had no previous engineering applications because of their scarcity and cost, are found to have favourable properties in some of the cases where choice is severely restricted and are emerging to an era of popularity.

QUESTIONS

1. Describe the structure of the rubidium atom of mass number 87 (Rb^{87}) in terms of the elementary particles.

This naturally-occurring isotope is radioactive, exhibiting β-decay and having a half-life of 6×10^{10} years. Explain this statement. What is the element formed in the decay process? [MST]

2. What isotope does ${}_{6}C^{14}$ become after radioactive emission of a β-particle? What isotope does ${}_{92}U^{238}$ become after emission of an α-particle?

3. Give a short account of the way in which the properties of an atom are dependent upon the number of protons and neutrons in its nucleus.

Describe **three** types of radiation which may result from nuclear instability and discuss some engineering applications of radioactivity. [MST]

4. Discuss the factors which determine the type of radioactive decay exhibited by a given nuclide.

The unit of radioactivity is defined as 1 curie $= 3{\cdot}7 \times 10^{10}$ disintegration s^{-1}. Show that one curie of radiation is produced by the decay of 1 g of the isotope ${}_{88}Ra^{226}$ which has a half-life of 1620 years. [MST]

5. ${}_{92}U^{238}$ will absorb a neutron and then emit a β-particle. What isotope does it become? If ${}_{92}U^{235}$ absorbs a neutron and then undergoes fission to form two atoms of the same isotope and two neutrons, what is the isotope?

6. Calculate the binding energy per nucleon of (*a*) ${}_{92}U^{238}$, and (*b*) ${}_{45}Rh^{103}$. These are the abundant isotopes and the atomic weights may be considered to be entirely due to them.

7. What energy is released by 1 g of ${}_{92}U^{235}$ undergoing fission into two approximately equal parts? (Binding energies change little for small changes in atomic number, and the results of question 6 should be used.)

8. In a test to determine the wear of piston rings a steel ring weighing 40 g is irradiated with neutrons until it has an activity of 10 microcuries due to the formation of the isotope Fe^{59} (half-life $\tau = 45{\cdot}1$ days). Six days later the ring is installed in an engine of sump capacity 6×10^{-3} m^3 and the engine is run for 30 days when the sump oil is found to have an activity of $12{\cdot}5 \times 10^6$ disintegrations min^{-1} m^{-3}. How much material has been worn off the ring?

[1 curie $= 3{\cdot}7 \times 10^{10}$ disintegrations s^{-1}.] [MST]

9. Define the terms *half-life* and *decay constant* as applied to the decay of radioactive material. What is the relationship between them?

Ra^{226} and C^{14} undergo radioactive decay, emitting α- and β-particles, respectively. What is the product in each case?

The activity due to the decay of the radioactive C^{14} present in timber from living trees is found to be 12·5 disintegrations per min per gramme of carbon. The half-life of C^{14} is 5568 years. What proportion of the carbon atoms are C^{14}?

If the corresponding activity for timber taken from a certain ancient Egyptian tomb is 5·7 disintegrations per min per gramme of carbon, estimate the age of the tomb. [MST]

10. A source of 2 MeV gamma rays is shielded by a 50 mm thick steel plate. Describe the methods by which the gamma rays can interact with the shielding material. Discuss the possible methods by which the original photon energy is degraded to thermal energy.

By what factor will the intensity of rays be reduced in the above case, if a beam of rays from such a source is reduced to half intensity by a 160 kg m^{-2} steel plate?

[Density of steel 7930 kg m^{-3}.] [MST]

APPENDIX I

Some Equilibrium Diagrams of Interest and Importance*

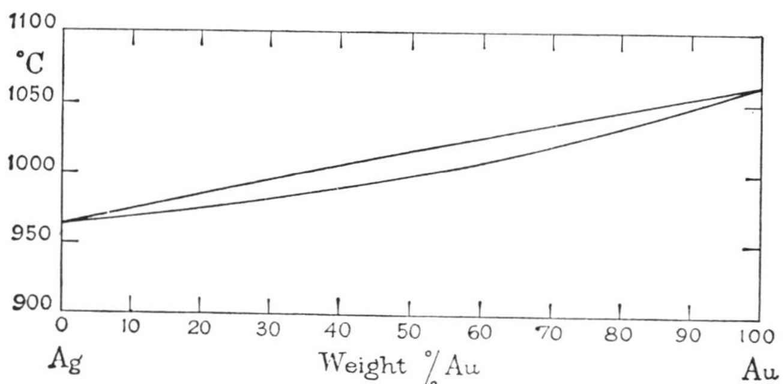

Fig. A.1.—Equilibrium diagram for silver-gold

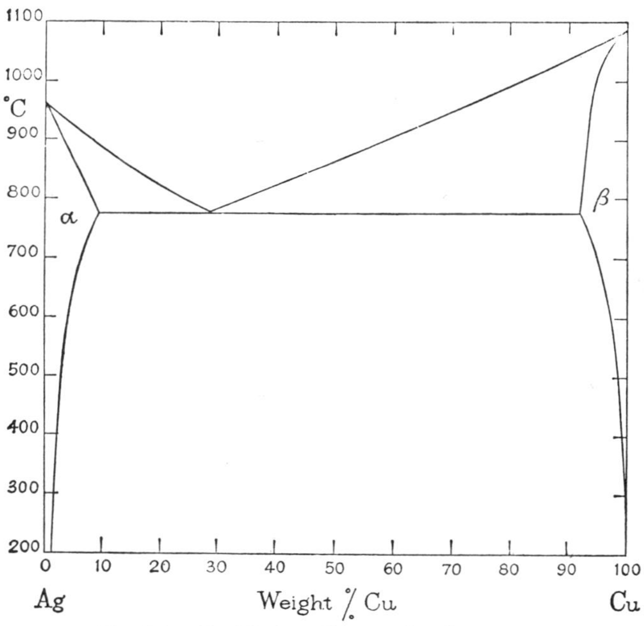

Fig. A.2.—Equilibrium diagram for silver-copper

* These are based on diagrams in the *Metals Reference Book*, second edition edited by Smithells, Butterworth.

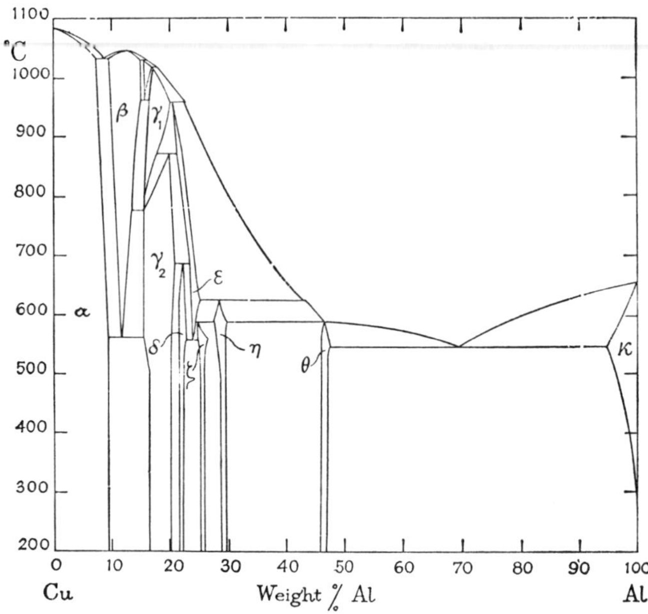

Fig. A.3.—Equilibrium diagram for aluminium-copper. (Aluminium bronze up to 15% Cu)

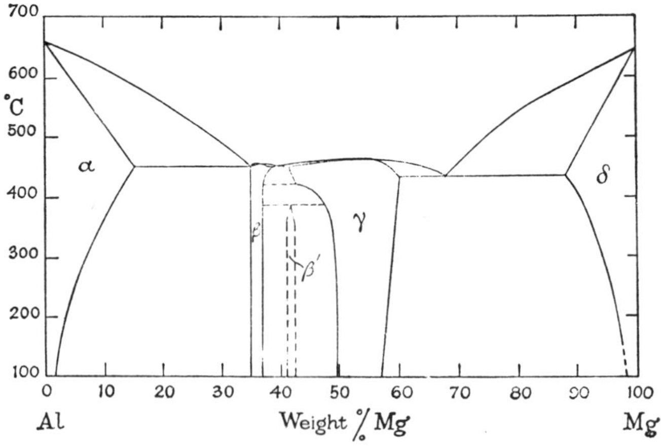

Fig. A.4.—Equilibrium diagram for aluminium-magnesium

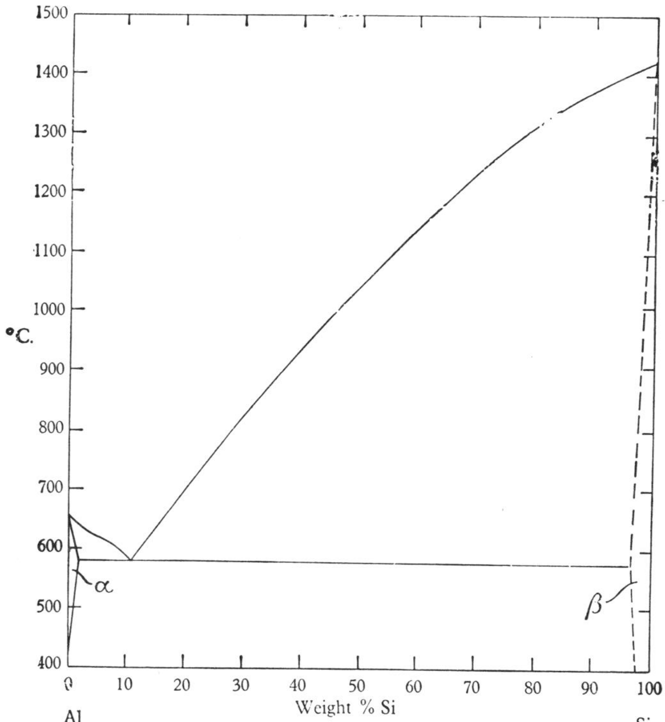

Fig. A.5.—Equilibrium diagram for aluminium-silicon

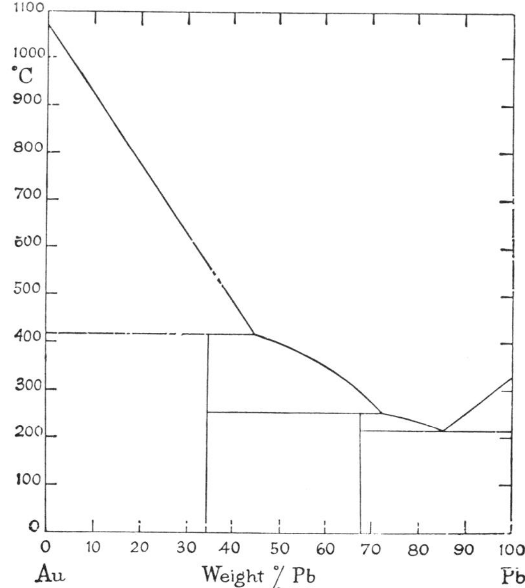

Fig. A.6.—Equilibrium diagram for gold-lead

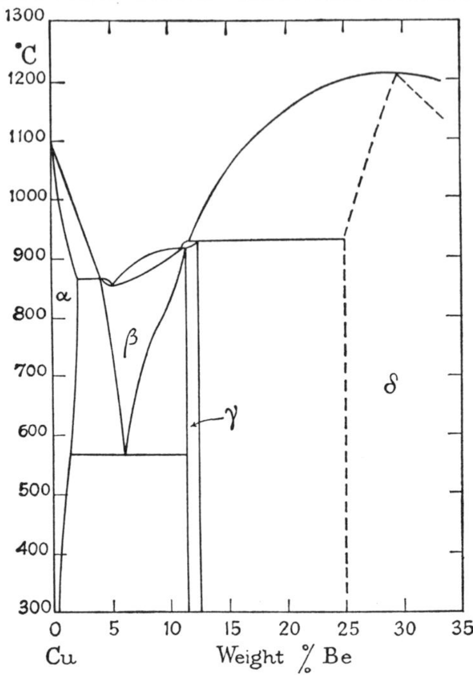

Fig. A.7.—Equilibrium diagram for copper-beryllium, copper-rich end (beryllium bronze)

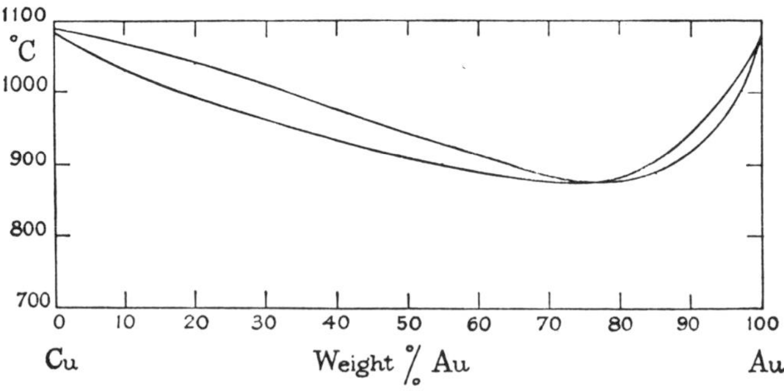

Fig. A.8.—Equilibrium diagram for copper-gold

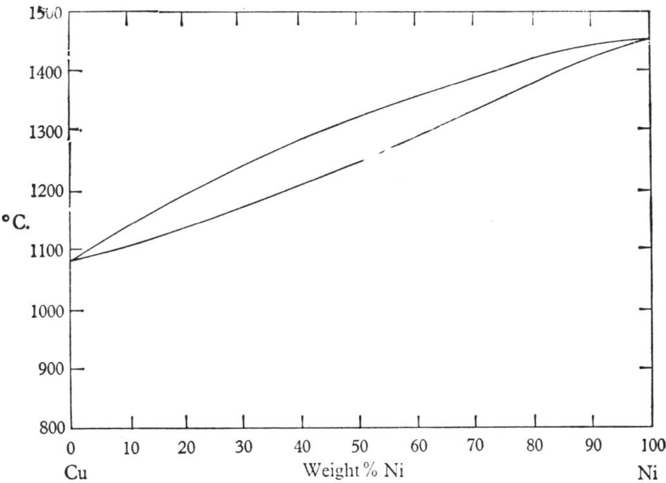

Fig. A.9.—Equilibrium diagram for copper-nickel

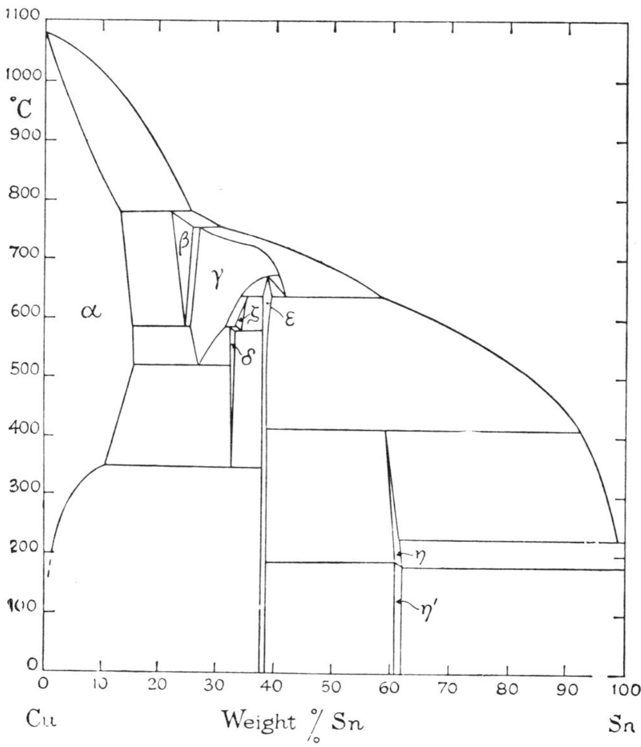

Fig. A.10.—Equilibrium diagram for copper-tin (tin bronzes)

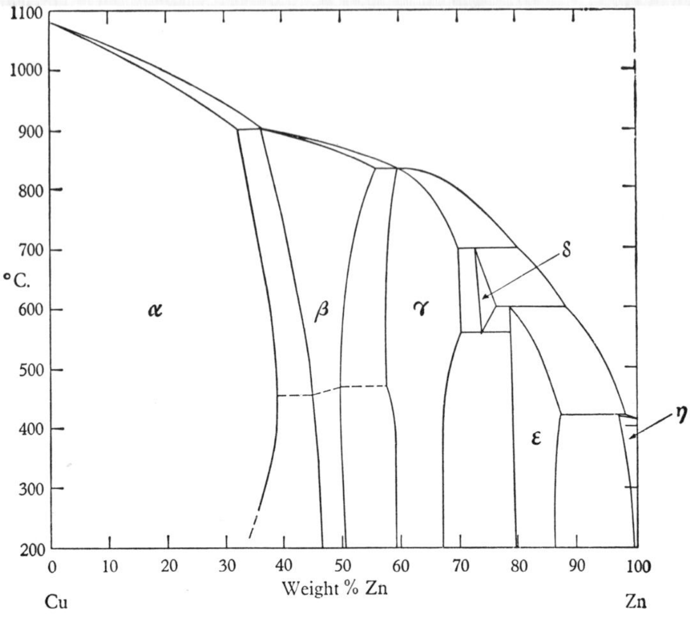

Fig. A.11.—Equilibrium diagram for copper-zinc (brasses)

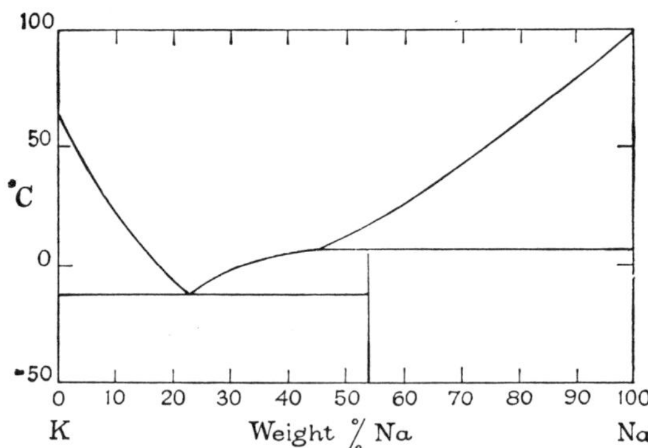

Fig. A.12.—Equilibrium diagram for potassium-sodium

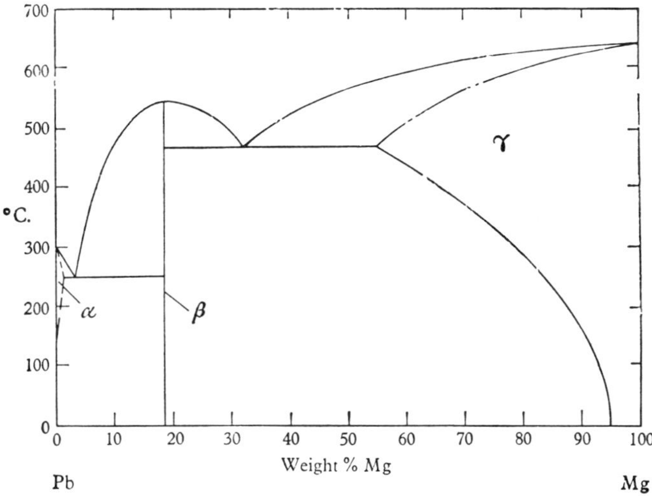

Fig. A.13.—Equilibrium diagram for magnesium-lead

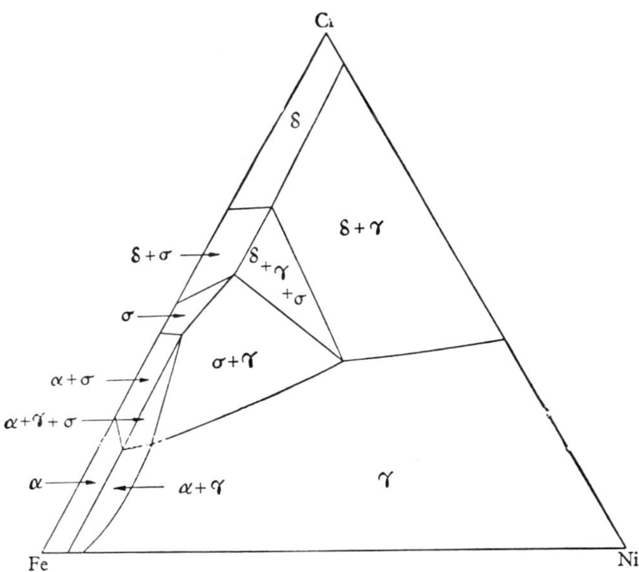

Fig. A.14.—Isothermal section at 650 °C of chromium-iron-nickel system

APPENDIX II

Equilibrium and Free Energy

As has been shown in various chapters in this book, materials can change their structure, phase or state due to changes in temperature and pressure. For any given values of these variables there will be a particular configuration which is the stable form, and at certain values two forms may exist in equilibrium. Of the forms possible for a system the stable one is that for which the free energy is least, and when two forms exist in equilibrium, their free energies are equal.

The concept of free energy will be outlined briefly here, but for a more complete treatment and proofs of the statements made, the reader is referred to textbooks on thermodynamics.

If a system, that is a portion of matter around which a hypothetical control surface may be drawn, has the following properties:

volume V
pressure p
absolute temperature T
internal energy U
entropy S

then the *Helmoltz free energy* F and the *Gibbs free energy* G are, respectively,

$$F = U - TS$$

and

$$G = U + pV - TS$$

The internal energy of a system is the sum of the energies of all the particles which make up the system, that is the kinetic energies of the particles which depend on their vibrational motions, etc., and the potential energies which are due to the forces between the particles (e.g., the electrostatic forces between ions and the van der Waals forces between atoms and molecules).

Entropy is defined as follows: If a system absorbs an infinitesimal quantity of heat dQ in a reversible manner, when at an absolute temperature T, then the increase of entropy of the system is

$$dS = dQ/T$$

If the change is irreversible, then

$$dS > dQ/T$$

By the principle of conservation of energy, the internal energy of a system can be altered only by giving or receiving energy in the form of heat or work. The First Law of Thermodynamics states that the change of internal energy dU is the difference of dQ, the heat supplied to the system, and dW, the work done by the system, i.e.,

$$dU = dQ - dW$$

If in any small change, the system changes its volume by dV, then it does an amount of work pdV against the surroundings:

$$dW = pdV$$

If a small change takes place in the state of a system, then

$$\begin{aligned} dF &= d(U-TS) \\ &= dU - TdS - SdT \\ &= dQ - dW - TdS - SdT \end{aligned}$$

For a change at constant temperature and volume

$$SdT = 0$$

$$dW = pdV = 0$$

Hence

$$dF = dQ - TdS$$

But, by the definition of entropy

$$TdS \geqq dQ$$

so that

$$dF \leqq 0$$

i.e., if any change takes place in a system at constant temperature and volume, the Helmholtz free energy either decreases or remains constant. If it is a change at conditions of equilibrium, then

$$dF = 0$$

and as this state can be approached only by a decrease of F, then F must have a minimum value under equilibrium conditions.

Similarly, for any small change,

$$\begin{aligned} dG &= d(U+pV-TS) \\ &= dQ - dW + pdV + Vdp - TdS - SdT \end{aligned}$$

For a change at constant temperature and pressure,

$$SdT = 0$$

and

$$Vdp = 0$$

Also, because,

$$dW = pdV$$

we have that

$$dG \leqq 0$$

Thus for equilibrium under these conditions either the Helmholtz free energy or the Gibbs free energy will remain constant for small changes of the remaining variables, and the actual value of F or G will be a minimum under such conditions.

The form of free energy to be considered is related to the conditions for equilibrium. In solids and liquids at atmospheric pressure, the pV term is sufficiently small compared with U and TS to be neglected, so that a condition of minimum F is adequate to define all equilibrium changes in these states.

APPENDIX III

SI Units

The units used in this book are almost all units of the *Système International d'Unités*. The seven basic units, from which all the others are derived, are:

length—metre (m)
mass – kilogramme (kg)
time—second (s)
thermodynamic temperature—kelvin (K)
electrical current—ampere (A)
luminous intensity—candela (cd)
amount of substance—mole (mol)

These are defined as follows:

metre—1 650 763·73 wavelengths *in vacuo* of the radiation corresponding to the transition between the energy levels $2p_{10}$ and $5d_5$ of the krypton-86 atom.

kilogramme—the mass of the international prototype which is in the custody of the *Bureau International des Poids et Mesures* (BIPM) at Sèvres, France.

second—the duration of 9 192 631 770 cycles of the radiation corresponding to the transition between the two hyperfine levels of the fundamental state of the caesium-133 atom.

kelvin—the fraction 1/273·16 of the thermodynamic temperature of the triple point of water.

ampere—that constant current which, if maintained in two parallel rectilinear conductors of infinite length, of negligible circular cross-section, and placed at a distance of 1 m apart in a vacuum, would produce between these conductors a force equal to 2×10^{-7} N m^{-1}.

candela—the luminous intensity, in the perpendicular direction, of a surface of 1/600 000 m^2 of a black body at the freezing temperature of platinum under a pressure of 101 325 Pa.

mole—amount of substance of a system which contains as many elementary entities as there are atoms in 0·012 kg of carbon 12.

The derived SI units which appear in this book are:

newton (N)—the force that will give a mass of 1 kilogramme an acceleration of 1 $m\,s^{-2}$.

pascal (Pa)—the pressure or stress equal to 1 $N\,m^{-2}$.

joule (J)—the energy or work due to a force of 1 N moving its point of application by 1 m.

volt (V)—the electric potential difference such that a current of 1 A at 1 V delivers energy at 1 $J\,s^{-1}$.

coulomb (C)—the quantity of electricity due to a current of 1 A flowing for 1 s.

farad (F)—the value of the capacitance that will store 1 C at a potential difference of 1 V.

SI multiplication factors used are:

Factor	prefix	symbol
10^{-9}	nano	n
10^{-6}	micro	μ
10^{-3}	milli	m
10^{3}	kilo	k
10^{6}	mega	M
10^{9}	giga	G

Other units which appear are:

electron volt (eV)—the energy acquired by an electron in traversing a potential difference of 1 V ($=1{\cdot}602\times10^{-19}$ J).

degree Celsius (°C)— $= T - 273{\cdot}13$, where T is temperature in K.

kilomole (kmol)—1 000 mol.

ANSWERS TO NUMERICAL QUESTIONS

Chapter 2

1. 2, 4, 2, 3, 5, 3, 2, 3, 6, 7
4. 121·3 nm; 1872 nm
6. $1{\cdot}82 \times 10^6$ m s^{-1}; 9·36 eV
7. 328 nm; 1·07 nm

Chapter 3

1. 86·9
2. 614 m s^{-1}
3. 1·53 Pa
5. 37·6

Chapter 4

3. 0·249 nm; $8{\cdot}91 \times 10^3$ kg m^{-3}
4. 0·2039 nm; 0·1442 nm; 0·2355 nm
6. 35° 20′; 55° 30′
7. $21{\cdot}0 \times 10^3$ kg m^{-3}; 109° 4′; 57° 35′
8. $(mC)^{1/(m-1)}$
9. 8·24 MJ / mol

Chapter 5

1. 1·87 nm; $2{\cdot}92 \times 10^{-18}$ J

Chapter 6

4. $LaSn_3$, La_2Sn_3, La_2Sn
5. 8·15
6. 0·414
7. 0·244 R

Chapter 7

1. Liquid (22% Ni) 47%; solid (37% Ni) 53%
2. $PbMg_2$; at 500 °C: 30Pb/70Mg solid solution;
at 300 °C: β-phase ($PbMg_2$) 19%
γ-phase (78% Mg) 81%
3. Na_2K; Liquid (32% Na) 14%, Na_2K 86%

6. 320 °C, 465 °C; Solid solution (11·3% A in B) 22%, A_2B 78% divided as follows:
   primary A_2B 34·5%
   eutectic A_2B 33·3%
   precipitated from eutectic solid solution 10·2%

8. (*a*)	One phase		30 Cr	30 Fe	40 Ni	
(*b*)	Two phases	γ	38 Cr	10 Fe	52 Ni	30%
		δ	83 Cr	10 Fe	7 Ni	70%
(*c*)	Three phases	γ	37 Cr	29 Fe	34 Ni	49%
		δ	66 Cr	27 Fe	7 Ni	45%
		σ	53 Cr	40 Fe	7 Ni	6%
9. (*a*)	Two phases	α	65 A	19B	16C	16%
		γ	23 A	8 B	69 C	84%
(*b*)	Three phases	α	60 A	25 B	15 C	27%
		β	10 A	80 B	10 C	20%
		γ	20 A	15 B	65 C	53%

10. $K = C_\alpha / C_E$

Chapter 8

2. −2·71%
3. +2.95% assuming true close packing [i.e. $c/a = 2\sqrt{(2/3)}$]
   +1·95% from actual c/a ratio
4. 0·124 nm; 0·350 nm
6. 0·50
7. 0·39
10. 0·48%

Chapter 9

3. (*a*) 0·0937; (*b*) 0·107
5. $2{\cdot}3 \times 10^8$ J (kmol of atoms)$^{-1}$
7. $6{\cdot}9 \times 10^{-20}$ J atom^{-1}
8. 120 MJ kmol^{-1}
9. 26×10^3; 34
10. $7{\cdot}84 \times 10^{-19}$ J (particle)$^{-1}$

Chapter 10

3. (*a*) 0·77 MPa
   (*b*) 16° 29′; 25° 9′
   (*c*) 1·17 MPa
4. 0·834 MPa

6. (*a*) 30°
   (*b*) 60°
   (*c*) 143 kPa
7. 6700
8. 11 nm; (110)
9. $1{\cdot}8 \times 10^{16}$ m^{-2}
10. $1{\cdot}66 \times 10^{7}$ m
11. 1·4 MPa
12. $5{\cdot}3 \times 10^{-7}$ m

Chapter 11

7. 13 days; 60 more days
8. −28 °C
9. 6 hours
10. Anneal, roll to 8 mm, anneal, roll to 4 mm, anneal, roll to 3 mm
11. 150 MPa
12. 0·126; 485 kPa

Chapter 12

1. (*a*) 158 MPa, (*b*) 132 MPa; 13·7%; 64%
2. 30%
3. $\sigma_t = 110\varepsilon^{0{\cdot}2}$ MPa; 67 MPa
4. 15·4
5. 22·6
6. 92 MPa; 28%
8. 142 kN
9. 0·415 HV
10. 130 kgf mm^{-2}; 0·175 kgf
11. 292
12. 163 kN; 91 kgf mm^{-2}

Chapter 13

4. 42%
5. HY 140; HY 180

Chapter 14

1. ± 182 MPa; ± 165 MPa
2. 98 MPa
3. 103; $5{\cdot}66 \times 10^{6}$; 9·8 μm cycle^{-1}; $1{\cdot}8 \times 10^{-13}$ m cycle^{-1}; ∼ 28000; ∼ $2{\cdot}8 \times 10^{6}$
4. Yes; $N_f = 4{\cdot}4 \times 10^{9}$ cycles

Chapter 15

2. 0·061 mm
3. $\varepsilon \propto \sigma^{5.2}$
4. $8{\cdot}9 \times 10^{-23}$ hr^{-1} $MPa^{-8.4}$; 8·4; 36·5 MPa
5. 15·4 MPa

Chapter 16

4. (*a*) 100% austenite
   (*b*) 8% ferrite, 92% austenite
   (*c*) 50% ferrite and pearlite, 50% austenite
   (*d*) 50% ferrite and pearlite, 50% martensite
   (*e*) 50% ferrite and pearlite, 50% tempered martensite
7. 16 mm, 474 kgf mm^{-2}

Chapter 17

3. 360, 470. No
4. 52 mm; 102 mm.

Chapter 18

2. Ca – 37% decrease, non-protective
   Cd – 18% increase, protective
5. 2·4 mm; 26·6 mm deep; 14·5 years
6. 6 days

Chapter 20

1. 0·732; 0·414; 0·381 nm
5. $3\ Al_2O_3.2\ SiO_2$
6. 339 kJ mol^{-1}; 230 kJ mol^{-1}

Chapter 21

5. (*a*) $7{\cdot}7 \times 10^{-6}$ s^{-1}; (*b*) 0·212
6. 283 K

Chapter 22

2. 134 GPa, 94×10^6 m^2 s^{-2}
3. 42%; 1·85 $N^{1/3}$ kg^{-1} $m^{7/3}$

Chapter 23

1. ${}_{38}\mathrm{Sr}^{87}$
2. ${}_{7}\mathrm{N}^{14}$, ${}_{90}\mathrm{Th}^{234}$
5. ${}_{93}\mathrm{Np}^{239}$, ${}_{46}\mathrm{Pd}^{117}$
6. (*a*) 7·6 MeV
   (*b*) 8·6 MeV
7. $9{\cdot}6 \times 10^{10}$ J
8. 0·235 g
9. ${}_{84}\mathrm{Rn}^{222}$; ${}_{7}\mathrm{N}^{14}$; $1{\cdot}05 \times 10^{-12}$; 6300 years
10. 5·65

INDEX

Half-lives as short as 10^{-9} s and as long as 10^{17} years have been observed.

Unstable nuclei do occur in the earth's crust, but the only ones now remaining are those which have a long half-life or are produced by decay of long-lived " parent " nuclei.

23.6. Interaction of radiation with matter

α and β rays lose their energies mainly by removing electrons from atoms in their paths, thereby creating ion-pairs. For every ion-pair produced in air, about 32·5 eV of energy are lost. Thus a particle with an initial energy of 0·5 to 5 MeV will have to cause many ionizations to lose its energy. An α-particle, being more massive and hence more slowly moving, causes far more ionizations than a β-particle in the same path length. Also due to the large mass compared with the electrons with which they collide, α-particles continue in fairly straight tracks. Hence, α-rays of a given energy have a fairly definite range in any given type of matter. Thus, for example, 4 MeV α-particles are stopped by 25 mm of air or 0·005 mm of lead.

Electrons are liable to large changes of direction on impact with other electrons, so that the paths of different β-rays will be quite different. 3 MeV β-rays are stopped by 13 m of air or 65 mm of concrete.

γ-rays do not cause ionization directly, but interact with material in three distinct ways. The photon energy may be used in ejecting an electron from an inner shell of an atom—the photoelectric effect—or the γ-ray may be scattered due to collision with electrons—the Compton effect. A γ-ray with more than 1·02 MeV energy may, if it passes very close to a nucleus, create an electron-positron pair. The 1·02 MeV is the rest-mass energy of the new particles. Any surplus energy becomes kinetic energy of the pair of particles.

The chance of a photon being stopped is constant along its path, so that the intensity of the γ-rays decreases in an exponential manner with distance. There are, however, secondary photons and electrons produced, so that the total radiation intensity does not fall off in an exact exponential manner. γ-rays cannot be screened off absolutely as can α- and β-rays, but the intensity can be reduced to tolerable limits. Some typical thicknesses required to reduce the intensity to 1/10 are

Energy (MeV)	Thickness of water	Thickness of lead
0.5	250 mm	12 mm
5	750 mm	50 mm

Ionization of living matter can occur due to any of these rays and an overdose is very unhealthy, so that effective screening is necessary for workers who handle radioactive materials.

23.7. Reaction cross-sections

Other unstable nuclei may be produced by bombarding "target" nuclei with "projectiles" which are other nuclei or neutrons. The projectile may be absorbed by the target nucleus forming a new nucleus which will normally be in a highly excited state. It will change rapidly from this state, either losing energy by γ-radiation and reverting to the ground state of the new nucleus, or forming a third type of nucleus or undergoing fission (see Section 23.3). The product nuclei in each case may be stable or may also decay in due course.

Two examples of the effect of neutron absorption are:

$${}_{27}\text{Co}^{59} + {}_{0}n^{1} \rightarrow {}_{27}\text{Co}^{60} \text{ excited}$$

$${}_{27}\text{Co}^{60} \text{ excited} \xrightarrow[\text{rapid}]{} {}_{27}\text{Co}^{60} \text{ ground state} + \text{energy}$$

$${}_{27}\text{Co}^{60} \text{ ground state} \xrightarrow[\text{half-life 5·25 years}]{} {}_{28}\text{Ni}^{60} + \beta + \gamma$$

and

$${}_{92}\text{U}^{235} + {}_{0}n^{1} \rightarrow {}_{92}\text{U}^{236} \text{ excited}$$

$${}_{92}\text{U}^{236} \text{ excited} \xrightarrow{\text{rapid}} \begin{cases} \text{either fission} \\ \text{or } {}_{92}\text{U}^{236} \text{ ground state} \end{cases}$$

$${}_{92}\text{U}^{236} \text{ ground state} \xrightarrow[\text{half-life } 2{\cdot}4 \times 10^{7} \text{ years}]{} {}_{90}\text{Th}^{232} + \alpha$$

Most fission products are β^- radioactive due to being neutron-rich.

The probability of interaction of a moving particle with a nucleus or other target depends on the kind of particle, its energy, and the target involved. It can be expressed quantitatively as the *cross-section* for interaction. In the remainder of this Section, discussion is restricted to the interaction of neutrons with nuclei.

Consider a sheet of material placed in the path of a stream of the moving particles. Suppose that the stream intensity (i.e., number of neutrons per unit area per second) decreases, on traversing a thickness dx, from I to $I+dI$ due to interactions of some particles with the targets.

If N_0 is the number of atoms per unit volume of the material, then a unit area of thickness dx of the sheet of the material contains N_0dx atoms. If each nucleus has an effective area σ, i.e., the area within which a neutron must pass if it is to interact with the nucleus, then

$$-\frac{dI}{I} = \sigma N_0 dx$$

σ is known as the nuclear cross-section for the particular reaction involved. The intensity I of a beam after traversing a sheet of thickness x will be given by

$$I = I_0 e^{-\sigma N_0 x}$$

where I_0 is the intensity of the incident stream.

Cross-sections are usually expressed in units of *barns*, one barn equalling 10^{-28} m^2. The total cross-section for the interaction of a neutron with a particular kind of nucleus may be divided in the proportions of the probabilities of the various types of interaction that may occur, that is into scattering cross-section and absorption cross-section, the latter being further subdivided into capture and fission cross-sections.

Scattering cross-sections are of the order of a few barns for most materials and are largely independent of the neutron energy. Absorption cross-sections vary greatly with the speed or energy of the neutron, being much greater for slow-moving than for fast neutrons, and also vary greatly from element to element.

$\Sigma = N_0\sigma$ is known as the *macroscopic cross-section.* It can be shown that $1/\Sigma$ is the corresponding mean free path (see p. 32) of the moving particle.

A fast neutron, such as one formed in the fission of a $_{92}U^{235}$ nucleus, has an energy of the order of 2 MeV. At each collision which causes scattering, there will be an energy interchange between the neutron and the particle with which it collides, resulting in a decrease of energy of the neutron. When the energy has decreased to the order of the thermal energy of the particles with which it comes in collision its mean energy (of the order of 0·025 eV) will not decrease further. The neutron is then known as a *thermal neutron.*

The fission cross-section of $_{92}U^{235}$ for thermal neutrons is much larger than for high-energy neutrons, whereas $_{92}U^{238}$ will undergo fission only after interaction with high-energy neutrons.

To make a fission process a self-sustaining chain reaction, as is necessary in a nuclear reactor, one of the neutrons produced per fission must be available to promote another fission. Although each fission produces, on average, 2·5 neutrons, capture and loss due to leakage from the system have to be considered.

Natural uranium, consisting of about 99·3% of U^{238} and 0·7% of U^{235}, has a relatively low fission cross-section for high-energy neutrons compared with the cross-sections for non-fission reactions, so that it is impossible to maintain a chain reaction with high-energy neutrons in natural uranium. By using a fuel enriched in U^{235} a chain reaction with fast neutrons can be achieved.

A chain reaction is possible in natural uranium if fission is mainly due to slow neutrons and if the neutrons are slowed down in a *moderator*, which is placed between the various fuel elements. The moderator must be of a material which has a high scattering cross-section and a relatively small absorption cross-section.

23.8. Radiation damage due to scattering

Owing to absorption and to scattering, various effects occur which alter the mechanical and physical properties of the material irradiated, such effects being known as *radiation damage.* A few of those effects which modify the mechanical properties will be discussed here.*

When fast-moving electrons are scattered, the atoms that cause the scattering are knocked out of place by the recoil and move with high energies which become dissipated in further collisions. The scattered atoms are known as *knock-ons.* Thermal neutrons do not have sufficient energy to cause any knock-ons.

Atoms are knocked out of place, leaving vacancies, and go into interstitial positions. The energy of a single vacancy in copper is about 1 eV and of an interstitial atom is about 4 eV, whereas 25 eV are needed to create a vacancy-interstitial pair. A fast-moving neutron can therefore create many such point defects while it is being slowed down. The excess energy between that needed to create a vacancy-interstitial pair and the energy remaining in the defects is released as heat, so that there is also intense heating in the zone in which the neutron is slowed down, a *thermal spike.* The temperature rise is sufficient to cause annihilation of many of the defects created by previous interactions by diffusion of the interstitials to vacancies, but a proportion remains.

Metals subjected to neutron irradiation become harder, and if they have a notch ductile-brittle transition (see p. 227), the transition temperature is raised (increases of 80 K in steel and 100 K in molybdenum have been reported). It is thought that the defects migrate to the dislocations already present and have an anchoring effect. There is less hardening due to neutron irradiation in a work-hardened metal, and even softening in a very heavily work-hardened material.

The damage can be restored by annealing, causing the interstitials and vacancies to migrate together and cancel. However, the stored energy is

* For a review, see A. H. Cottrell, " Effects of Neutron Irradiation on Metals and Alloys," *Met. Rev.*, Vol. 1, 1956, p. 479. Also A. H. Cottrell, Thomas Hawkesley Memorial Lecture, 1959, Inst. Mech. Engrs,